PERIODIC STRUCTURES

PERIODIC STRUCTURES

MODE-MATCHING APPROACH AND APPLICATIONS IN ELECTROMAGNETIC ENGINEERING

Ruey-Bing (Raybeam) Hwang

Department of Electrical Engineering
National Chiao-Tung University
Hsinchu, Taiwan

IEEE

WILEY

This edition first published 2013

Registered office
John Wiley & Sons Singapore Pte. Ltd., 1 Fusionopolis Walk, #07-01 Solaris South Tower, Singapore 138628

For details of our global editorial offices, for customer services and for information about how to apply for permission to reuse the copyright material in this book please see our website at www.wiley.com.

Library of Congress Cataloging-in-Publication Data

Hwang, Ruey-Bing.
Periodic structures : mode-matching approach and applications in electromagnetic engineering / Ruey-Bing (Raybeam) Hwang.
pages cm
Includes bibliographical references and index.
ISBN 978-1-118-18803-3 (hardback)
1. Electric filters. 2. Optoelectronic devices. 3. Wave guides.
4. Antennas (Electronics) 5. Photonic crystals. 6. Crystal lattices–Electric properties.
7. Electromagnetic waves. I. Title.
TK7872.F5H93 2013
621.3–dc23

2012027877

A catalogue record for this book is available from the Library of Congress.

ISBN: 9781118188033

Typeset in 11/13pt Times by Thomson Digital, Noida, India
Printed and bound in Singapore by Markono Print Media Pte Ltd

Contents

Preface

My objective in writing this book is twofold. The first objective is to build up a firm and rigorous mathematical framework, namely a mode-matching approach and transmission-line network representation, for analyzing the typical problem of wave processes involved in periodic structures ranging from the one-dimensional to the three-dimensional. The second objective is to allow the reader to understand that most of the interesting phenomena occurring in contemporary periodic structures can be clarified using existing classical electromagnetism, such as coupled-mode theory, phase-matching condition, and so on. I believe that some people will question the mode-matching method in handling periodic structures in regard to two disadvantages: the slow convergence rate for metallic structures and the infinite structure under consideration. I confess that how to solve the electromagnetic fields in a finite periodic structure is not my major concern in this book; after all, there are many well-developed commercial software packages available nowadays based on several numerical methods (e.g., the finite-difference time (frequency) domain, the finite-element method, the integral equation with moment method, the finite-integration method) for dealing with real-world electromagnetic problems in the microwave and optical communities. As to the convergence problem, the modal transmission-line approach to be elucidated in this book can tackle the task. The mode-matching method has its own advantages in facilitating the understanding of electromagnetic fields using the concept of the *modal solution*; for example, the *eigenwave solution* in a periodic medium can reveal information concerning the mode phase- and dispersion-relation; the mode dispersion relation in a gratings-assisted waveguide can be directly determined by solving the generalized eigenvalue equation rather than by extracting from electromagnetic fields.

Some of the subject matter in this book has been presented for several years as a one-semester course in the Graduate School of National Chiao-Tung University, Hsinchu, Taiwan. The prerequisites for the course are a knowledge of linear algebra and electromagnetic theory. I have not attempted the task of referring to all relevant publications. The lists of books and journal and conference articles in the reference sections at the ends of each chapter are representative, but are by no means exhaustive.

I would like to thank Professor Song-Tsuen Peng (Professor Emeritus at National Chiao-Tung University) who guided me to this exciting field during my PhD study.

I appreciate Prof. Fung-Yuel Chang for polishing my English writing and encouraging me to pursue independent research when he was a visiting Professor in National Chiao-Tung University.

I wish to acknowledge with thanks the consistent encouragement and support I received from Professor Jan-Dong Tseng (National Chin-Yi University of Technology).

I would like to express my gratitude to Professor Chang-Yu Wu, Professor Ching-Wen Hsue at National Taiwan University of Science and Technology, and Professor Raj Mittra at Pennsylvania State University for their constructive suggestions. The following warrant particular mention: Dr. Cheng-Chi Hsiao, Professors T. Tamir and K. M. Leung at Polytechnic University, Brooklyn, New York, my colleagues at National Chiao-Tung University, Professor Sin-Horng Chen (Dean), Professor Li-Chun Wang (Chairman), Professor Jen-Tsai Kuo, Professor Lin-Kun Wu, Professor Kuan-Kin Chan, Professor Yu-De Lin, Professor Shyh-Jong Chung, Professor Yi-Chiu, Professor Malcolm Ng, and Professor Edward-Yi Chang (Dean of Office of Research and Development, NCTU), Professor Ching-Cheng Tien at Chung Hua University, Professor Chien-Jen Wang at National University of Tainan, Professor Chih-Wen Kuo and Professor Ken-Huang Lin at National Sun Yat-Sen University, and Professor Kitazawa Toshihide at Ritsumeikan University, Professors Ming-Shing Lin and Chung-I G. Hsu at National Yunlin University of Science and Technology, Professor Shyue-Win Wei at National Chi-Nan University, Professor Dau-Chyrh Chang at Oriental Institute of Technology.

I would convey special acknowledgement to James Murphy (Publisher at John Wiley & Sons) for his full and consistent support.

Last but not the least, a special word of thanks is also due my wife, Belinda, who took care of everything, including me, during the writing and preparation of this book.

Ruey-Bing (Raybeam) Hwang
Hsinchu, Taiwan
March 2012

1

Introduction

1.1 Historical Perspective on the Research in Periodic Structures

The class of periodic structures has been a subject of continuing interest in the literature. The main effort in the past has been on the scattering and guiding of waves by one-dimensional (1D) periodic structures. In particular, the microwave field has employed periodic structures in many different applications, of which a few examples are linear accelerators, slow-wave structures in microwave tubes, filters, artificial dielectrics, slot arrays, phase-array antennas, frequency-selective structures, leaky-wave antennas, and so on. On the other hand, the 1D periodic structure also has its own applications in optical engineering; for example, in dielectric gratings used in integrated-optics applications (Tamir 1975, 1979) (diffraction gratings for beam splitting, grating couplers, and leaky-wave structures).

In recent years, considerable attention has been focused on the numerical and experimental studies of wave phenomena associated with two-dimensional (2D) and three-dimensional (3D) structures, and many applications have been demonstrated. Most of the potential applications were found and developed in the optics community, such as the *photonic crystal*. A photonic crystal contains dielectric or metallic inclusions periodically arranged in a 2D or 3D lattice pattern; these mimic a natural crystal with a small and basic building block of atoms or molecules repeated in space. The periodic nature of the dielectric function results in the simultaneous reflection of waves from each period, producing a stop-band where the wave propagation is forbidden. Such behavior is analogous with the electronic *band gap* in electronic materials, which is caused by introducing a gap into the energy band structure of a crystal so that electrons are forbidden to propagate with a certain energy in a certain direction (Kittle 1986). A semiconductor is the best representative having a complete band gap between

Periodic Structures: Mode-Matching Approach and Applications in Electromagnetic Engineering, First Edition.
Ruey-Bing (Raybeam) Hwang.

the valence and conduction energy bands. Therefore, engineering an artificial crystal with a *complete* stop-band, which can extend its gap to all possible directions, becomes a hot spot of research interest. With the complete stop-band (or photonic band gap), we may design a photonic crystal waveguide to guide light in a channel surrounded by photonic crystals, which are operated in the stop-band or below-cutoff condition, even in a tight corner (Joannopoulos *et al.* 1995). Additionally, the complete stop-band can also be employed to design a planar and linear defect. By tailoring the size of defect in a photonic crystal, the single mode can be pinned to the defect, producing the so-called *photonic crystal microcavities* (Joannopoulos *et al.* 1995; Soukoulis 2001).

More recently, researchers found that through engineering the commonly used dielectric materials – for example, constructing resonators and wire arrays (with *subwavelength* period) made of metallic strip lines printed on a dielectric substrate (Shelby *et al.* 2001; Smith *et al.* 2000) – one may obtain artificial materials that have properties that may not be found in nature, such as having simultaneously negative $\varepsilon_{\text{eff}}(\omega)$ and $\mu_{\text{eff}}(\omega)$ in a certain frequency band. Such an artificial material is called a *metamaterial*, which gains its electromagnetic properties from the structure instead of the chemical composition. Interestingly, the "unusual" electromagnetic property of negative refraction (negative refractive index) caused by the simultaneous existence of negative permittivity and permeability were demonstrated numerically (Engheta and Ziolkowski 1964) or experimentally (Eleftheriades and Balmain 2005).

Regarding the mathematical method for analyzing the electromagnetic field in periodic structures (e.g., microwave periodic structures, optical gratings, photonic crystals, and metamaterials), several numerical methods are developed in the literature. Sigalas *et al.* (2001) calculated the transmission and absorption of electromagnetic waves in 2D and 3D metallic band-gap structures by using the transfer-matrix method. The finite-difference method was employed by Smith *et al.* (1995) to study the defect-mode resonant cavity in a 2D metal photonic lattice. The plane-wave expansion of the fields and dielectric function is commonly employed to calculate the band structure within an infinite structure (Sakoda 2004). However, its convergence becomes poor when the contrast of the dielectric constant is large, particularly for the metallic system. Nicorovici and McPhedran (1994) and Chin *et al.* (1994) employed an analytical formulation using Green's function based on lattice sums to calculate the scattering characteristics of 2D photonic crystals consisting of an array of circular metallic cylinders. They used the cylindrical harmonics, which inherently satisfy the boundary conditions at the interfaces, to expand the fields in the periodic structure. Thus, highly accurate results with relatively low computation resources are permitted to investigate the large structure. Moreover, the analytical, numerical, and computational methods used in the modeling of scattering and guiding problems in modeling photonic crystals, which include the scattering matrix methods, multipole theory, mode-matching technique, the finite-difference frequency-domain mode solution, and the finite-difference time-domain method, were reported by Yasumoto (2005).

1.2 From 1D Periodic Stratified Medium to 3D Photonic Crystals: An Overview of this Book

As mentioned in the Preface of this book, I do not attempt to teach you any novel methods to solve the electromagnetic fields in complex periodic structures or to improve the numerical accuracy and convergence rate. Contrarily, the structures under consideration in each chapter are simple but essential; they can be regarded as building blocks in periodic structures. In fact, for a structure with a complex unit cell pattern, the staircase approach can be exploited to partition them into a cascade of multiple 1D or 2D periodic layers.

On the basis of the building-block approach used in this book, it is worth noting that a structure to be discussed in an ensuing chapter is, in fact, evolved from that being demonstrated in the previous chapter; for instance, the 2D periodic structure under the situation of *in-plane propagation* depicted in Chapter 4 can be regarded as a stack of 1D dielectric grating layers illustrated in Chapter 3, and so on. A brief introduction to each chapter is now given in the following subsections.

1.2.1 Chapter 2: Wave Propagation in Multiple Dielectric Layers

The field solutions in a uniform dielectric medium will be introduced first. If the wave propagation direction is designated as the z-axis, then the two bases in the vector space, which can expand an arbitrary electric and magnetic field (eigenfunction expansion), will be rigorously proved those relate to the propagation vector of a plane wave on the transverse (x–y) plane and are independent of the spatial position (x and y). Such a complete set is in the so-called *eigen coordinate* system rather than the spatial coordinate system that we are familiar with. The electromagnetic fields are expressed in terms of the superposition of the above-mentioned basis along the transverse plane; their expansion coefficients are z-dependent functions. Furthermore, the expansion coefficient is referred to as the voltage-wave function for the tangential electric field, while the current-wave function is for the tangential magnetic field (Felsen and Marcuvitz 1973). As a consequence, the transmission-line network representation is developed for expressing the general field solution of electromagnetic fields in a uniform medium (Oliner 1963).

From the electromagnetic boundary condition, the tangential components of the electric and magnetic fields are continuous at the interface between two dielectric media, resulting in the voltage and current waves being continuous at the interface. In doing so, the electromagnetic fields in a structure consisting of multiple parallel dielectric layers can be expressed in terms of a cascade of transmission-line sections (Figure 1.1). The scattering characteristics of a multilayered structure can be easily determined by successively cascading the input–output relation (*recursive-impedance method*, *transfer-matrix method* and *scattering-matrix method*) of each layer. Moreover, the *transverse resonance technique* will also be introduced to solve the mode

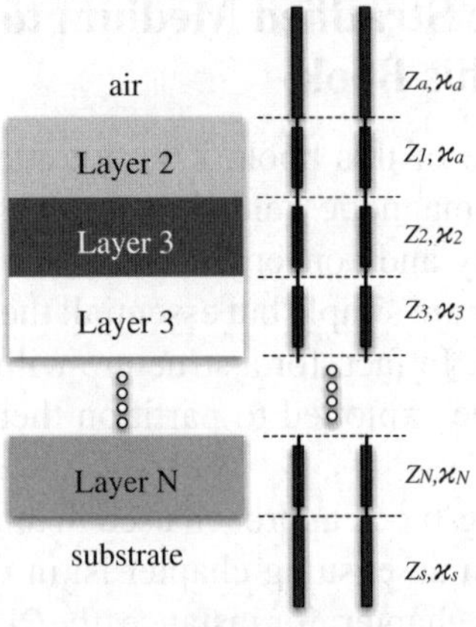

Figure 1.1 A structure consisting of multiple parallel dielectric layers and its transmission-line network representation

dispersion relation of an open or closed[1] wave-guiding structure consisting of multiple parallel dielectric layers.

1.2.2 Chapter 3: One-Dimensional Periodic Medium

Specifically, the 1D periodic structure with the unit cell containing several parallel dielectric layers can be regarded as a special case in the above-mentioned problem category. As will become clear later on, the reflection due to the stop-band in a 1D periodic structure with finite periods will be demonstrated.

The Bloch–Floquet (periodic) boundary condition for an infinite periodic medium will then be introduced. The electromagnetic fields problem in an infinite 1D dielectric periodic medium, which contains a holographic grating with the dielectric function $\varepsilon(x) = \varepsilon_g\{1 + 2\delta\cos[(2\pi x)/d)]\}$ and the parallel dielectric layers shown in Figure 1.2, subject to the periodic boundary condition, will be converted into an electromagnetic boundary-value problem presented in the form of an eigenvalue equation. Furthermore, the eigenvalue and its associated eigenvector stand for the propagation constant[2] of the wave supported in the periodic medium and the source-free electromagnetic fields in the unit cell, respectively. Regarding the mathematical details for resolving the

[1] An open structure means that the cross-section (or transverse plane), which is perpendicular to the wave-guiding (longitudinal) direction, is unbound, while the closed structure is bound.

[2] The propagation constant mentioned here may be that of the Bloch wave along the periodicity direction or that along the tangential (interface) direction.

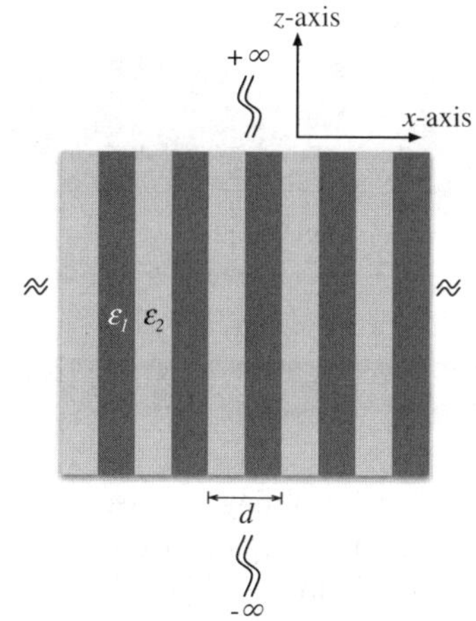

Figure 1.2 A 1D periodic medium consisting of two parallel dielectric layers in a unit cell; the direction perpendicular to the periodicity is assumed to be infinite in extent

eigenvalue problem mentioned previously, two commonly used approaches[3] including the *modal transmission-line* (Peng 1989) and *Fourier-series expansion* (Moharam and Gaylord 1981) will be illustrated. Both approaches have their respective pros and cons; for instance, for the periodic medium having a extremely large dielectric contrast in the constituent media or consisting of negative permittivity or permeability, the modal transmission-line can obtain the exact eigenvalues; however, the Fourier-series-based approach needs a considerably large number of Fourier bases for expanding the periodic dielectric function but obtaining a solution with some errors. Although the error can be reduced by increasing the number of Fourier bases, it certainly needs much more computation resources in dealing with the problem and inevitably degrades the computational efficiency. Contrarily, the eigenvalue in the Fourier-series-based approach can be directly obtained by solving the matrix eigenvalue equation without resorting to complex roots searching; however, it is inevitable in the modal transmission-line approach. For the commonly seen structure in optical engineering, the dielectric contrast is moderate (excluding the case with a plasma medium). Therefore, the Fourier-series-based approach can handle most of the problems.

Additionally, the *phase diagram*, which draws the relation among the propagation constants along the x- and z-axes and k_o, is demonstrated for understanding the coupling between space harmonics, and also the propagation characteristics concerning

[3] The rigorous coupled-wave analysis approach was developed by Moharam and Gaylord. The modal-transmission line approach was developed by Peng *et al.* (1975). More than a decade later, Peng extended the original mathematical formulation, which can only deal with the single polarization problem, into one able to handle the oblique incident of plane wave, simultaneously considering the hybrid TE–TM modes.

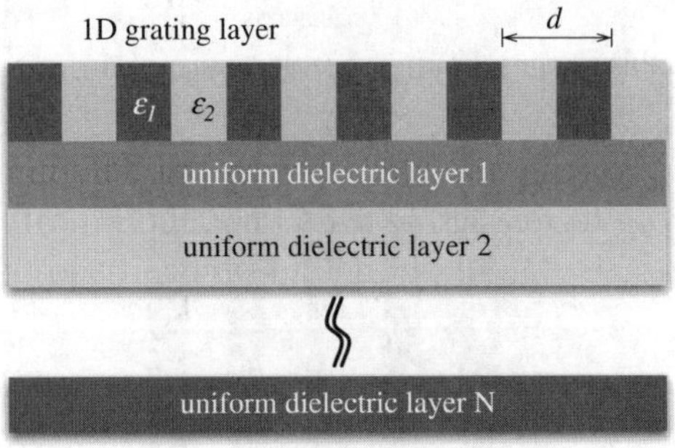

Figure 1.3 A 1D grating mounted on a multilayered structure consisting of parallel dielectric layers; the finite thickness 1D grating can be regarded as the finite truncation of the 1D periodic medium along the z-axis sketched in Figure 1.2

the phase and group velocities of the wave in a 1D periodic medium. Another tremendously useful plot, the *Brillouin diagram* (also termed a kd versus βd diagram), is also introduced to allow us to understand the electromagnetic coupling between space harmonics in the frequency domain. Specifically, the stop-band (band-gap or forbidden band), wherein the wave experiences a strong reflection, caused by the contra-flow interaction, will be addressed.

In Figure 1.2, the structure is assumed to be infinite in extent and, therefore, the general field solutions are obtained by solving the eigenvalue problem. If we truncate the original structure into a finite thickness along the z-axis but retain the infinite periods along the x-axis, the finite-thickness periodic structure will become a so-called *grating*, which is commonly used in optical engineering. The grating is usually mounted on the top of a multilayered structure, as depicted in Figure 1.3. Such a structure can be employed as a diffraction grating, which splits and diffracts the incident wave into several beams traveling in different directions, or serve as a grating-assisted waveguide, which is able to couple the incident light into the dielectric waveguide beneath the grating later, as shown in this figure; reciprocally, the waveguide mode, such as a surface wave, supported in the uniform dielectric layer can be converted into the space (leaky) wave radiating into the surrounding medium. Additionally, the reflector or deflector can also be designed based on the grating waveguide. In this book, the general electromagnetic-field solutions (hybrid TE–TM modes) will be taken into account and the electromagnetic coupling between the two polarizations will also be addressed.

1.2.3 Chapter 4: Two- and Three-Dimensional Periodic Structures

We stack up and interleave the finite-thickness 1D periodic structure (or grating) – which is on the top of the multiple dielectric layers depicted in Figure 1.3 – and uniform separator to form a 2D periodic structure,[4] as shown in Figure 1.4. If the

[4] Here, the 2D periodic structure means that it has periodicities along two directions.

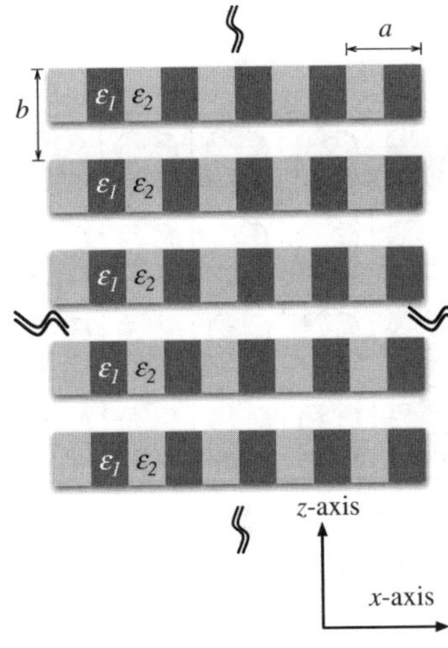

Figure 1.4 By stacking up and interleaving a 1D grating layer (which is sketched in Figure 1.3) and a uniform separator, a 2D periodic structure can be achieved. In this figure, the periods along the horizontal and vertical directions are denoted as a and b, respectively. The period along the horizontal direction is assumed to be infinity. The number of 1D grating layers along the vertical direction is also infinity

in-plane propagation is considered,[5] the input–output relation of the unit cell consisting of a 1D periodic layer and a uniform separator can be readily determined by cascading the generalized scattering matrices (Hall *et al.* 1988) of the 1D grating and uniform dielectric layer. Furthermore, for an infinite stack of 1D periodic layers (becomes a 2D periodic medium), we can find the eigenwave (source-free) solution supported in such a medium of infinite extent. By imposing the the periodic boundary condition (along the z-axis) on the input–output relation of the unit cell, we can obtain a generalized eigenvalue equation with the eigenvalue relating to the propagation constant along the z-axis. As usual, the phase diagram and dispersion (Brillouin) diagram will be demonstrated. Notably, in comparison with the 1D periodic case, the mode coupling in a 2D periodic structure is much more complex, since the space harmonics along the two directions, say the x- and z-axes, have to be taken into account, at the same time. Specifically, the *slanted* stop-band, which is caused by the coupling between the space harmonics along the two periodicities, will be presented and elucidated.

For the above-mentioned 2D infinite periodic structure, the wave propagation direction is limited to the x–z plane. The mathematical formulation demonstrated there cannot treat the problem that the energy propagates along the y-axis; namely, the

[5] In-plane indicates that the wave is propagating along the x–z plane.

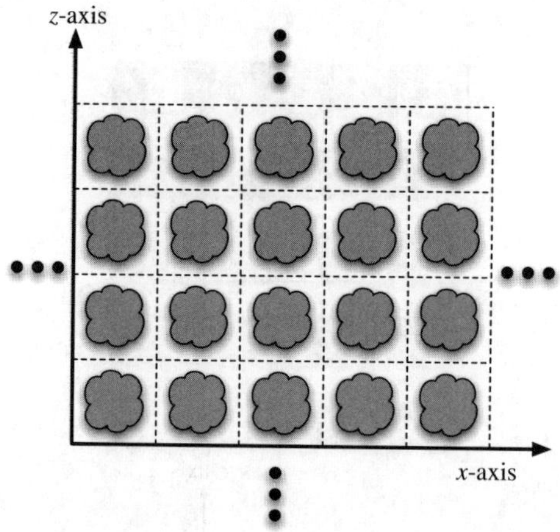

Figure 1.5 A 2D periodic medium has periods along the x- and z-axes. For example, you may consider this as a 2D dielectric columns array immersed in a uniform medium

out-of-plane propagation. Let us now redraw the 2D periodic structure as shown in Figure 1.5; the structure is infinite in extent along both x- and z- directions. The dimension along the y-axis can be a finite number. Under this situation, the propagation constant along the y-axis, k_y, does not vanish. If we consider a plane wave is incident from the top of the structure, the energy propagation along the y-axis shall be of considerable concern. Moreover, since the propagation constants along the transverse plane, k_x and k_z, are prescribed by the incident plane wave, the problem will be one of how to figure out k_y. In view of the problem description, the mathematical procedures will be totally different from those mentioned previously. To this end, the double Fourier-series expansion approach will be developed. The electric and magnetic fields in a 2D periodic medium will be decomposed into the superposition of plane-wave solutions invoking the periodicities along the two axes; therefore, the infinite number of space harmonics along the x- and z-axes must be taken into account. The input–output relation for a finite-thickness 2D periodic layer (let us imagine the structure shown in Figure 1.5 is confined along the y-axis) along the y-axis, as usual, will be presented in the form of generalized scattering matrix. As a consequence, we are able to evaluate the scattering properties of a 2D grating-assisted waveguide, as well as its guidance characteristics, by invoking the transverse resonance technique.

Regarding the 3D periodic structure shown in Figure 1.6, it can be considered as a stack of 2D periodic structures. Since we have built up the input–output relation of a finite-thickness 2D periodic structure, the scattering characteristics of a 3D periodic structure consisting of finite 2D periodic layers can be readily determined by successively cascading the generalized scattering matrix of each layer.

Additionally, if the layer number along the y-axis is infinity, we have a 3D periodic medium of infinite extent. Therefore, how to obtain the source-free solution (eigen-

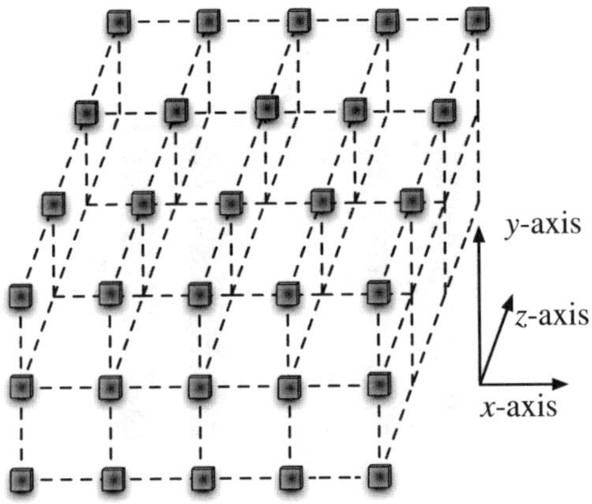

Figure 1.6 A 3D periodic medium has periods along the x-, y-, and z-axes. For example, you may stack up a 2D periodic structure along the y-axis. The 2D periodic structure may have the pattern shown in Figure 1.5

value problem) will be a major concern. As usual, by imposing the periodic boundary condition along the y-axis over a period, we can set up a generalized eigenvalue equation; however, the matrix size is much larger than that in the 2D periodic structure. The 3D phase diagram, which plots the relation among k_x, k_y, k_z for a given wavelength (or k_o), will be conducted; the dispersion diagram will also be figured out. Specifically, the stop-band behavior caused by the cross-polarization coupling between the *TE-like* and *TM-like* space harmonics presented in the dispersion diagram will be carefully examined and elucidated by the reflection property of the corresponding 3D periodic structure, however, with a finite thickness along the z-axis.

1.2.4 Chapter 5: Introducing Defects into Periodic Structures

In this part we aim at studying the types of modes supported in a 1D periodic medium with defects. As shown in Figure 1.7, several uniform layers are removed from a complete 1D periodic medium. The structure is a so-called the 1D periodic structure with plane (y–z plane) defect. Initially, such a structure was designed to serve as a waveguide for guiding the microwave signal. Significantly, the two semi-infinite periodic stratified media are operated in the stop-band or below-cutoff condition, so that the mode is bouncing back and forth between the two 1D periodic side walls. The transverse resonance technique will be employed to calculate the mode dispersion relation. The two types of modes with sinusoidal and parabolic variation along the transverse direction (x-axis) within the defect channel are obtained; the former is the *parallel-plate-waveguide-like mode* and the latter is the *surface state* with the electric field decaying in both directions.

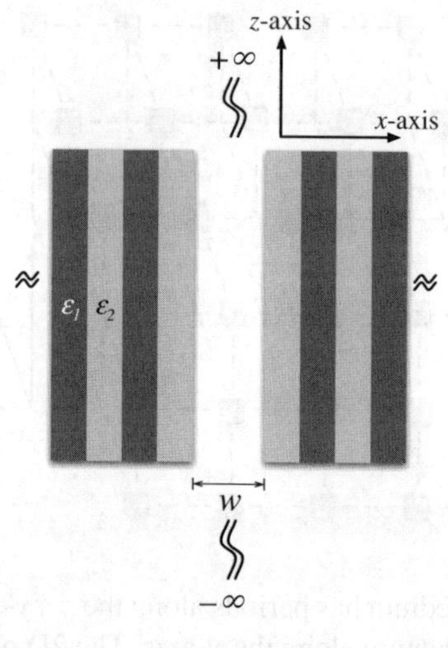

Figure 1.7 We carve out several uniform layers from the 1D periodic medium shown in Figure 1.2 to obtain a parallel-plane waveguide (defect region) sandwiched by a pair of 1D periodic structures

When one or more 1D periodic layers are carved out from the 2D periodic medium shown in Figure 1.4, the structure becomes the one depicted in Figure 1.8. Alternatively, the structure can also be regarded as a parallel-plane channel sandwiched by a pair of 2D periodic structures. As usual, the scattering properties of the structure can be determined by successively cascading the input–output relation of the 1D periodic layer and uniform dielectric slab.

In addition, the guidance properties of the waveguide mode supported in the parallel-plate-waveguide-like structure are also taken into account. In view of the symmetry of the structure with respect to the central plane in the defect channel, the open-circuit (terminated by a perfect magnetic conductor (PMC)) and short-circuit (terminated by a perfect electric conductor (PEC)) bisection, as depicted in Figure 1.9, can be applied to simplify the mathematical analysis. Through the numerical calculation, the mode dispersion relations, which generally are complex numbers with the phase and attenuation constants respectively denoted as β and α, will be obtained. The attenuation of the wave during propagation means power leakage into the surrounding medium, resulting in a *leaky wave*. The leaky-wave and beam-steering phenomena will be systematically examined in detail.

Since such a structure is similar to a Fabry–Perot filter with the 2D periodic layers served as the partially reflection mirror, the wavelength-selective transmission is in presence behaving as creating narrow passbands in a forbidden band. In this book, we will build up a framework for relating the scattering and guiding characteristics of the defected periodic structure. Specifically, the phase-match

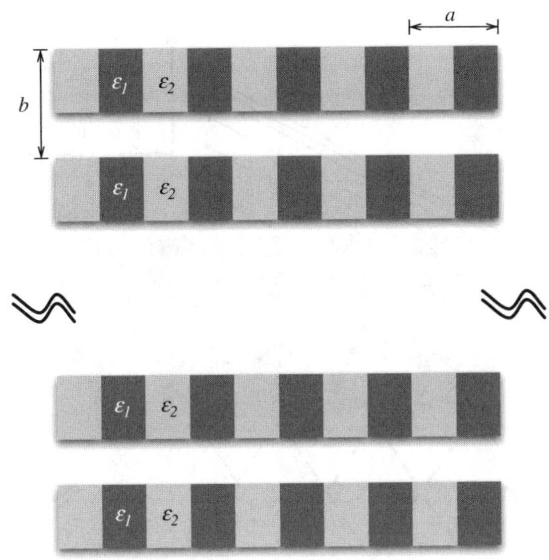

Figure 1.8 We carve out a row of a 1D periodic layer from a 2D periodic medium shown in Figure 1.4 to obtain a parallel-plane waveguide (defect region) sandwiched by a pair of 2D periodic structures

condition between the tangential-component of the phase constant of an incident plane wave and the phase constant of the defect-channel mode is carefully inspected for predicting the wavelengths of resonance transmission, which is attributed to the re-radiation of the leaky-mode excited by the incident plane wave due to phase-match condition.

1.2.5 Chapter 6: Periodic Impedance Surface

The Huygens principle states that the field solution in a region V' is completely determined by the tangential fields specified on the surface S' enclosing V'. Let us imagine a periodic structure with a period along the x-axis, as depicted in Figure 1.10; the structure is assumed to be infinite in extent along the x-axis. Such a periodic structure consists of metallic corrugation and metal plate in a unit cell, where the metallic corrugation can be regarded as the *ppwg* short-circuited at its bottom end. Since the metallic periodic structure under consideration is infinite in extent, the scattering and guidance waves only exist in the upper half-space. The Huygens principle mentioned previously can be extended to the *impedance boundary condition (IBC)* prescribed on an artificial surface above the structure under consideration. Referring to Figure 1.10, the IBC is defined on the top of the 1D metallic periodic structure. We will show that if the input impedance looking into the slot (the region between two metallic corrugations shown in Figure 1.11) is employed to approximate the *periodic IBC* (or

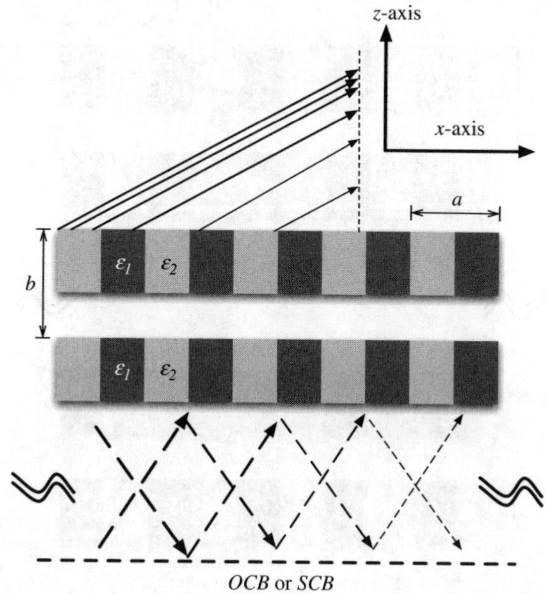

Figure 1.9 Owing to the symmetry of the structure with respect to the central plane in the defect channel shown in Figure 1.8, the open-circuit bisection (OCB) or short-circuit bisection (SCB) can be applied to simplify the structure redrawn in this figure. In this figure, the zigzag pattern inside half of the defect channel denotes the waveguide mode is bouncing back and forth between the two interfaces; the termination plane and waveguide side wall are made of 2D periodic structures. The wave leakage into the surrounding medium means that the mode has a complex propagation constant, which is decaying along the guiding axis while the amount of attenuating power is radiating into the air

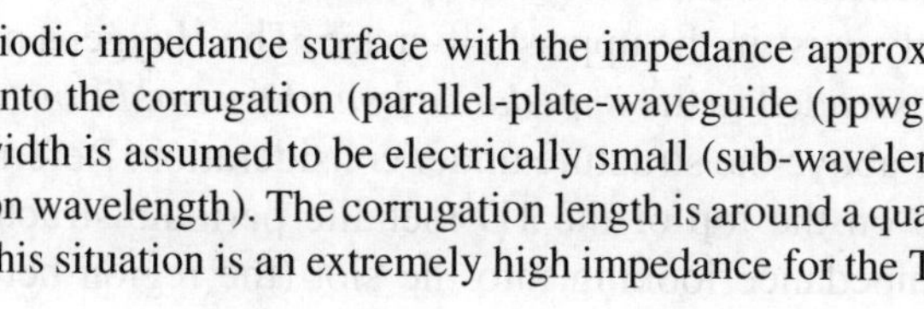

Figure 1.10 A periodic impedance surface with the impedance approximated by the input impedance looking into the corrugation (parallel-plate-waveguide (ppwg)-like structure); the width of the ppwg width is assumed to be electrically small (sub-wavelength width, which is smaller than operation wavelength). The corrugation length is around a quarter wavelength; the input impedance in this situation is an extremely high impedance for the TM wave considered in this example

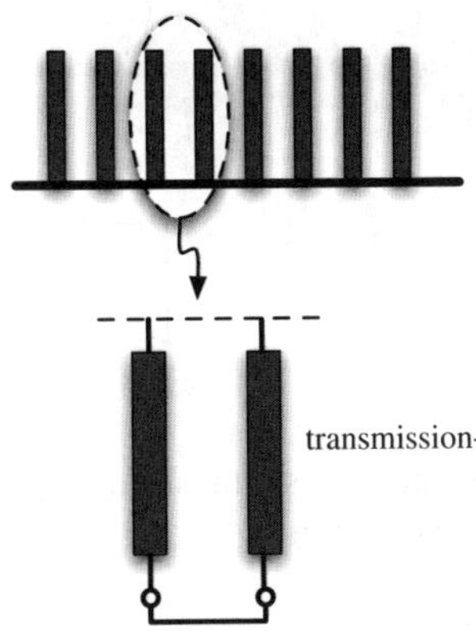

Figure 1.11 A corrugated metal surface and its transmission-line representation: the region between two corrugations is the ppwg region with sub-wavelength width; the input impedance looking into each slot is equal to $\mathrm{j}Z_0 \tan k_0 t$ for TM mode, where k_0 is the free-space wavenumber and t is the length of the corrugation

periodic impedance surface), their scattering results can approach the exact solutions obtained by the rigorous mode-matching approach (MMA). I would emphasize that the approach of the periodic impedance surface is rigorous; the errors are due to the approximation error of the periodic impedance surface. Additionally, the scattering analysis concerning a 2D dyadic periodic impedance surface is also considered, as the method to obtain the 2D dyadic periodic impedance surface from a real structure is beyond the scope of this book.

1.2.6 Chapter 7: Exotic Dielectrics Made of Periodic Structures

With the mathematical background for resolving those problems that contain eigenwaves (source-free solution) in a periodic medium and the scattering and guidance characteristics of a periodic structure ranging from one to three dimensions having been elucidated in the previous six chapters, no more new mathematical approaches will be presented in this chapter.

From the literature, we know that most metamaterials are made of a periodic structure with *sub-wavelength* periods. In microwave applications, the unit cell is usually composed of metallic strips printed on a dielectric substrate. In optics, the unit cell is made by engineering a dielectric material; the fabrication process is analogous with that of a photonic crystal. As is well known, the study of artificial dielectrics is not a new one (Kock 1948). The artificial dielectric indeed is a 2D or 3D periodic structure; however, the method of theoretical analysis only involves that of the 1D periodic

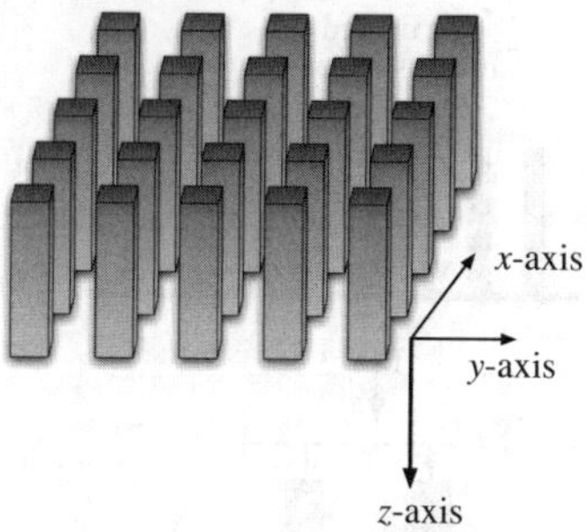

Figure 1.12 An exotic (artificial) dielectric synthesized by putting a 2D dielectric (or metallic) rods array in a uniform dielectric medium

structure. As shown in Figure 1.12, an artificial dielectric (or metamaterial) made of a 2D dielectric or metallic columns array immersed in a uniform dielectric medium is presented. We will apply the well-developed Mode-Matching Approach (MMA) for investigating the wave process involved in such a medium made of periodic structures. Specifically, the *essential* phase relation of an eigenwave will be employed to understand the group (or energy) velocity of the Bloch wave propagating in such an *effective* medium. With the phase relation, we will know that the *effective refractive index* depends not only on frequency, but also on the angle. As will become clear later on, we may *engineer the phase relation* of an exotic material to manipulate the energy flow inside that material. Finally, an analytical expression regarding the effective refractive index of an artificial dielectric made of a 1D periodic structure will be illustrated.

References

Chin, S. K., Nicorovici, N. A., and McPhedran, R. C. (1994) Green's function and lattice sums for electromagnetic scattering by a square array of perfectly conducting cylinders. *Physical Review E, Statistical Physics, Plasmas, Fluids, and Related Interdisciplinary Topics* **49**, 4590–4602.

Eleftheriades, G. V. and Balmain, K. G. (eds) (2005) *Negative Refraction Metamaterials: Fundamental Properties and Applications*, John Wiley & Sons.

Engheta, N. and Ziolkowski, R. W. (1964) *Metamaterials: Physics and Engineering Explorations*, John Wiley & Sons.

Felsen, L. B. and Marcuvitz, N. (1973) *Radiation and Scattering of Waves*, Prentice-Hall, Englewood Cliffs, NJ.

Hall, R. C., Mittra, R., and Mitzner, K. M. 1988 Scattering from finite thickness resistive strip gratings. *IEEE Transactions on Antennas and Propagation* **36**, 504–510.

Joannopoulos, J. D., Meade, R. D., and Winn, J. N. (1995) *Photonic Crystals: Modeling the Flow of Light*, Princeton University Press, Princeton, NJ.

Kittle, C. (1986) *Solid-State Physics*, John Wiley & Sons, Inc., New York, NY.

Kock, W. E. (1948) Metallic delay lenses. *Bell Systems Technical Journal* **27**, 58–82.

Moharam, M. G. and Gaylord, T. K. (1981) Rigorous coupled-wave analysis of planar-grating diffraction. *Journal of the Optical Society of America* **71**, 811–818.

Nicorovici, N. A. and McPhedran, R. C. (1994) Experimental and theoretical results for a two-dimensional metal photonic band-gap cavity. *Physical Review E, Statistical Physics, Plasmas, Fluids, and Related Interdisciplinary Topics* **50**, 3143–3160.

Oliner, A. A. (1963) *1963 Short Course on "Microwave Field and Network Techniques": Radiating Periodic Structures: Analysis in Terms of k vs. β Diagrams*, Electrophysics Department, Polytechnic Institute of Brooklyn, New York, NY.

Peng, S.-T., Tamir, T., and Bertoni, H. L. (1975) Theory of dielectric grating waveguides. *IEEE Transactions on Microwave Theory and Techniques* **23**, 123–133.

Peng S.-T. 1989 Rigorous formulation of scattering and guidance by dielectric grating waveguides: general case of oblique incidence. *Journal of the Optical Society of America* **6**, 1869–1883.

Sakoda, K. (2004) *Optical Properties of Photonic Crystals*, Springer.

Shelby, R. A., Smith, D. R., Nemat-Nasser, S. C., and Schultz, S. (2001) Microwave transmission through a two-dimensional, isotropic, left-handed metamaterial. *Applied Physics Letters* **4**, 489–491.

Sigalas, M. M., Chan, C. T., Ho, K. M., and Soukoulis, C. M. (2001) Metallic photonic band-gap materials. *Physical Review B, Condensed Matter* **52**, 11744–11751.

Smith, D. R., Schultz, S., Kroll, N. *et al.* (1995) Experimental and theoretical results for a two-dimensional metal photonic band-gap cavity. *Appl. Phys. Lett.* **65**, 645–647 .

Smith, D. R., Padilla, W. J., Vier, D. C. *et al.* (2000) Composite medium with simultaneously negative permeability and permittivity. *Physical Review Letters* **84**, 4184–4187.

Soukoulis, C. M. (ed.) (2001) *Photonic Crystals and Light Localizaton in the 21st Century*, Kluwer Academic Publishers.

Tamir, T. (1975) Leaky waves in planar optical waveguides. *Nouvelle Revue Française d'Optique* **5**, 273–284.

Tamir, T. (ed.) (1979) *Integrated Optics: Topics in Applied Physics*, Springer, Berlin.

Yasumoto, K. (2006) *Electromagnetic Theory and Applications for Photonic Crystals*, CRC Press, Boca Raton, FL.

Further Readings

Born, M. and Wolf, E. (1964) *Principles of Optics*, Macmillan, New York, NY.

Hunsperger, R. G. (1985) *Integrated Optics: Theory and Technology*, 2nd edn, Springer-Verlag.

Marcuse, D. (1974) *Theory of Dielectric Optical Waveguides*, Academic Press, New York, NY.

Oliner, A. A. (1999) Periodic structures and photonic-band-gap terminology: historical perspectives. In Proceedings of the 29th European Microwave Conference, vol. I, pp. 295–298.

2

Wave Propagation in Multiple Dielectric Layers

As is well known from the textbooks in electromagnetics, the plane-wave solutions in a uniform medium can be decomposed into two polarized modes. Therefore, each one can be analyzed with relative ease. However, in the presence of a discontinuity between periodic structures and a uniform medium, the single polarization mode can no longer satisfy the electromagnetic boundary condition when the plane wave is incident obliquely (out-of-plane propagation). The superposition of the two polarized modes is needed to determine the general solution. Consequently, our main effort in this section is to introduce a rigorous and systematic way for putting the two polarization modes back together. By invoking the mathematics of dyadic and eigenvalue problems, we will prove that the two orthogonal eigenbases correspond to TE and TM waves and the two associated eigenvalues relate to the wave (or characteristic) impedances of the two polarizations. Moreover, the wave propagating along the longitudinal direction can be interpreted using the commonly used transmission-line equation. Namely, through the rigorous mathematical process described in this chapter, the electromagnetic field problem in a uniform layer can be transformed into the electric circuit analogy using the transmission-line representation.

2.1 Plane-Wave Solutions in a Uniform Dielectric Medium

We begin with the Maxwell equations in the frequency domain, where μ and ϵ are the permeability and permittivity of the dielectric medium. Moreover, the electrical

Periodic Structures: Mode-Matching Approach and Applications in Electromagnetic Engineering, First Edition.
Ruey-Bing (Raybeam) Hwang.

charge and current are assumed to be absent in the medium. Thus, the four equations are given below.

$$\nabla \times \underline{E} = -\mathrm{j}\omega\mu\underline{H} \tag{2.1}$$

$$\nabla \times \underline{H} = +\mathrm{j}\omega\varepsilon\underline{E} \tag{2.2}$$

$$\nabla \cdot \underline{E} = 0 \tag{2.3}$$

$$\nabla \cdot \underline{B} = 0 \tag{2.4}$$

Taking the curl operator of both sides of Equation (2.1) and substituting Equation (2.2) into its right-hand side, we obtain

$$\nabla \times \nabla \times \underline{E} = k^2\underline{E} \tag{2.5}$$

where $k^2 = \omega^2\mu\varepsilon$.

From the vector identity

$$\nabla \times \nabla \times \underline{E} = \nabla\nabla \cdot \underline{E} - \nabla^2\underline{E} \tag{2.6}$$

Equation (2.5) becomes

$$\nabla^2\underline{E} + k^2\underline{E} = 0 \tag{2.7}$$

Equation (2.7) contains three partial differential equations with the electric field along the x-, y- and z-axis. The general solutions to Equation (2.7) can be written as

$$E_i(x, y, z) = (A\,\mathrm{e}^{\mathrm{j}k_x x} + B\,\mathrm{e}^{-\mathrm{j}k_x x})(C\,\mathrm{e}^{\mathrm{j}k_y y} + D\,\mathrm{e}^{-\mathrm{j}k_y y})(E\,\mathrm{e}^{\mathrm{j}k_z z} + F\,\mathrm{e}^{-\mathrm{j}k_z z}) \tag{2.8}$$

where i represents x, y, or z. Parameters k_x , k_y, and k_z are respectively the propagation constants along the x-, y-, and z-axes, which satisfy the following relation:

$$k_x^2 + k_y^2 + k_z^2 = k^2 \tag{2.9}$$

Moreover, parameters A, B, C, D, E, and F are constants to be determined by the electromagnetic boundary conditions. In a multilayered environment, since the energy transmission across the layers is of primary concern, it is natural to define the z-axis as the longitudinal direction, while the x–y plane is designated as the transverse plane perpendicular to the z-axis. Additionally, owing to the phase-matching condition, the tangential components of the propagation vector in each layer remain the same as those of the incident plane wave. Here, we assume that the phase constants of the incident plane wave along the transverse plane are k_x and k_y, respectively. Thus, the electric and magnetic field components can be written as

$$E_i(x, y, z) = E_i(z)\exp(-\mathrm{j}k_x x)\exp(-\mathrm{j}k_y y) \tag{2.10}$$

$$H_i(x, y, z) = H_i(z)\exp(-\mathrm{j}k_x x)\exp(-\mathrm{j}k_y y) \tag{2.11}$$

where $E_i(z)$ and $H_i(z)$ take into account the waves bouncing back and forth along the $+z$- and $-z$-axes both.

It is noted that we consider the plane wave propagating in a uniform medium with finite thickness; therefore, the wave function along the z-axis should contain both the forward and backward propagating terms. Contrarily, only the forward or backward propagating term is present if the structure is infinite in extent along the z-axis. For simplicity, we define the propagation vector along the transverse plane as

$$\underline{k}_t = k_x \underline{x}_o + k_y \underline{y}_o \tag{2.12}$$

Therefore, the electric and magnetic field vectors can be rewritten as follows:

$$\underline{E}(x, y, z) = [\underline{x}_o E_x(z) + \underline{y}_o E_y(z) + \underline{z}_o E_z(z)] \exp(-\mathrm{j}\underline{k}_t \cdot \underline{\rho}) \tag{2.13}$$

$$\underline{H}(x, y, z) = [\underline{x}_o H_x(z) + \underline{y}_o H_y(z) + \underline{z}_o H_z(z)] \exp(-\mathrm{j}\underline{k}_t \cdot \underline{\rho}) \tag{2.14}$$

where $\underline{\rho} = x\underline{x}_o + y\underline{y}_o$ is the position vector along the transverse plane.

The above equation can be further decomposed into a 2D transverse vector and a longitudinal component as

$$\underline{E}(x, y, z) = [\underline{E}_t(z) + \underline{z}_o E_z(z)] \exp(-\mathrm{j}\underline{k}_t \cdot \underline{\rho}) \tag{2.15}$$

$$\underline{H}(x, y, z) = [\underline{H}_t(z) + \underline{z}_o H_z(z)] \exp(-\mathrm{j}\underline{k}_t \cdot \underline{\rho}) \tag{2.16}$$

Similarly, the curl operator can also be decomposed into the components along the transverse and longitudinal directions:

$$\nabla = \underline{x}_o \frac{\partial}{\partial x} + \underline{y}_o \frac{\partial}{\partial y} + \underline{z}_o \frac{\partial}{\partial z} = \nabla_t + \underline{z}_o \frac{\partial}{\partial z} \tag{2.17}$$

Substituting Equations (2.15)–(2.17) into Equations (2.1) and (2.2), we obtain the following four equations:

$$\underline{k}_t \times \underline{E}_t(z) = \omega\mu \underline{H}_z(z) \tag{2.18}$$

$$\frac{\mathrm{d}}{\mathrm{d}z} \underline{z}_o \times \underline{E}_t(z) = -\mathrm{j}\omega\mu \underline{H}_t(z) + \mathrm{j}\underline{k}_t \times E_z(z) \underline{z}_o \tag{2.19}$$

$$\underline{k}_t \times \underline{H}_t(z) = -\omega\varepsilon \underline{E}_z(z) \tag{2.20}$$

$$\frac{\mathrm{d}}{\mathrm{d}z} \underline{z}_o \times \underline{H}_t y(z) = \mathrm{j}\omega\varepsilon \underline{E}_t(z) + \mathrm{j} H_z(z) \underline{k}_t \times \underline{z}_o \tag{2.21}$$

Substituting Equation (2.18) in Equation (2.21) to eliminate H_z results in the following linear relationship between the transverse field vectors $\underline{z}_o \times \underline{E}_t(z)$ and $\underline{H}_t(z)$:

$$\frac{\mathrm{d}}{\mathrm{d}z} \underline{H}_t(z) = \overline{\overline{\xi}} \cdot \underline{z}_o \times \underline{E}_t(z) \tag{2.22}$$

$$\overline{\overline{\xi}} = -\mathrm{j} \frac{k^2 \overline{\overline{I}} - \underline{k}_t \underline{k}_t}{\omega\mu} \tag{2.23}$$

where $\overline{\overline{\xi}}$ is a 2D dyadic and $\overline{\overline{I}}$ is the 2D unit dyadic. Notably, the definition of the dyadic and its properties are shown in Appendix A. Similarly, the elimination of E_z from Equations (2.19) and (2.20) yields

$$\frac{\mathrm{d}}{\mathrm{d}z}\underline{z}_\mathrm{o} \times \underline{E}_\mathrm{t}(z) = \overline{\overline{\chi}} \cdot \underline{H}_\mathrm{t}(z) \tag{2.24}$$

$$\overline{\overline{\chi}} = -\mathrm{j}\frac{\underline{k}_\mathrm{t}\underline{k}_\mathrm{t} + k_z^2\overline{\overline{I}}}{\omega\varepsilon} \tag{2.25}$$

where $\overline{\overline{\alpha}}$ is a 2D dyadic.

So far, we have reduced the 3D vector differential equations in Equations (2.1) and (2.2) into two 2D vector differential equations, Equations (2.22) and (2.24). Notably, although $\overline{\overline{\xi}}$ and $\overline{\overline{\alpha}}$ are expressed in terms of the 2D dyadic form, they can in fact also be regarded as 2×2 matrices. Their eigenvector and eigenvalues can be readily resolved. After some mathematical procedures, we may determine the two eigenvalues υ and τ of dyadic $\overline{\overline{\alpha}}$, and their associated eigenvector, are $\underline{\alpha}$ and $\underline{\beta}$, respectively. Significantly, the 2D dyadic $\overline{\overline{\alpha}}$ shares the same eigenvectors with $\overline{\overline{\xi}}$, and their two eigenvalues are denoted as ζ and ϱ.

$$\overline{\overline{\chi}} \cdot \underline{\alpha} = \upsilon\underline{\alpha} \tag{2.26}$$

$$\overline{\overline{\chi}} \cdot \underline{\beta} = \tau\underline{\beta} \tag{2.27}$$

$$\upsilon = -\mathrm{j}\omega\mu \tag{2.28}$$

$$\tau = -\mathrm{j}\frac{k_z^2}{\omega\varepsilon} \tag{2.29}$$

$$\overline{\overline{\xi}} \cdot \underline{\alpha} = \zeta\underline{\alpha} \tag{2.30}$$

$$\overline{\overline{\xi}} \cdot \underline{\beta} = \varrho\underline{\beta} \tag{2.31}$$

$$\zeta = \mathrm{j}\frac{k_z^2}{\omega\mu} \tag{2.32}$$

$$\varrho = -\mathrm{j}\omega\varepsilon \tag{2.33}$$

$$\underline{\alpha} = \frac{\underline{k}_\mathrm{t}}{k_\mathrm{t}} \tag{2.34}$$

$$\underline{\beta} = \underline{z}_\mathrm{o} \times \underline{\alpha} \tag{2.35}$$

where k_t is the magnitude of $\underline{k}_\mathrm{t}$, the transverse propagation vector. Apparently, the two eigenvectors in Equations (2.34) and (2.35) are perpendicular to each other and they are both perpendicular to the z-axis. Thus, the three vectors $\underline{\alpha}$, $\underline{\beta}$, and $\underline{z}_\mathrm{o}$ form a set of orthonormal basis vectors in a rectangular coordinate system, which are different from the structure coordinate system with x-, y- and z-axes.

Once the eigenvectors and eigenvalues of the two dyadics are obtained, the two dyadics can be expanded as follows:

$$\overline{\overline{\chi}} = \upsilon\underline{\alpha\alpha} + \tau\underline{\beta\beta} \tag{2.36}$$

$$\overline{\overline{\xi}} = \zeta\underline{\alpha\alpha} + \varrho\underline{\beta\beta} \tag{2.37}$$

2.2 Transmission-Line Network Representation of a Dielectric Layer of Finite Thickness

In the previous section, the linear relationship between the tangential electric and magnetic field components is represented by a 2D dyadic. Moreover, the 2D dyadic in Equations (2.22) and (2.24) are expanded in terms of the two orthogonal bases in Equations (2.36) and (2.37). From matrix theory, the transverse vectors in Equations (2.22) and (2.24) can also be expressed in terms of the superposition of the two bases:

$$\underline{z}_{\rm o} \times \underline{E}_{\rm t}(z) = V'(z)\underline{\alpha} + V''(z)\underline{\beta} \tag{2.38}$$

$$\underline{H}_{\rm t}(z) = I'(z)\underline{\alpha} + I''(z)\underline{\beta} \tag{2.39}$$

where $V'(z)$, $V''(z)$ are the expanding coefficients for transverse electric field, while $I'(z)$, and $I''(z)$ are those for transverse magnetic field. Notably, those expanding coefficients are function of z, because that the eigenbases are constant vectors independent of the position variable, z. The above equations are the so-called transmission-line network representation (Felsen and Marcuvitz 1973).

Substituting Equations (2.38) and (2.39) into Equations (2.22) and (2.24) accompanied by Equations (2.36) and (2.37), we obtain the following four first-order differential equations:

$$\frac{\rm d}{{\rm d}z}V'(z) = \upsilon I'(z) \tag{2.40}$$

$$\frac{\rm d}{{\rm d}z}V''(z) = \tau I''(z) \tag{2.41}$$

$$\frac{\rm d}{{\rm d}z}I'(z) = \zeta V'(z) \tag{2.42}$$

$$\frac{\rm d}{{\rm d}z}I''(z) = \varrho V''(z) \tag{2.43}$$

These four equations are expressed in the form of traditionally used transmission-line equations. In comparison with the transmission-line equations given in

Equations (2.44) and (2.45), we may define the characteristic impedance (and admittance) of the TE and TM modes in Equations (2.46) and (2.47), respectively:

$$\frac{\mathrm{d}}{\mathrm{d}z}V(z) = -\mathrm{j}k_z ZI(z) \tag{2.44}$$

$$\frac{\mathrm{d}}{\mathrm{d}z}I(z) = -\mathrm{j}k_z YV(z) \tag{2.45}$$

$$Z = \begin{cases} \omega\mu/k_z; & \text{TE} \\ k_z/\omega\varepsilon; & \text{TM} \end{cases} \tag{2.46}$$

$$Y = \begin{cases} k_z/\omega\mu; & \text{TE} \\ \omega\varepsilon/k_z; & \text{TM} \end{cases} \tag{2.47}$$

$$k_z = \sqrt{k_\mathrm{o}^2\mu\epsilon - k_\mathrm{t}^2} \tag{2.48}$$

It is noted that the impedance and admittance are the inverse of each other for both the TE and TM modes . By solving the differential equations in Equations (2.44) and (2.45), it is easy to determine the voltage and current function as follows:

$$V(z) = V_+\,\mathrm{e}^{-\mathrm{j}k_z z} + V_-\,\mathrm{e}^{+\mathrm{j}k_z z} \tag{2.49}$$

$$I(z) = Y[V_+\,\mathrm{e}^{-\mathrm{j}k_z z} - V_-\,\mathrm{e}^{+\mathrm{j}k_z z}] \tag{2.50}$$

In a uniform dielectric layer of finite thickness, the voltage and current waves bounce back and forth along the longitudinal direction (the z-axis). In Equations (2.49) and (2.50), the first and second terms correspond respectively to the forward- and backward-propagating waves.

For easy reference, we rearrange the electric- and magnetic-field components obtained previously and list them in Table 2.1 for TE and TM waves.

Table 2.1 Electric and magnetic components for TE and TM waves

TE($V'' = 0$, $I'' = 0$)	TM($V' = 0$, $I' = 0$)
$E_x = V'(z)k_y/k_\mathrm{t}$	$E_x = V''(z)k_x/k_\mathrm{t}$
$E_y = -V'(z)k_x/k_\mathrm{t}$	$E_y = V''(z)k_y/k_\mathrm{t}$
$E_z = 0$	$E_z = -I''(z)k_\mathrm{t}/\omega\varepsilon$
$H_x = I'(z)k_x/k_\mathrm{t}$	$H_x = -I''(z)k_y/k_\mathrm{t}$
$H_y = I'(z)k_y/k_\mathrm{t}$	$H_y = I''(z)k_x/k_\mathrm{t}$
$H_z = -V'(z)k_\mathrm{t}/\omega\mu$	$H_z = 0$

From Equation (2.38) and Table 2.1, the x and y components of an arbitrary electric field can be written as

$$E_x(x, y, z) = \left[V'(z)\frac{k_y}{k_t} + V''(z)\frac{k_x}{k_t}\right]\exp(-\mathrm{j}\underline{k}_t \cdot \underline{\rho}) \tag{2.51}$$

$$-E_y(x, y, z) = \left[V'(z)\frac{k_x}{k_t} - V''(z)\frac{k_y}{k_t}\right]\exp(-\mathrm{j}\underline{k}_t \cdot \underline{\rho}) \tag{2.52}$$

If we put the above two tangential electric-field components in a vector, Equations (2.51) and (2.52) can be expressed in terms of the following vector-matrix form:

$$\begin{bmatrix} -E_y \\ E_x \end{bmatrix} = \mathbb{T}\begin{bmatrix} V'(z) \\ V''(z) \end{bmatrix} \mathrm{e}^{-\mathrm{j}\underline{k}_t \cdot \underline{\rho}} \tag{2.53}$$

$$\mathbb{T} = \begin{bmatrix} \frac{k_x}{k_t} & -\frac{k_y}{k_t} \\ \frac{k_y}{k_t} & \frac{k_x}{k_t} \end{bmatrix} \tag{2.54}$$

Similarly, the general tangential magnetic-field vector can also be expressed in terms of the vector-matrix form:

$$\begin{bmatrix} H_x \\ H_y \end{bmatrix} = \mathbb{T}\begin{bmatrix} I'(z) \\ I''(z) \end{bmatrix} \mathrm{e}^{-\mathrm{j}\underline{k}_t \cdot \underline{\rho}} \tag{2.55}$$

It is noted that the voltage and current waves, $V(z)$ and $I(z)$, satisfy the transmission-line equations in Equations (2.44) and (2.45). Their voltage and current solutions of the TE and TM modes can be put together and filled into vectors given as follows:

$$\underline{V}(z) = \mathrm{e}^{-\mathrm{j}k_z z}\underline{V}_+ + \mathrm{e}^{+\mathrm{j}k_z z}\underline{V}_- \tag{2.56}$$

$$\underline{I}(z) = \mathbb{Y}[\underline{V}_+ \mathrm{e}^{-\mathrm{j}k_z z} - \underline{V}_- \mathrm{e}^{+\mathrm{j}k_z z}] \tag{2.57}$$

$$\underline{V}(z) = \begin{bmatrix} V'(z) \\ V''(z) \end{bmatrix} \tag{2.58}$$

$$\underline{I}(z) = \begin{bmatrix} I'(z) \\ I''(z) \end{bmatrix} \tag{2.59}$$

$$\underline{V}_+ = \begin{bmatrix} V'_+ \\ V''_+ \end{bmatrix} \tag{2.60}$$

$$\underline{V}_- = \begin{bmatrix} V'_- \\ V''_- \end{bmatrix} \tag{2.61}$$

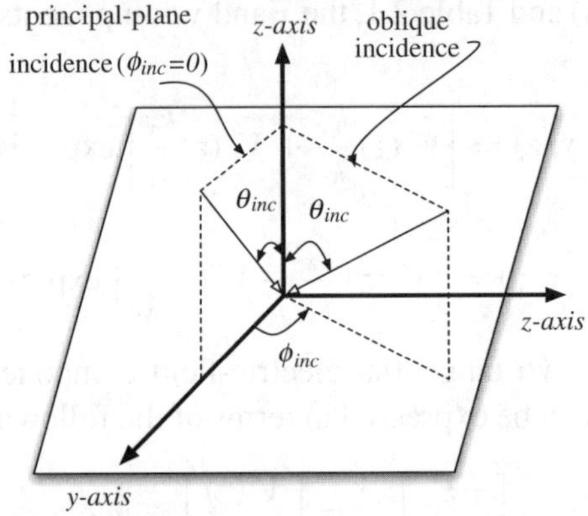

Figure 2.1 Principal and oblique incidence

Equations (2.53) and (2.55) indicate that arbitrary electric and magnetic fields can be expressed in terms of the superposition of the TE and TM modes. For a uniform dielectric layer with an incident arbitrary plane wave, we may first individually treat the scattering characteristics subject to TE- and TM-polarized waves. Once the transmission-line voltage and current waves are obtained, we apply the transformation formula in Equations (2.53) and (2.55) for determining the hybrid electric and magnetic field. This method will allow us to tackle the scalar electromagnetic fields rather than to directly deal with the 3D vector electromagnetic fields.

Example 2.1

A plane wave is propagating in free space. The elevation angle $\theta_{\rm inc}$ and azimuthal angle $\phi_{\rm inc}$ are defined in the spherical coordinate system shown in Figure 2.1. Write down the transformation matrix given in Equation (2.54).

Solution

The transverse component of the propagation vector along the x- and y-axis is written as

$$k_x = k_{\rm o} \sin\theta \cos\varphi$$
$$k_y = k_{\rm o} \sin\theta \sin\varphi$$
$$k_{\rm t} = k_{\rm o} \sin\theta$$

Substituting k_x and k_y into Equation (2.54), we may determine the following transformation matrix:

$$\mathbb{T} = \begin{bmatrix} \cos\varphi & -\sin\varphi \\ \sin\varphi & \cos\varphi \end{bmatrix}$$

We may find that the transformation matrix written previously, in fact, is a coordinate transformation matrix transforming the coordinate system from the structure coordinate system by rotating an angle counter-clockwise about the origin. The new coordinate system is an eigen-coordinate system (Figure 2.2) that is totally different from the structure coordinate system, (x, y, z).

2.2.1 Wave Propagating in Regular and Exotic Mediums

Recently, many researchers have developed some artificial materials made from traditionally used materials, such as a dielectric medium containing metallic inclusions. By properly designing the unit-cell configuration with periods that are usually much smaller than the operational wavelength, the effective permittivity and permeability may have very different properties compared with the commonly used materials (Engheta and Ziolkowski 1964; Smith *et al.* 2004). For example, we may have the so-called double negative (DNG) medium simultaneously exhibiting negative permeability and permittivity, an artificial magnetic conductor (AMC) without ferromagnetic composition, a near-zero refractive index medium, and a plasma-like medium with positive permeability and negative permittivity. In this section, we will discuss the wave-propagating characteristics in the four mediums with the following combinations of permittivity and permeability:

1. positive permittivity and positive permeability;
2. negative permittivity and positive permeability;

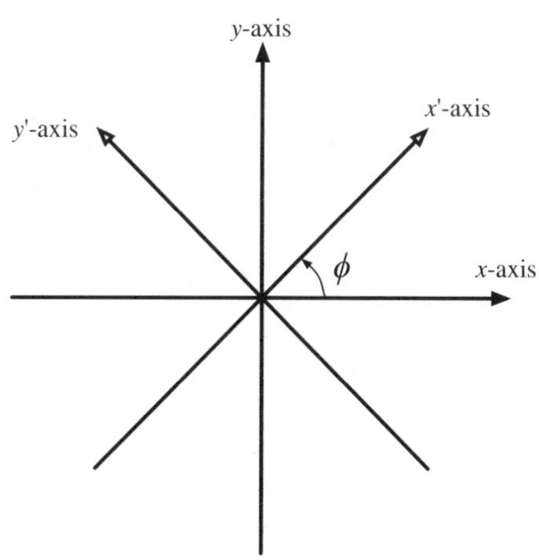

Figure 2.2 Transformation between space- and eigen-coordinate systems

3. negative permittivity and negative permeability;
4. positive permittivity and negative permeability.

With the given permittivity and permeability, the propagation constant along the z-axis can be obtained from Equation (2.48). After substituting the above four combinations of electrical properties into Equation (2.48), we obtained their z-direction propagation constants given from Equations (2.62)–(2.65):

$$\mu > 0 \quad \text{and} \quad \varepsilon > 0 \xrightarrow{\text{yields}} k_z = \sqrt{\omega^2 \mu \varepsilon - k_t^2} \tag{2.62}$$

$$\mu > 0 \quad \text{and} \quad \varepsilon < 0 \xrightarrow{\text{yields}} k_z = \sqrt{-\omega^2 \mu \left|\varepsilon\right| - k_t^2} = -\mathrm{j}\alpha_z \tag{2.63}$$

$$\mu < 0 \quad \text{and} \quad \varepsilon < 0 \xrightarrow{\text{yields}} k_z = \pm\sqrt{\omega^2 \left|\mu\right| \left|\varepsilon\right| - k_t^2} \tag{2.64}$$

$$\mu < 0 \quad \text{and} \quad \varepsilon > 0 \xrightarrow{\text{yields}} k_z = \sqrt{-\omega^2 \left|\mu\right| \varepsilon - k_t^2} = -\mathrm{j}\alpha_z \tag{2.65}$$

The first medium is a commonly used medium. The second medium is a plasma medium able to be found in nature. For instance, a noble metal in visible light exhibits the plasma effect. Contrarily, the third and fourth mediums have, so far, not been found in nature, but can be synthesized using traditionally used metallic and dielectric materials. Hence, they are usually called artificial dielectrics. Considering the limiting case that a plane wave is normally incident into the four mediums, respectively, the propagation constants in a commonly used dielectric medium, shown in Equation (2.62), are real numbers. The z-direction propagation constants in Equations (2.63) and (2.65) are pure imaginary numbers, which imply that the waves are attenuating along the z-axis in those mediums. In Equation (2.64), since both the permeability and permittivity are negative numbers, their product is still a positive number; the square root may be a positive or a negative number. However, by considering that the real part of the wave impedance in Equation (2.46) must be positive (negative resistance is not acceptable for a passive medium), we have to choose the minus sign of the z-axis propagation constant k_z. The propagation characteristics of waves in the above four mediums are plotted in Figure 2.3 for ease of reference. In quadrants 1 and 3, the plane wave propagates without attenuation, while the propagation vector of the wave in quadrant 3 (DNG medium) is opposite to that in quadrant 1. The waves in quadrants 2 and 4 decay along the propagating direction.

In the following example, we will demonstrate the negative refraction of a plane wave propagating from a regular medium into an exotic medium while simultaneously having negative permittivity and permeability.

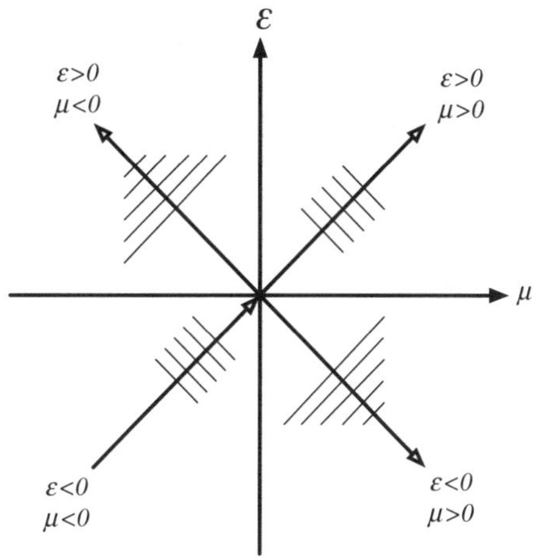

Figure 2.3 Wave propagation in the four media

Example 2.2

Consider a TE-polarized plane wave incident from air into a DNG medium with permittivity and permeability given as: $\varepsilon = -1.2\varepsilon_o$ and $\mu = -1\mu_o$. If the incident angle is 60° counted from the normal direction in the air region and the wavelength is 2 mm, plot the electric-field contour and explain the physical picture of the wave process involved in the structure under consideration.

Solution

To plot the electric field contour, we write a computer code using MATLAB®. Since the second medium is a DNG medium, the propagation constant along the z-axis must be

$$k_z^{(2)} = \sqrt{k_o^2(-1)(-1.2) - k_o^2(\sin 60°)^2} = \pm\sqrt{0.45}k_o$$

From Equation (2.46), the characteristic impedance of the wave in the DNG medium is $-\omega\mu_o/k_z^{(2)}$. To have positive characteristic impedance, we must choose the minus sign of $k_z^{(2)}$. Therefore, the characteristic impedance is equal to $Z_o = 120\pi/\sqrt{0.45}$.

Moreover, the electric fields in the respective regions are obtained by solving the voltage function of the cascade of the two semi-infinite transmission lines.

Figure 2.4 shows the contour distribution of the $E_y(x, z)$ component in the structure consisting of the air region ($z < 0$) and the DNG medium ($z > 0$). The direction normal to the wave front in each region represents the wave propagation direction. Because of the mismatch between the characteristic impedance of the waves in the two media, reflection occurs. Thus, the wave front in the air region is no longer a plane. Notably, in the DNG medium, owing to the minus sign of k_z, the propagating vector is directed inwards, as shown in Figure 2.4. For ease of reference, the wave transmission into a regular dielectric medium with $\varepsilon = 1.2\varepsilon_o$ and $\mu = \mu_o$ is also plotted for realizing the difference between the two cases. As shown in Figure 2.5, the direction of refraction follows Snell's law. Notably, the refracted wave in the DNG medium is in the direction opposite to that in the regular dielectric medium, and is called negative refraction.

2.3 Scattering Characteristics of Plane Wave by Multiple Dielectric Layers

So far, by invoking the mathematics of the dyadic and the eigenvalue problem we have determined a set of orthonormal bases for expanding an arbitrary electromagnetic field as the energy propagation is along the z-axis (longitudinal direction). The expanding coefficients of the tangential electric and magnetic fields respectively correspond to the voltage and current waves. Specifically, the voltage and current waves satisfy the transmission-line equations. Consequently, for the structure consisting of multiple uniform dielectric layers, the energy propagation along the longitudinal direction per-

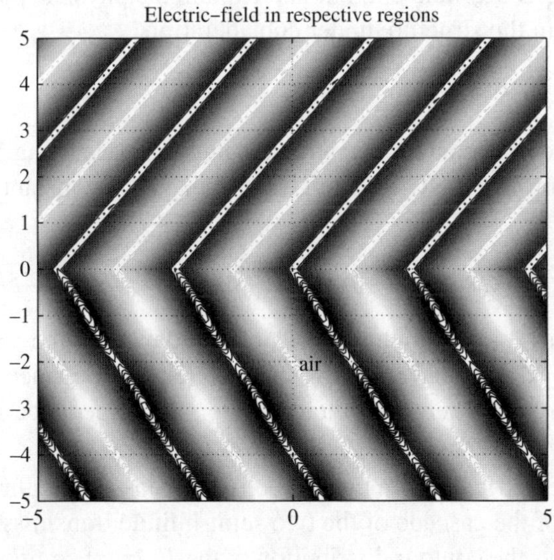

Figure 2.4 The contour of $E_y(x, z)$ in air and the DNG medium. The vertical and horizontal axes respectively denote the z- and x-axis; the region of $z \leq 0$ is the air region, while the region of $z > 0$ is the DNG medium with $\varepsilon_r = -1.2$ and $\mu_r = -1$. The plane wave is incident from the air region

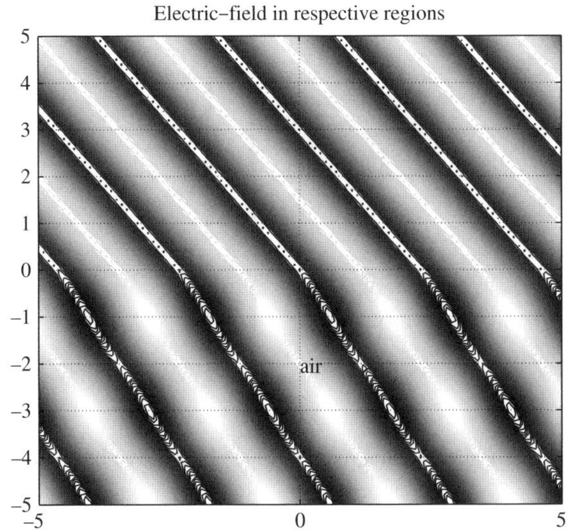

Figure 2.5 The contour of $E_y(x, z)$ in air and the dielectric medium. The vertical and horizontal axes respectively denote the z- and x-axes; the region of $z \leq 0$ is the air region, while the region of $z > 0$ is the dielectric medium with $\varepsilon_r = 1.2$ and $\mu_r = 1$. The plane wave is incident from the air region

pendicular to the surface of the layers can be figured out by evaluating the cascade of finite-length transmission lines, while the eigenfunction, shown in Equations (2.34) and (2.35), along the transverse plane remains the same for each layer because of the phase-match condition.

In this section we will discuss the scattering characteristics of plane waves by multilayered structures. Three approaches will be introduced: the recursive-impedance method, the transfer-matrix method, and the scattering-matrix method . Moreover, these approaches can also be extended to treat the problem of a periodic structure; that is, we may repeat the multilayer periodically along the thickness direction to achieve a 1D periodic structure.

Figure 2.6 shows a structure consisting of multiple uniform dielectric layers, each of which is uniform in its permittivity and permeability and is infinite in extent along the transverse (x–y) plane; however, the thickness is finite along the longitudinal (z) direction. A plane wave is incident on this planar structure with elevation angle θ, which is counted from the z-axis, and azimuthal angle ϕ, which is counted from the x-axis, as shown in Figure 2.1. We will employ three different approaches for determining the scattering characteristics of such a conical structure.

First of all, as we have derived previously, for an arbitrary plane-wave incidence problem, we may decompose them into the superposition of TE- and TM-polarization waves. Through the respective scattering analysis, the general solution can be obtained

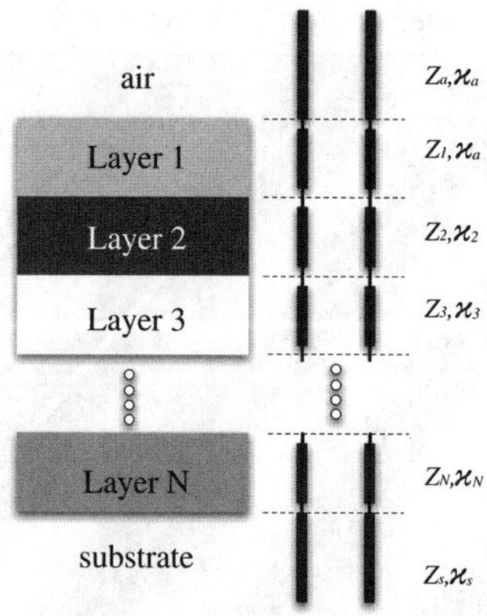

Figure 2.6 Reflection and transmission of a plane wave by a dielectric medium

by putting them together. What we have to do now is to formulate the scattering characteristics of the multilayered structure under the TE- and TM-polarized wave incidence, respectively.

In the previous section, we established the one-to-one correspondence between the electromagnetic fields and transmission-line voltage and current waves. Moreover, based on the boundary condition, the tangential electric and magnetic fields must be continuous at the interface between the dielectric layers. From Table 2.1, this implies that the voltage and current waves must also be continuous at each layer interface. In doing so, the electromagnetic field in a multilayered environment can be converted into a cascade of a transmission-line network composed of finite-length transmission-line sections.

2.3.1 *Recursive-Impedance Method*

We first consider a finite-length transmission line having the propagation constant κ and characteristic impedance Z_c. The start and stop positions of the transmission line individually are z_s and $z_s + t$. Such a transmission line is terminated by impedance Z_l. The voltage- and current-wave functions of the transmission line are written as

$$V(z) = V_+ \mathrm{e}^{-\mathrm{j}\kappa(z-z_s)} + V_- \mathrm{e}^{+\mathrm{j}\kappa(z-z_s)} \tag{2.66}$$

$$I(z) = Y_c[V_+ \mathrm{e}^{-\mathrm{j}\kappa(z-z_s)} - V_- \mathrm{e}^{+\mathrm{j}\kappa(z-z_s)}] \tag{2.67}$$

where V_+ and V_- represent the forward- and backward-propagating voltages, respectively.

The termination condition is given as follows by relating the voltage to current at the end of the line via the load impedance:

$$V(z_s + t) = Z_l I(z_s + t) \tag{2.68}$$

The input impedance looking into the transmission line and the voltage transfer function can be determined as follows through some mathematical procedures:

$$Z_{in} = Z_c \frac{1 + \Gamma}{1 - \Gamma} \tag{2.69}$$

$$\Gamma = \Gamma_l \exp(-j2\kappa t) \tag{2.70}$$

$$\Gamma_l = \frac{Z_l - Z_c}{Z_l + Z_c} \tag{2.71}$$

$$T(z; z_s) = \frac{V(z)}{V(z_s)} = e^{-j\kappa(z-z_s)} \frac{1 + \Gamma_l e^{-j2(t+z_s-z)}}{1 + \Gamma} \tag{2.72}$$

Let us start with the last section of the transmission line (Nth section) shown in Figure 2.6. Because of the presence of load impedance (the intrinsic impedance of the surrounding medium below the Nth layer), the input impedance at the input terminal of this line can be obtained through Equation (2.69). Moreover, such an input impedance can also be regarded as the load impedance of the $(N - 1)$th transmission-line section. Again, we may employ Equation (2.69) to determine the input impedance of this line at its input terminal. By exploiting the same procedure recursively, we may finally obtain the input impedance looking into the input terminal of the overall transmission-line network, which is denoted as $Z_{in}^{(total)}$.

Here, we assume that the intrinsic impedance of the surrounding medium connecting to the first layer is characterized as Z_i. Therefore, the voltage and current at the input terminal of the transmission-line network are written as

$$V(0) = (1 + \Gamma_i)a \tag{2.73}$$

$$I(0) = Y_i(1 - \Gamma_i)a \tag{2.74}$$

$$\Gamma_i = \frac{Z_{in}^{(total)} - Z_i}{Z_{in}^{(total)} + Z_i} \tag{2.75}$$

where a is the amplitude of the incident wave in the input region.

Once the voltage at the input terminal is determined, the voltage at any position along any section of transmission line can be readily determined by applying the voltage transfer function given in Equation (2.72), as long as the voltage at the input terminal of the transmission line is known.

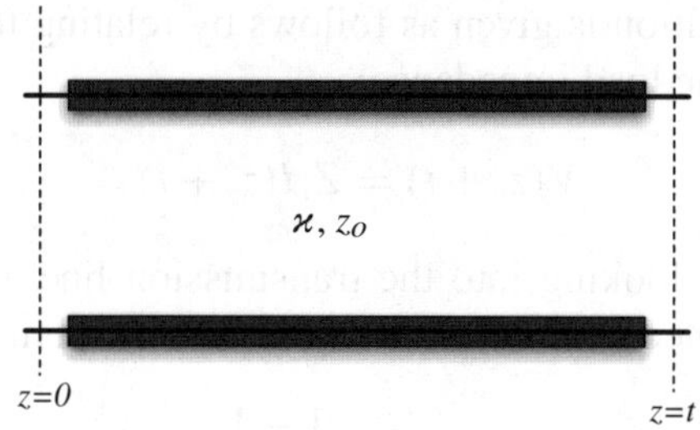

Figure 2.7 A finite-length transmission line

2.3.2 *Transfer-Matrix Method*

The second approach is the transfer-matrix method that is extensively employed in microwave engineering. We begin with the same finite-length transmission line shown in Figure 2.7. The voltage and current at the input terminal are as follows:

$$V(z_s) = V_+ + V_- \tag{2.76}$$

$$I(z_s) = Y_c(V_+ - V_-) \tag{2.77}$$

Solving the forward- and backward-propagating voltages, V_+ and V_-, in the last two equations, we may express them in terms of the voltage and current at the input terminal. After substituting them into the transmission-line solutions in Equations (2.66) and (2.67), we obtain the voltage and current functions everywhere along the transmission line in terms of those at the input terminal, which are given as

$$\begin{bmatrix} V(z) \\ I(z) \end{bmatrix} = \begin{bmatrix} \cos\kappa(z-z_s) & -jZ_c \sin\kappa(z-z_s) \\ -jY_c \sin\kappa(z-z_s) & \cos\kappa(z-z_s) \end{bmatrix} \begin{bmatrix} V(z_s) \\ I(z_s) \end{bmatrix} \tag{2.78}$$

Therefore, the input–output relation of the voltage-current vector of the transmission line can be written as

$$\begin{bmatrix} V(z_s+t) \\ I(z_s+t) \end{bmatrix} = \mathbb{T}(\kappa t) \begin{bmatrix} V(z_s) \\ I(z_s) \end{bmatrix} \tag{2.79}$$

$$\mathbb{T}(\kappa t) = \begin{bmatrix} \cos\kappa t & -jZ_c \sin\kappa t \\ -jY_c \sin\kappa t & \cos\kappa t \end{bmatrix} \tag{2.80}$$

Matrix $\mathbb{T}$ is called the transfer-matrix or ABCD-matrix relating the voltage and current at the input and output terminals.

Additionally, we may also present the above results as the following form:

$$\begin{bmatrix} V(z_s) \\ I(z_s) \end{bmatrix} = \mathbb{R}(\kappa t) \begin{bmatrix} V(z_s + t) \\ I(z_s + t) \end{bmatrix} \tag{2.81}$$

$$\mathbb{R}(\kappa t) = \begin{bmatrix} \cos \kappa t & jZ_c \sin \kappa t \\ jY_c \sin \kappa t & \cos \kappa t \end{bmatrix} \tag{2.82}$$

The above expression indicates that the voltage–current vector at the input terminal can also be obtained by multiplying the transfer matrix $\mathbb{R}$ with the voltage–current vector at the output terminal. Notably, the matrix $\mathbb{R}$ is the inversion of matrix $\mathbb{T}$. Returning to the transmission-line network shown in Figure 2.4, each section of the transmission line has the input–output relation given in Equation (2.80) or (2.82). Consequently, we can readily determine the input–output relation of the overall network by sequentially multiplying the transfer matrices:

$$\begin{bmatrix} V_{out} \\ I_{out} \end{bmatrix} = \mathbb{T} \begin{bmatrix} V_{inp} \\ I_{inp} \end{bmatrix} \tag{2.83}$$

$$\mathbb{T} = \prod_{i=N}^{1} \mathbb{T}_i(\kappa_i t_i) \tag{2.84}$$

$$\begin{bmatrix} V_{inp} \\ I_{inp} \end{bmatrix} = \mathbb{R} \begin{bmatrix} V_{out} \\ I_{out} \end{bmatrix} \tag{2.85}$$

$$\mathbb{R} = \prod_{i=1}^{N} \mathbb{R}_i(\kappa_i t_i) \tag{2.86}$$

$$\kappa_i = \sqrt{\omega^2 \mu_i \varepsilon_i - k_t^2} \tag{2.87}$$

From the termination condition $V_{out} = Z_{out} I_{out}$ at the last section of transmission line, the input impedance looking into the first section of the transmission line can be written as

$$Z_{in}^{(total)} = \frac{V_{inp}}{I_{inp}} = \frac{r_{11} Z_{out} + r_{12}}{r_{21} Z_{out} + r_{22}} \tag{2.88}$$

$$\mathbb{R} = \begin{bmatrix} r_{11} & r_{12} \\ r_{21} & r_{22} \end{bmatrix} \tag{2.89}$$

Problem 2.1

Prove that if the $\mathbb{T}$ matrix is given and with the entries denoted as t_{11}, t_{21}, t_{12}, and t_{22}, respectively, the $\mathbb{R}$ matrix can be written as follows:

$$\mathbb{R} = \begin{pmatrix} t_{22} & -t_{12} \\ -t_{21} & t_{11} \end{pmatrix}$$

If the incident wave amplitude in the surrounding medium is denoted as a, the voltage and current at the input interface are written as

$$V_{\text{inp}} = (1 + \Gamma_{\text{i}})a \tag{2.90}$$

$$I_{\text{inp}} = Y_{\text{i}}(1 - \Gamma_{\text{i}})a \tag{2.91}$$

$$\Gamma_{\text{i}} = \frac{Z_{\text{in}}^{(\text{total})} - Z_{\text{i}}}{Z_{\text{in}}^{(\text{total})} + Z_{\text{i}}} \tag{2.92}$$

where Y_{i} is the characteristic admittance of the surrounding medium.

Therefore, the voltage–current vector at the output interface could be obtained by substituting Equations (2.91) and (2.92) into Equation (2.83). The transmission coefficient can then be determined by evaluating the following equation:

$$\tau = \frac{V_{\text{out}}}{V_{\text{inp}}} \tag{2.93}$$

2.3.2.1 Eigenwave in a 1D Periodic Medium Consisting of Two Dielectric Mediums in a Unit Cell

Certainly, we may extend the above formulation to 1D periodic structures consisting of a stack of unit cells (periods). We consider a 1D periodic structure composed of N periods, each of which has the same transfer matrix $\mathbb{T}$. From the transfer-matrix theory described previously, the input–output relation of the overall structure can be written as $\mathbb{T}^N$. From eigen-decomposition of a matrix, the 2×2 matrix $\mathbb{T}$ can be factorized as

$$\mathbb{T} = \mathbb{Q}\mathbb{D}\mathbb{Q}^{-1}$$

where $\mathbb{Q}$ is the square matrix whose first and second columns are the eigenvectors of matrix $\mathbb{T}$. Parameter $\mathbb{D}$ is a diagonal matrix whose diagonal elements are the corresponding eigenvalues. Therefore, the matrix $\mathbb{T}^N$ can be immediately obtained and written as

$$\mathbb{T}^N = \mathbb{Q}\mathbb{D}^N\mathbb{Q}^{-1}$$

The above equation reveals that the property of the matrix $\mathbb{T}^N$, in fact, is determined by that of the diagonal matrix $\mathbb{D}$. Strictly speaking, the two eigenvalues of the matrix $\mathbb{D}$ are the dominant factors.

Now we return to the detail of the transfer-matrix of the unit cell consisting of two dielectric layers. From Equation (2.80), the transfer matrix of the unit cell is written as

$$\mathbb{T} = \begin{bmatrix} \cos\kappa_2 t_2 & -jZ_2 \sin\kappa_2 t_2 \\ -jY_2 \sin\kappa_2 t_2 & \cos\kappa_2 t_2 \end{bmatrix} \begin{bmatrix} \cos\kappa_1 t_1 & -jZ_1 \sin\kappa_1 t_1 \\ -jY_1 \sin\kappa_1 t_1 & \cos\kappa_1 t_1 \end{bmatrix} \tag{2.94}$$

where κ_i, t_i, and Z_i ($i = 1$ or 2) are the propagating constants along the periodicity, thickness, and characteristic impedance in each layer.

If the eigenvalue of the transfer-matrix $\mathbb{T}$ is designated as λ, the characteristic equation is written as

$$\det(\mathbb{T} - \lambda\mathbb{I}) = 0 \tag{2.95}$$

$$\lambda^2 - \text{trace}(\mathbb{T})\lambda + \det(\mathbb{T}) = 0 \tag{2.96}$$

From Equation (2.96), we may find that the two eigenvalues λ_1 and λ_2 satisfy the following equations:

$$\lambda_1 + \lambda_2 = \text{trace}(\mathbb{T}) \tag{2.97}$$

$$\lambda_1 \cdot \lambda_2 = \det(\mathbb{T}) = 1 \tag{2.98}$$

Equation (2.98) tells us that the two eigenvalues are the inverse of each other. Therefore, we may assume that the two eigenvalues are

$$\lambda_1 = e^{-j\theta} \tag{2.99}$$

$$\lambda_2 = e^{+j\theta} \tag{2.100}$$

This means that the voltage- and current-wave experience a phase shift with phase angle θ once it is propagating through a unit cell. Substituting Equations (2.99) and (2.100) into Equation (2.97), we may obtain the following equation:

$$\cos\theta = \cos\kappa_1 t_1 \cos\kappa_2 t_2 - Q \sin\kappa_1 t_1 \sin\kappa_2 t_2 \tag{2.101}$$

$$Q = \frac{1}{2}\left(\frac{Z_1}{Z_2} + \frac{Z_2}{Z_1}\right) \tag{2.102}$$

Looking at the right-hand side of Equation (2.101), the term may be an arbitrary real number (consider lossless dielectric media); however, its value affects the propagating characteristics significantly. If it is less than or equal to one, the angle θ is a real number. Contrarily, if it is greater than unity, the angle θ must be a complex number. Phase angle θ is determined by the product of the average propagation constant in the unit cell along the periodicity direction, denoted by $\bar{\kappa}$, and the period d. For a real number of average propagation constant, the wave is propagating through the unit cell; however, the wave is decaying in the unit cell when the average propagation constant is a complex number, such as $\bar{\kappa} = \beta - j\alpha$, where β and α are the phase and attenuation constant. Notably, the minus and plus signs in Equations (2.99) and (2.100) stand for the forward- and backward-propagating waves, respectively. The backward-propagating wave can be regarded as the reflection of the forward-propagating one, so that if $\bar{\kappa}$ is a complex number, both the waves are decaying along their propagating direction.

Example 2.3

A structure is composed of two uniform layers with permittivity and permeability constant: $(10\mu_o,10\varepsilon_o)$ and $(-10\mu_o,-10\varepsilon_o)$. Notably, the medium with negative permeability and permittivity is a so-called the DNG medium. When the propagation constant along the transverse plane, which is perpendicular to the periodicity, is assumed to be zero, describe the propagating behavior of the eigenwaves.

Solution

Owing to the vanishing of the transverse phase constants, the z-axis propagation constant in the DNG medium (designated as the second layer) is written as

$$\kappa_2 = \pm 10\sqrt{\omega^2 \mu_o \varepsilon_o} \tag{2.103}$$

We have two choices concerning the sign of k_z in the DNG medium. As mentioned previously, we may choose the correct one by considering that the real part of the wave impedance in the DNG medium is a positive number (a negative resistor is not allowed in a passive medium). Consequently, we have the solution $\kappa_2 = -10k_o$ and the impedances are positive real numbers:

$$Z_2 = \begin{cases} \dfrac{-10\omega\mu_o}{\kappa_2}; & \text{TE} \\ \dfrac{\kappa_2}{-10\omega\varepsilon_o}; & \text{TM} \end{cases} \tag{2.104}$$

Moreover, the impedance and propagation constant in the regular dielectric medium (designated as the first layer) are written as follows:

$$Z_1 = \begin{cases} \dfrac{10\omega\mu_o}{\kappa_1}; & \text{TE} \\ \dfrac{\kappa_1}{10\omega\varepsilon_o}; & \text{TM} \end{cases} \tag{2.105}$$

$$\kappa_1 = 10k_o \tag{2.106}$$

Since the two characteristic impedances are the same, $Q = 1$ in Equation (2.102). Therefore, Equation (2.101) becomes

$$\cos\theta = \cos(\kappa_1 t_1 + \kappa_2 t_2) \tag{2.107}$$

$$\cos\theta = \cos(10k_o t_1 - 10k_o t_2) \tag{2.108}$$

As mentioned previously, θ is the phase angle of the wave propagating through the two-layered structure. Interestingly, if the two layers share the same thickness, the phase-delay angle is zero. This is explained by the wave experiencing a phase lag in the first layer ; however, in the second layer it experiences a phase advance. Since the phase-angle of delay and advance are equal, the total delay angle vanishes. As a consequence, the average propagation constant is zero. This means that the phase delay angle is independent of the length of the structure having such a composite material consisting of DNG and conventional dielectric medium.

2.3.3 Scattering-Matrix Method

As shown in Figure 2.8, we consider here a cascade of two semi-infinite transmission lines, which form a junction at $z = 0$, having propagation constants κ_1 and κ_2 and characteristic impedances Z_1 and Z_2, respectively. The incident and reflected waves in line 1 are respectively a_1 and b_1, while those in line 2 are a_2 and b_2. Owing to the boundary condition that the voltage and current at the interface between the two lines must be continuous, we have the following relations:

$$a_1 + b_1 = b_2 + a_2 \tag{2.109}$$

$$Y_1(a_1 - b_1) = Y_2(b_2 - a_2) \tag{2.110}$$

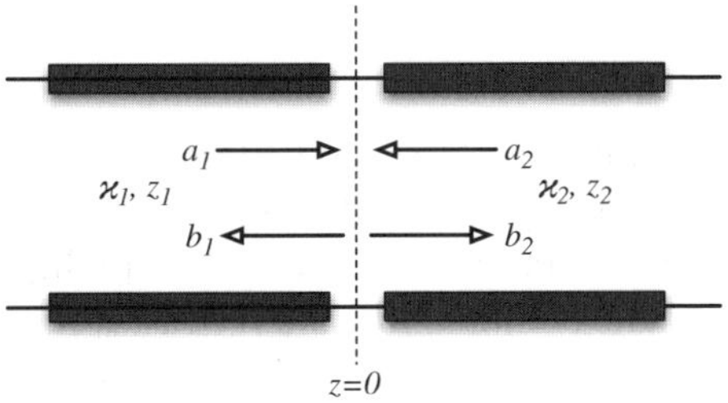

Figure 2.8 Cascade of the two semi-infinite transmission lines

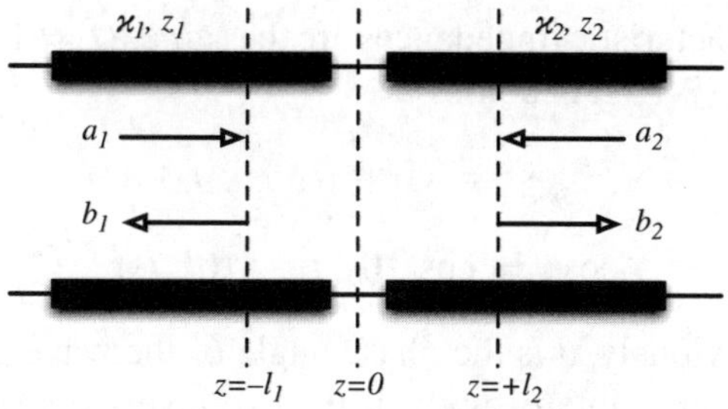

Figure 2.9 Cascade of the two semi-infinite transmission lines

After some mathematical procedures, we obtain the relationship between the incident and reflected wave vectors written as

$$\begin{bmatrix} b_1 \\ b_2 \end{bmatrix} = \mathbb{S} \begin{bmatrix} a_1 \\ a_2 \end{bmatrix} \tag{2.111}$$

$$\mathbb{S} = \begin{bmatrix} \frac{Y_1-Y_2}{Y_1+Y_2} & \frac{2Y_2}{Y_1+Y_2} \\ \frac{2Y_1}{Y_1+Y_2} & \frac{Y_2-Y_1}{Y_1+Y_2} \end{bmatrix} \tag{2.112}$$

where $\mathbb{S}$ is called the scattering matrix.

It is noted that the scattering matrix we obtained is defined at the discontinuity. If the scattering matrix were to be defined at two different interfaces with distances l_1 and l_2 away from $z = 0$, as shown in Figure 2.9, then the original scattering matrix has to be multiplied by the two transition matrices at its left- and right-hand sides, which is given as

$$\begin{bmatrix} b_1(-l_1) \\ b_2(+l_2) \end{bmatrix} = \mathbb{TST} \begin{bmatrix} a_1(-l_1) \\ a_2(+l_2) \end{bmatrix} \tag{2.113}$$

$$\mathbb{T} = \begin{bmatrix} \mathrm{e}^{-\mathrm{j}\kappa_1 l_1} & 0 \\ 0 & \mathrm{e}^{-\mathrm{j}\kappa_2 l_2} \end{bmatrix} \tag{2.114}$$

Problem 2.2

Prove the equations in Equations (2.113) and (2.114).

The cascade of scattering matrices is not as simple as that of the transfer-matrix. For the cascade of the transfer-matrix, what we need is to multiply each of the matrices sequentially. Nevertheless, the combination rule for cascading two scattering matrices is relatively complicated. The proof of the combination rule can be seen in the textbook by Pozar (2005) and we will not repeat it again. We just list the results for your reference.

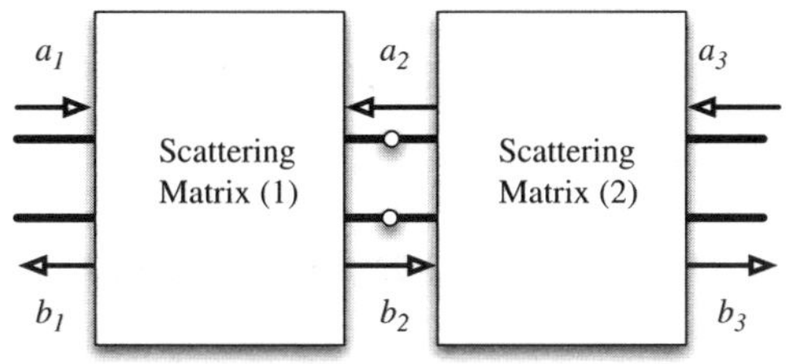

Figure 2.10 Cascade of the two scattering matrices

As shown in Figure 2.10, two scattering matrices, denoted $\mathbb{S}^{(1)}$ and $\mathbb{S}^{(2)}$, with the elements of the scattering matrices given in Equations (2.116) and (2.117), connect to each other. Although not shown here, the voltage and current must be continuous at the interface between the two scattering matrices. After some mathematical procedures, the scattering matrix of the overall structure (or combination rule) is given by

$$\begin{bmatrix} b_1 \\ b_3 \end{bmatrix} = \begin{bmatrix} s_{11}^{(1)} + \frac{s_{12}^{(1)} s_{11}^{(2)} s_{21}^{(1)}}{1 - s_{11}^{(2)} s_{22}^{(1)}} & \frac{s_{12}^{(1)} s_{12}^{(2)}}{1 - s_{11}^{(2)} s_{22}^{(1)}} \\ \frac{s_{21}^{(2)} s_{21}^{(1)}}{1 - s_{22}^{(1)} s_{11}^{(2)}} & s_{22}^{(2)} + \frac{s_{21}^{(2)} s_{22}^{(1)} s_{12}^{(2)}}{1 - s_{22}^{(1)} s_{11}^{(2)}} \end{bmatrix} \begin{bmatrix} a_1 \\ a_3 \end{bmatrix} \tag{2.115}$$

$$\begin{bmatrix} b_1 \\ b_2 \end{bmatrix} = \begin{bmatrix} s_{11}^{(1)} & s_{12}^{(1)} \\ s_{21}^{(1)} & s_{22}^{(1)} \end{bmatrix} \begin{bmatrix} a_1 \\ a_2 \end{bmatrix} \tag{2.116}$$

$$\begin{bmatrix} a_2 \\ b_3 \end{bmatrix} = \begin{bmatrix} s_{11}^{(2)} & s_{12}^{(2)} \\ s_{21}^{(2)} & s_{22}^{(2)} \end{bmatrix} \begin{bmatrix} b_2 \\ a_3 \end{bmatrix} \tag{2.117}$$

Problem 2.3

Derive the combination formula of the two scattering matrices shown in Equation (2.115).

For example, the transmission-line network depicted in Figure 2.6 contains $N - 1$ discontinuities, which are located at the interface between two adjacent dielectric layers. As usual, we first derive the scattering matrix at the discontinuity. Then the building-block approach will be exploited to cascade each of the scattering matrices for determining the scattering matrix of the overall transmission-line network.

Once the combination rule of the cascade of two scattering matrices is understood, we can move on to deal with the scattering matrix of a complex structure like that in Figure 2.6. Here, we redraw the transmission-line network of Figure 2.6 as shown in Figure 2.11. Each scattering matrix contains an interface between two adjacent

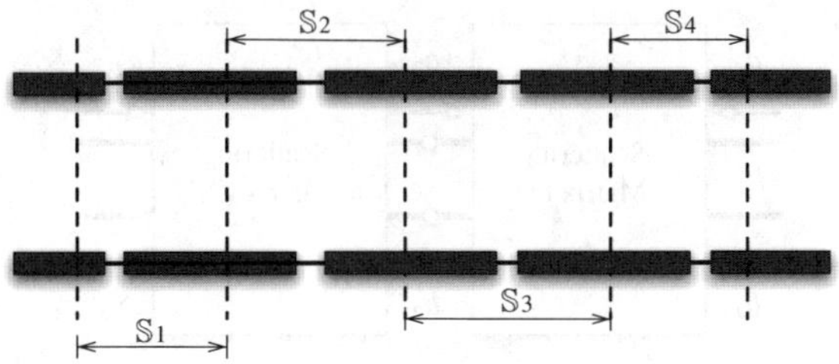

Figure 2.11 Cascade of the scattering matrices

transmission lines. Therefore, the overall scattering matrix can be readily obtained by recursively performing the combination rule in Equation (2.115). The mathematical notation of the scattering matrix of the network shown in Figure 2.11 can be expressed as follows:

$$\begin{bmatrix} b_{\mathrm{i}} \\ b_{\mathrm{o}} \end{bmatrix} = \mathbb{S}^{(1)} \otimes \mathbb{S}^{(2)} \otimes \mathbb{S}^{(3)} \otimes \mathbb{S}^{(4)} \ldots \mathbb{S}^{(N+1)} \begin{bmatrix} a_{\mathrm{i}} \\ a_{\mathrm{o}} \end{bmatrix} \tag{2.118}$$

where the $\otimes$ symbol represents the combination operation between two scattering matrices given in Equation (2.115). Parameters b_{i} and b_{o} are the reflected waves at the input and output interfaces, respectively, while a_{i} and a_{o} individually are the incident waves at the input and output interfaces. Moreover, the reflection and transmission coefficients can then be determined by the same procedures described in the previous two approaches.

2.3.3.1 Scattering Analysis for 1D Periodic Layers of Finite Thickness

In this subsection we will investigate the scattering characteristic of a 1D periodic structure, which is shown in Figure 2.12, consisting of a finite number of unit cells, each of which includes two dielectric layers with relative permittivity, permeability, and thickness, given respectively as $(\varepsilon_1, \mu_1, t_1)$ and $(\varepsilon_2, \mu_2, t_2)$. Such a structure is a typical one commonly used in microwave and optics engineering for serving as a band-stop filter. Here, we exploited the above-mentioned approaches of recursive-impedance, the transfer-matrix, and the scattering-matrix. Although not shown here, the same results were obtained; however, the computation resources are very different for the three approaches.

Considering the first example with the structure and incident parameters given in the caption of Figure 2.13, the top figure shows the reflectance and transmittance versus wavelength. The reflectance and transmittance are defined as the reflection power and transmitted power normalized to the incident power; therefore, they satisfy the power conservation law; that is, their sum should be equal to unity. As shown in this figure, there are three bands of total reflection. As is well known, they are due to the stop-band of the 1D periodic structure. Nevertheless, the physical insight of the total reflection

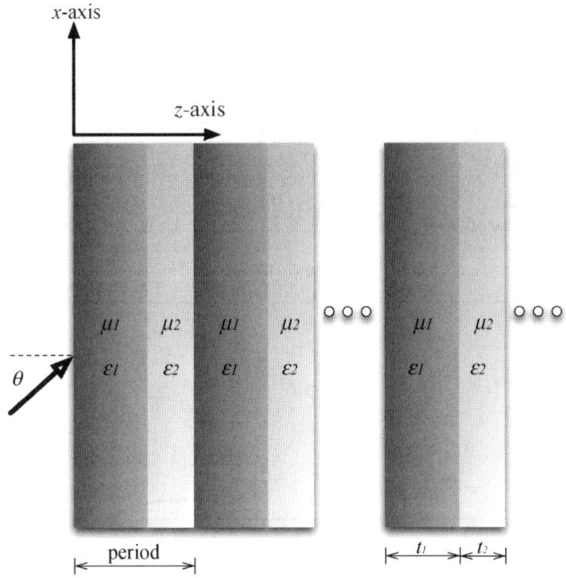

Figure 2.12 A 1D periodic structure

(or stop-band) occurring in these frequency bands is the point that we would discuss in this section.

As was described in the preceding subsection, since the 1D periodic structure is made up of a cascade of unit cells, the scattering characteristic of the overall structure relates to that of the unit cell. The eigenvalues in Equations (2.99) and (2.100) determine the propagation constants of the wave traveling therein. Additionally, their associated eigenvectors determine the characteristic impedances:

$$\mathbb{T}\begin{bmatrix} V_{\text{inp}}^{(1)} \\ I_{\text{inp}}^{(1)} \end{bmatrix} = \lambda_1 \begin{bmatrix} V_{\text{inp}}^{(1)} \\ I_{\text{inp}}^{(1)} \end{bmatrix} \tag{2.119}$$

$$\mathbb{T}\begin{bmatrix} V_{\text{inp}}^{(2)} \\ I_{\text{inp}}^{(2)} \end{bmatrix} = \lambda_2 \begin{bmatrix} V_{\text{inp}}^{(2)} \\ I_{\text{inp}}^{(2)} \end{bmatrix} \tag{2.120}$$

$$Z_{\text{c}}^{(1)} = \frac{V_{\text{inp}}^{(1)}}{I_{\text{inp}}^{(1)}} \tag{2.121}$$

$$Z_{\text{c}}^{(2)} = \frac{V_{\text{inp}}^{(2)}}{I_{\text{inp}}^{(2)}} \tag{2.122}$$

The characteristic impedance is chosen based on the rule that the real part of the impedance should be a positive number when the wave is propagating along the $+z$-axis.

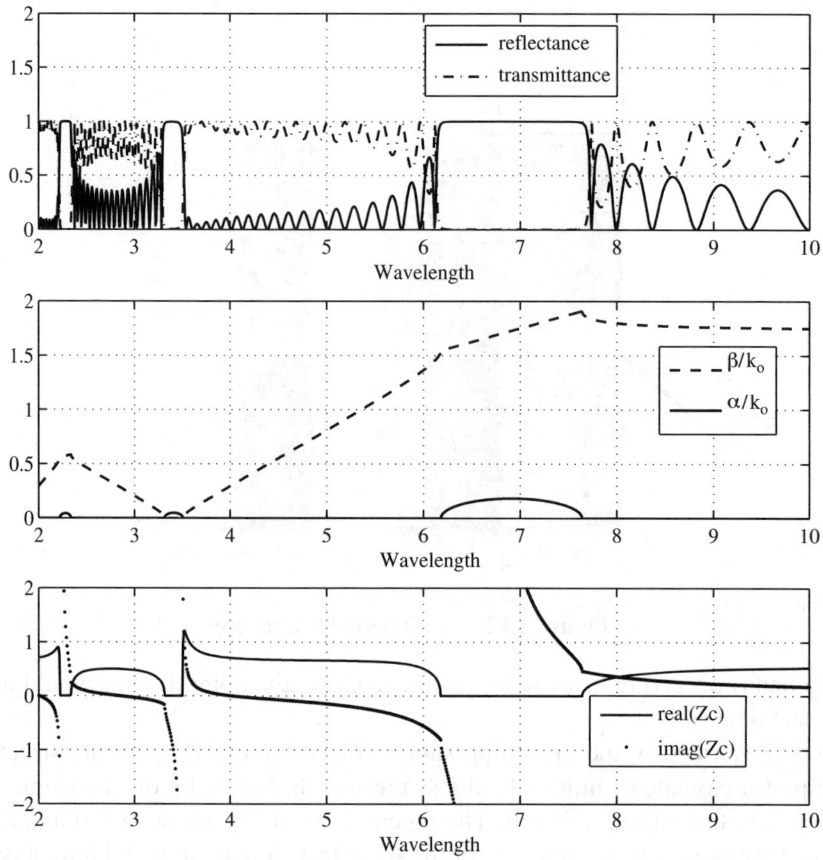

Figure 2.13 Scattering characteristics of a two-tone 1D periodic structure composed of 20 periods, each of which consists of two dielectric mediums with respective relative permittivities, permeabilities, and thicknesses of ($\varepsilon_1 = 2$, $\mu_1 = 1$, $t_1 = 1$) and ($\varepsilon_2 = -4$, $\mu_2 = -1$, $t_2 = 1$). The input and output regions are both air. The TE-polarized (E_y mode) plane with incident angle $\theta = 0°$ is impinging on the structure

The propagation constant (including the phase and attenuation constants) versus wavelength obtained from the eigenvalues is also drawn in Figure 2.13 (middle), where the dashed and solid lines individually represent the normalized phase and attenuation constants. Obviously, the nonzero attenuation constant in the three bands indicates that the wave is decaying, enabling a strong reflection when it propagates through 20 unit cells. That is the reason why it exhibits total reflection in the three stop-band regions. Notably, the regions with the nonzero attenuation and the vanished attenuation are the so-called stop-band and pass-band regions, respectively.

Incidentally, in the long-wavelength (low-frequency) and pass-band region, the effective refractive index (defined as β/k_o) is almost constant, since the period is smaller than that of the wavelength. Contrarily, in the high-frequency (short-wavelength) region, the dispersive effect is significant, as shown in the figure.

In addition to the propagation constant, the characteristic impedance also affects the scattering characteristics because that the reflection coefficient relates to the impedance mismatch. From the bottom figure in Figure 2.13, the stop-band regions exhibit reactive impedance, which is inductive or capacitive one; thus, the incident energy is reflected back to the input region.

Furthermore, we have carried out the scattering analysis for the 1D periodic structure composed of conventional dielectric material and a DNG medium. The negative relative permittivity and permeability are attached in the caption of Figure 2.14. Distinct from Figure 2.13, although the two mediums have very different electrical properties, the stop-band phenomenon is not obvious. Interestingly, the normalized attenuation constant is zero; the normalized phase constant is lower than that in the conventional

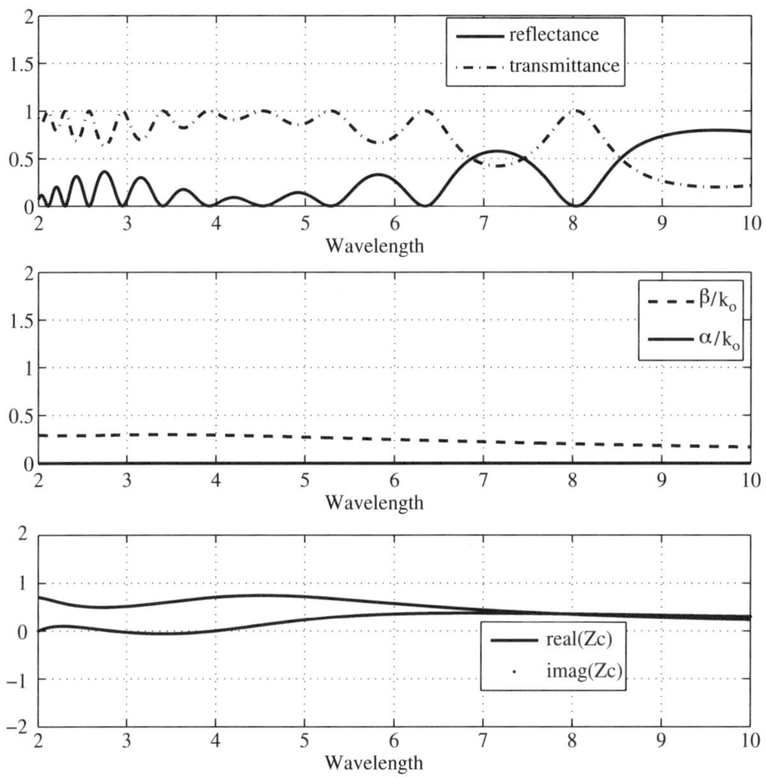

Figure 2.14 Scattering characteristics of a two-tone 1D periodic structure composed of 20 periods, each of which consists of two dielectric mediums with respective relative permittivities, permeabilities, and thicknesses of ($\varepsilon_1 = 2$, $\mu_1 = 1$, $t_1 = 1$) and ($\varepsilon_2 = -4$, $\mu_2 = -1$, $t_2 = 1$). The input and output regions are both air. The TE-polarized (E_y mode) plane with incident angle $\theta = 0°$ is impinging on the structure

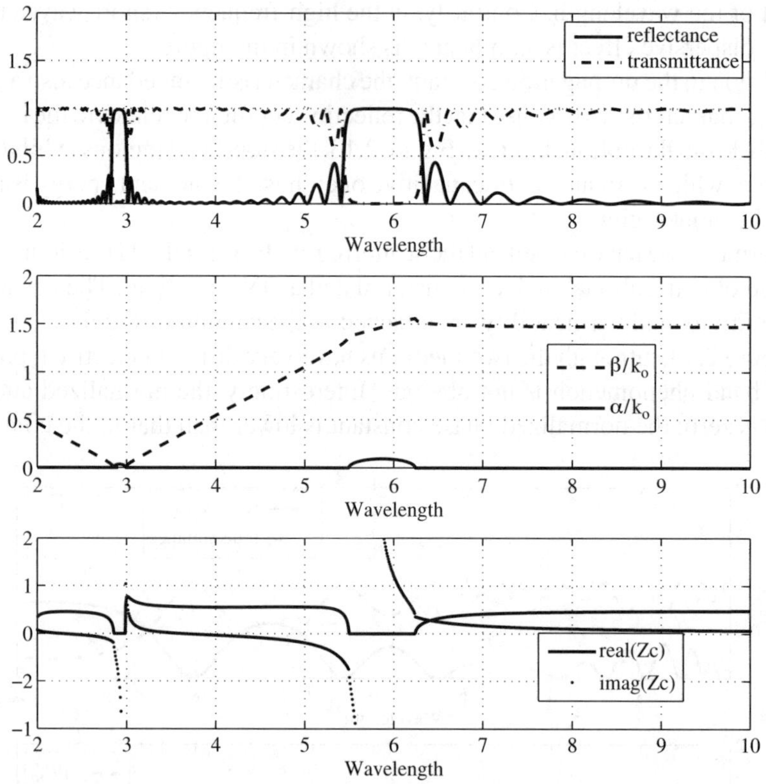

Figure 2.15 Scattering characteristics of a two-tone 1D periodic structure composed of 20 periods, each of which consists of two dielectric mediums with respective relative permittivities, permeabilities, and thicknesses of ($\varepsilon_1 = 2$, $\mu_1 = 1$, $t_1 = 1$) and ($\varepsilon_2 = -4$, $\mu_2 = -1$, $t_2 = 1$). The input and output regions are both air. The TM-polarized (H_y mode) plane with incident angle $\theta = 60°$ is impinging on the structure

dielectric medium due to the partial cancellation contributed by the negative propagation constant in the DNG medium. It is instructive to note that the stop-band is not obvious, since the characteristic impedance of the unit cell does not show a pure reactive one.

In addition to the TE-polarized incident wave, the scattering properties of the 1D periodic structure illuminated by a TM-polarized plane wave are also obtained and the results are shown in Figure 2.15. Although the TM-polarized wave has the same propagation constant in each layer compared with those in Figure 2.13, the different characteristic impedances in the two layers cause the different positions of stop-band regions. The regions with nonzero attenuation constant and reactive impedance shown in the middle and bottom panels can indeed be utilized to predict the occurrence of the stop-band.

Additionally, we choose the composite conventional dielectric material and DNG medium to let the wave have nearly the same magnitude of the propagation constant

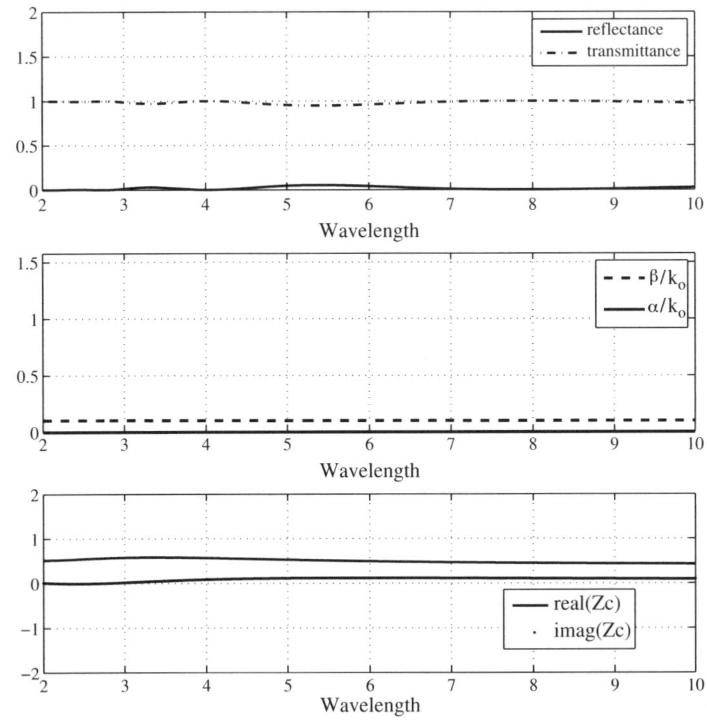

Figure 2.16 Scattering characteristics of a two-tone 1D periodic structure composed of 20 periods, each of which consists of two dielectric mediums with respective relative permittivities, permeabilities, and thicknesses of $(\varepsilon_1 = 2, \mu_1 = 1, t_1 = 1)$ and $(\varepsilon_2 = -2.5, \mu_2 = -1, t_2 = 1)$. The input and output regions are both air. The TM-polarized (H_y mode) plane with incident angle $\theta = 60^\circ$ is impinging on the structure

but opposite sign in the two regions. The results, including the scattering properties, normalized phase and attenuation constants, and the characteristic impedances, are shown in Figure 2.16. Interestingly, the almost zero effective refractive index indicates the near-zero phase delay angle when the wave is propagating through such a periodic structure. Moreover, owing to the good impedance match between the characteristic impedance of the unit cell and those in the input and output regions, the reflectance is inconsiderable in the scattering analysis.

2.4 Transverse Resonance Technique for Determining the Guiding Characteristics of Waves in Multiple Dielectric Layers

2.4.1 Transverse Resonance Technique

In addition to the scattering characteristics, in this section we will discuss the guiding characteristics of waveguide modes supported in a dielectric waveguide consisting of multiple uniform dielectric layers, which is commonly used in the optics community

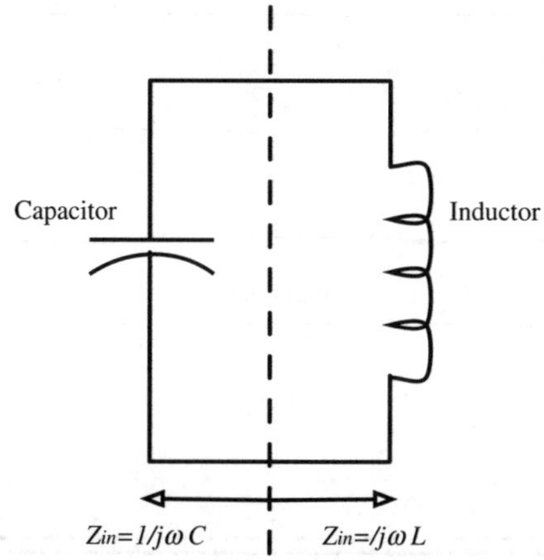

Figure 2.17 A resonant circuit

(Hunsperger 1985; Marcuse 1974; Tamir 1979). Before dealing with the guided-wave problem in a multilayered environment, we first revisit a commonly used electronic circuit: a resonant circuit containing an inductor and capacitor.

As shown in Figure 2.17, owing to the lossless lumped element and metal wire, the interchange of electric and magnetic energy keeps the total energy constant in this resonant circuit. Consequently, the sum of the input impedance looking into the left- and right-hand sides must vanish, resulting in the following equation:

$$\overleftarrow{Z}_{\text{in}} + \overrightarrow{Z}_{\text{in}} = 0 \tag{2.123}$$

$$\frac{1}{\text{j}\omega C} + \text{j}\omega L = 0 \tag{2.124}$$

From Equation (2.124), the resonant angular frequency is given as

$$\omega = \frac{1}{\sqrt{LC}} \tag{2.125}$$

Returning to the scattering characteristics of a multilayered structure shown in Figure 2.6, the relationship between the incident and reflected waves is

$$b = \frac{Z_{\text{in}} - Z_{\text{a}}}{Z_{\text{in}} + Z_{\text{a}}} a \tag{2.126}$$

where Z_{in} and Z_a are the input impedance looking into the structure and characteristic impedance in the surrounding medium, respectively. Parameters a and b are the incident- and reflected-wave amplitudes. For the guiding problem, there is no incident wave impinging from the surrounding medium; that is, $a = 0$. From the fundamental

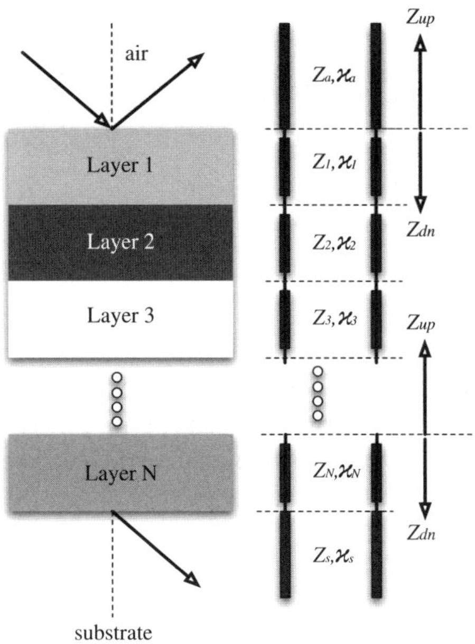

Figure 2.18 Transverse resonance condition in a multilayered structure

mathematics, the necessary and sufficient condition for the existence of reflected wave b must be

$$Z_{\text{in}} + Z_{\text{a}} = 0 \tag{2.127}$$

Equation (2.127) is termed the *transverse resonance equation*, which states that the energy is stored on the transverse plane perpendicular to the transmission-line direction, which is similar to the resonant circuit analogy depicted previously. More precisely, Equation (2.127) can be rewritten as

$$Z_{\text{up}} + Z_{\text{dn}} = 0 \tag{2.128}$$

where Z_{up} and Z_{dn} individually represent the input impedances looking into the upward and downward directions from the interface between the multilayered structure and surrounding medium. Furthermore, one can prove that the interface for implementing the transverse resonance equation can be arbitrarily chosen to be at any surface on the transverse plane within the multilayered structure, as shown in Figure 2.18.

2.4.2 Will Surface Waves be Supported in a Single Interface Environment?

Consider a single interface between two semi-infinite spaces, one filled with dielectric medium and the other with air, depicted in Figure 2.19. Is it possible to support a

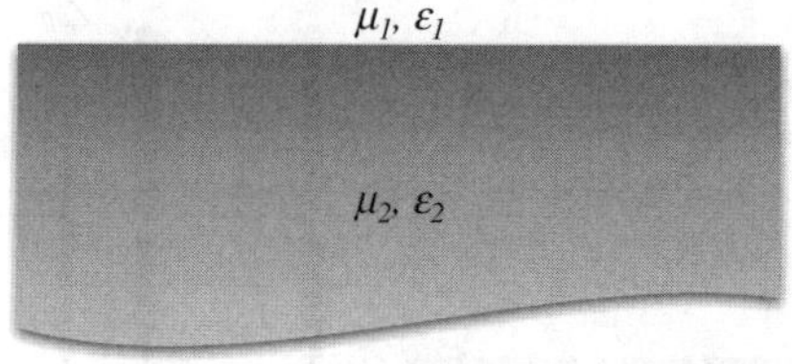

Figure 2.19 A single interface between two semi-infinite mediums

guided wave along the interface? In the following, we will discuss this problem based on the transverse resonance technique described previously.

First of all, we assume that the permittivity and permeability of the dielectric medium are $\kappa\varepsilon_o$ and μ_o, respectively, where κ is the relative dielectric constant. The characteristic impedances for the TE and TM waves are

$$Z_{dn} = \begin{cases} \frac{\omega\mu_o}{\kappa_d}; & \text{TE} \\ \frac{\kappa_d}{\omega\varepsilon_o\kappa}; & \text{TM} \end{cases} \tag{2.129}$$

$$Z_{up} = \begin{cases} \frac{\omega\mu_o}{\kappa_a}; & \text{TE} \\ \frac{\kappa_a}{\omega\varepsilon_o}; & \text{TM} \end{cases} \tag{2.130}$$

where

$$\kappa_a = \sqrt{k_o^2 - k_t^2} \tag{2.131}$$

$$\kappa_d = \sqrt{k_o^2\kappa - k_t^2} \tag{2.132}$$

and k_t is the propagation constant along the transverse plane, which is continuous across the two media. Parameters κ_a and κ_d are the propagation constants along the z-axis in the air and dielectric medium, respectively.

We first solve the transverse resonance equation for determining the TE-polarized guided-wave propagation constant k_t. Apparently, the solution does not exist. This means that no TE guided wave can be supported in the structure. On the other hand, there a solution exists when we evaluate the transverse resonance equation for TM polarization, and the solution is given as

$$k_t = k_o\sqrt{\frac{\kappa}{1+\kappa}} \tag{2.133}$$

For a commonly used dielectric medium, the relative dielectric constant is greater than unity. Therefore, the propagation constant on the transverse plane, k_t, is smaller than the wave number in free-space k_o. This indicates that the wave cannot be bound to the interface because the z-axis components of the propagation constant in air and dielectric medium are both real numbers. The necessary and sufficient conditions for the existence of guided waves in this structure are that κ and $1 + \kappa$ are both negative numbers. If this is the case, then k_t must be greater than k_o. For a medium with relative dielectric constant satisfying the condition $\kappa < -1$, it is a plasma medium. Here, such a guided wave is called a surface wave, since it decays along the z-axis in both semi-infinite spaces.

So, we know that the TM surface wave can be supported at the interface between two semi-infinite planes made up of plasma medium and air.

2.4.3 Single Dielectric Layer Backed with a PEC or PMC

Let us start with a simple but commonly used structure in microwave engineering. Figure 2.20 shows a conductor-backed dielectric layer (or grounded substrate) with relative dielectric constant ε_s and thickness t. The surrounding medium is assumed to be air. Such a structure is a typical one used extensively as a printed-circuit board. One may print the micro-strip line or the other planar transmission lines on the grounded substrate for transmitting the electromagnetic signal. However, we will not tackle the guided-wave problem of the planar transmission-line modes in this subsection. We will focus ourselves on the waveguide modes supported in the grounded-substrate structure. Such a waveguide mode is called a surface wave or substrate mode.

Additionally, Figure 2.21 shows the same dielectric slab; however, we replace the PEC by a PMC. The boundary condition of a PMC states that the tangential magnetic field vanishes on the surface, which is different from the PEC boundary condition with vanishing tangential electric field on the surface. Although PMCs do not exist in nature, the above case is actually commonly used in dealing with the dispersion relation of waveguide modes in a symmetric structure. For example, considering a dielectric slab immersed in a uniform medium, such as air; owing to the symmetry of the structure the electric field therein can be decomposed into two types: even and odd modes. For the even mode, the electric field is symmetric with respect to the bisection plane; thus, an open-circuit boundary condition is placed on the bisection plane. On

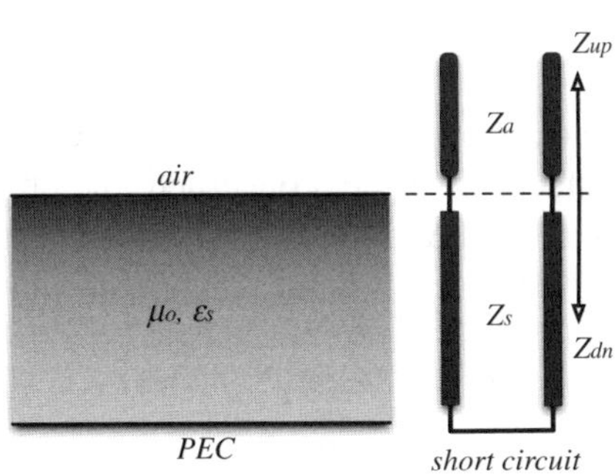

Figure 2.20 A dielectric slab backed with a PEC

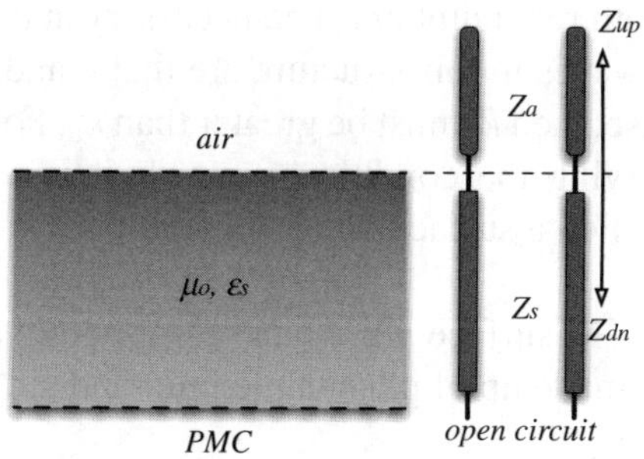

Figure 2.21 A dielectric slab backed with a PMC

the other hand, the short-circuit boundary condition is applied on the bisection plane to obtain the anti-symmetry field distribution. In Figures 2.20 and 2.21, we draw the structure together with its transmission-line network representation for SCB and OCB, respectively.

Returning to Figures 2.20 and 2.21, the schematic diagram of the transmission-line network representation for evaluating the transverse resonance equation is shown. The transmission-line representation of the dielectric layer is short-circuit or open-circuit at its bottom end and is connecting to the wave impedance in air characterized by Z_a at its top end. Therefore, the transverse resonance equation is written as

$$Z_{up} + Z_{dn} = 0 \tag{2.134}$$

$$Z_{up} = Z_a \tag{2.135}$$

$$Z_{dn} = jZ_s \tan \kappa_s t; \quad \text{short circuit} \tag{2.136}$$

$$Z_{dn} = -jZ_s \cot \kappa_s t; \quad \text{open circuit} \tag{2.137}$$

$$\kappa_s = k_o\sqrt{\varepsilon_s - \varepsilon_{eff}} \tag{2.138}$$

$$\varepsilon_{eff} \xrightarrow{\text{def}} \left(\frac{k_t}{k_o}\right)^2 \tag{2.139}$$

In Equation (2.138) we define a new parameter, ε_{eff}, called the effective dielectric constant. As far as the surface waves are concerned, the effective dielectric constant must be greater than ε_a and smaller than ε_s. Consequently, the propagation constant along the z-axis in the air region, κ_a, is a pure imaginary number written as

$$\kappa_a = -jk_o\sqrt{|\varepsilon_a - \varepsilon_{eff}|} \tag{2.140}$$

Since the wave is attenuating along the z-axis in the air region. Contrarily, the z-axis component of the propagation constant κ_s is a real number in the substrate region. This means that the waves in the substrate region form a standing wave pattern along the transverse plane perpendicular to the direction of surface-wave propagation. The dispersion equations for the TE and TM surface waves are as follows:

$$\tan\frac{2\pi t}{\lambda}\sqrt{\varepsilon_s-\varepsilon_{eff}}+\sqrt{\frac{\varepsilon_s-\varepsilon_{eff}}{\varepsilon_{eff}-\varepsilon_a}}=0 \quad \text{TE-SCB} \tag{2.141}$$

$$\tan\frac{2\pi t}{\lambda}\sqrt{\varepsilon_s-\varepsilon_{eff}}-\frac{\varepsilon_s}{\varepsilon_a}\sqrt{\frac{\varepsilon_{eff}-\varepsilon_a}{\varepsilon_s-\varepsilon_{eff}}}=0 \quad \text{TM-SCB} \tag{2.142}$$

$$\tan\frac{2\pi t}{\lambda}\sqrt{\varepsilon_s-\varepsilon_{eff}}-\sqrt{\frac{\varepsilon_{eff}-\varepsilon_a}{\varepsilon_s-\varepsilon_{eff}}}=0 \quad \text{TE-OCB} \tag{2.143}$$

$$\tan\frac{2\pi t}{\lambda}\sqrt{\varepsilon_s-\varepsilon_{eff}}+\frac{\varepsilon_a}{\varepsilon_s}\sqrt{\frac{\varepsilon_s-\varepsilon_{eff}}{\varepsilon_{eff}-\varepsilon_a}}=0 \quad \text{TM-OCB} \tag{2.144}$$

Since the effective dielectric constant is between ε_a and ε_s, we may solve Equations (2.141)–(2.144) using the Newton–Raphson method for searching the roots of the nonlinear equations. Alternatively, we may consider the normalized thickness t/λ as the unknown to be solved and directly take the inverse of the trigonometric function for a given effective dielectric constant. Thus, we may obtain the normalized thickness expressed as a function of the effective dielectric constant. Moreover, the cutoff condition ($\varepsilon_{eff}=\varepsilon_a$) of the TE- and TM-polarized surface modes can be determined as follows:

$$\left.\frac{t}{\lambda}\right|_{\text{cutoff}}^{\text{TE-SCB}}=\frac{2n-1}{4\sqrt{\Delta\varepsilon}} \quad n=1,2,3,\ldots\infty \tag{2.145}$$

$$\left.\frac{t}{\lambda}\right|_{\text{cutoff}}^{\text{TM-SCB}}=\frac{n}{2\sqrt{\Delta\varepsilon}} \quad n=0,1,2,3,\ldots\infty \tag{2.146}$$

$$\left.\frac{t}{\lambda}\right|_{\text{cutoff}}^{\text{TE-OCB}}=\frac{n}{2\sqrt{\Delta\varepsilon}} \quad n=0,1,2,3,\ldots\infty \tag{2.147}$$

$$\left.\frac{t}{\lambda}\right|_{\text{cutoff}}^{\text{TM-OCB}}=\frac{2n-1}{4\sqrt{\Delta\varepsilon}} \quad n=1,2,3,\ldots\infty \tag{2.148}$$

$$\Delta\varepsilon=\varepsilon_s-\varepsilon_a \tag{2.149}$$

Problem 2.4

Prove Equations (2.141)–(2.144).

Figure 2.22 shows the dispersion relation of the surface waves supported in the structures shown in Figures 2.20 and 2.21. The vertical axis represents the effective dielectric constant, which is defined as the ratio of k_t/k_o, while the horizontal axis is

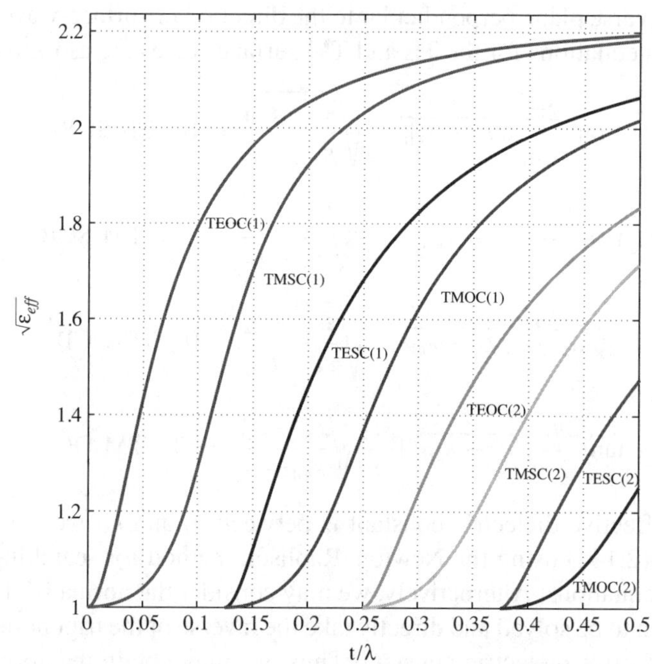

Figure 2.22 Dispersion relation of the surface waves supported in the PEC- and PMC-backed dielectric slab with relative dielectric constant $\varepsilon_s = 5.0$

the normalized thickness. defined as the ratio of substrate thickness to the operational wavelength. In fact, they have an infinite number of surface modes; however, we just plot some of them to illustrate their physical consequences. From this figure, we know that TM-SCB and TE-OCB surface waves have zero cutoff frequency; namely, it can propagate at a very low frequency down to dc (direct current). It may be conjectured that once the TM-polarized surface waves are excited via the planar transmission-line, micro-strip antenna, or via-hole in a grounded-substrate environment, they can conduct from dc to any frequency. Therefore, they propagate in the substrate and result in electromagnetic coupling between circuits, causing electromagnetic interference problems. Additionally, this figure allows us to know how many and what types of surface modes are excited at a certain frequency. Namely, by drawing a vertical line at the given normalized thickness, you can determine the number of surface modes by counting the number of intersection points between this vertical line and the dispersion curves in Figure 2.22.

To understand the effect of ε_s on the dispersion relation of surface modes, we calculate the second case with a high relative dielectric constant, $\varepsilon_s = 26$. As shown in Figure 2.23, compared with the dispersion curves shown in Figure 2.22, the dispersion

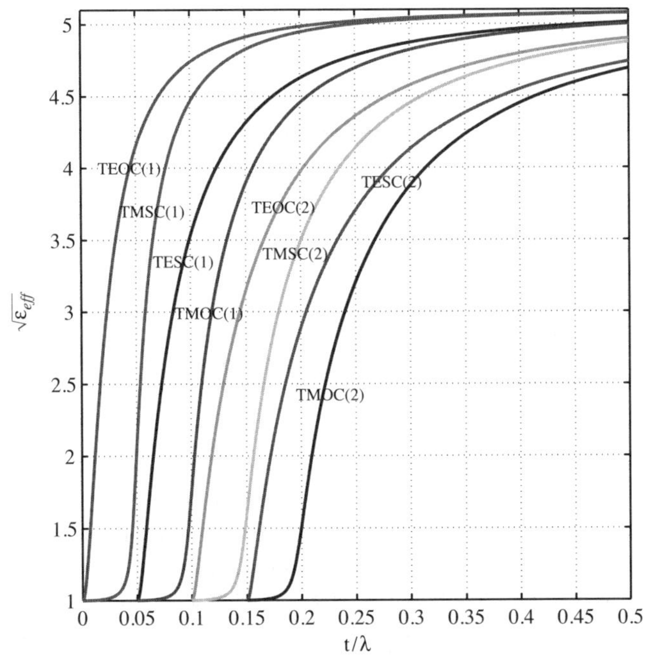

Figure 2.23 Dispersion relation of the surface waves supported in the PEC- and PMC-backed dielectric slab with relative dielectric constant $\varepsilon_s = 26$

curves seem to squeeze together. Therefore, for the same operation frequency, a large number of surface modes are excited.

2.4.4 *Mode Dispersion Relation of a Closed Structure Consisting of Dielectric Layers*

In addition to resolving the mode dispersion of multiple dielectric layers in an open structure, the transverse resonance technique can also be employed to solve the mode dispersion relation for a closed structure consisting of dielectric layers, such as stratified mediums in a parallel-plate-waveguide (ppwg). In this section, the ppwg filled with multiple dielectric layers will be taken as an example to discuss the mode dispersion relations and their eigenfunction profile.

Investigation of a cloaking structure that is able to make a metallic object invisible under the illumination by microwaves or millimeter waves in a waveguide environment has attracted considerable attention recently. Some of the cloaking structures made up of anisotropic or non-homogeneous artificial materials need specific material properties, such as negative permittivity and permeability, which are hard to obtain in practice. Although some so-called metamaterials made of periodic structures were

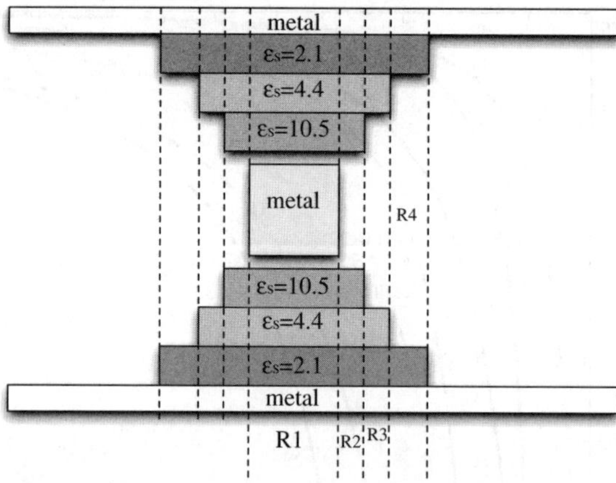

Figure 2.24 A cloaking structure consisting of composite dielectrics: the top view of the structure configuration. The metal block is the object to be cloaked, while the composite dielectric consisting of three dielectric layers attached to the waveguide sidewall is termed the cloaking structure. The two metal plates outside the cloaking structure as well as the metal block are the side walls of a rectangular waveguide, which is commonly used in microwave engineering. Here, we assumed that the field has no variation along the height direction of the waveguide, so that the ppwg filled with the dielectric layers in a 2D environment is considered in the theoretical analysis. For the mode dispersion analysis, each constituent region, which is composed of parallel metallic plates and parallel dielectric layers, is denoted by Ri, where $i = 1, 2, 3, 4$. In order to understand the mode dispersion relation in each constituent region, the waveguide filled with dielectric layers is infinite in extent along the guiding axis. Figures 2.25–2.28 show the structure configuration (top view) of the constituent regions

reported to be successful in achieving the goal, they usually need a complex fabrication process and are operated in a very narrow band. In the following, I will review a novel broadband cloaking structure consisting of composite dielectrics and use it as an example to calculate the dispersion characteristics of waveguide modes (Hwang and Chin 2011). Nevertheless, we will not target the scattering analysis for the overall structure, since the theory and numerical technique for dealing with the discontinuity using the modal expansion method (or mode-matching method) is not yet delivered.

Figure 2.24 shows the structure configuration (top view) of the metallic block and the cloaking structure put inside a metallic rectangular waveguide. Here, because the structure and medium along the waveguide height direction are uniform, the field variation along that direction can be neglected for the waveguide operated in its fundamental mode, say TE_{10} mode. Consequently, the 3D vector boundary-value problem can be reduced to that of the 2D scalar one.

As was indicated in the Figure 2.24, the structure was partitioned into four regions, characterized by R1, R2, R3, and R4, each of which consists of a stack of dielectric layers. With regard to the structures in Figures 2.25–2.28, since the modes supported

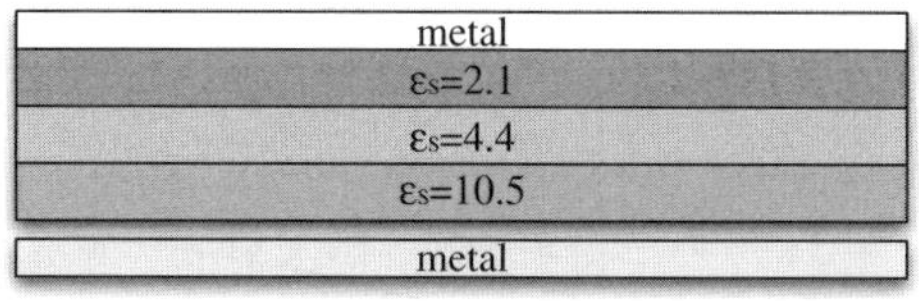

Figure 2.25 Structure configuration in R1

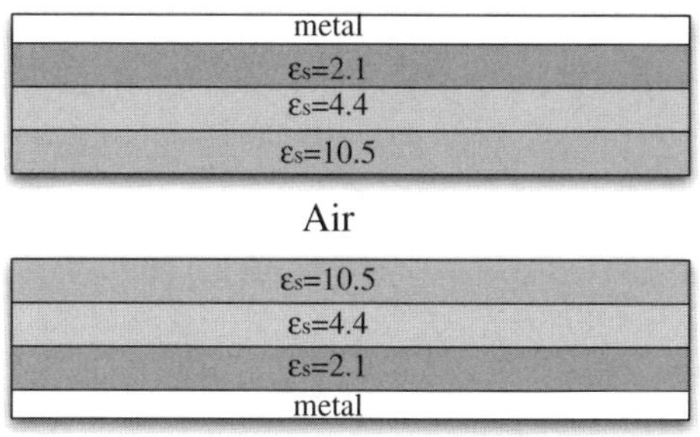

Figure 2.26 Structure configuration in R2

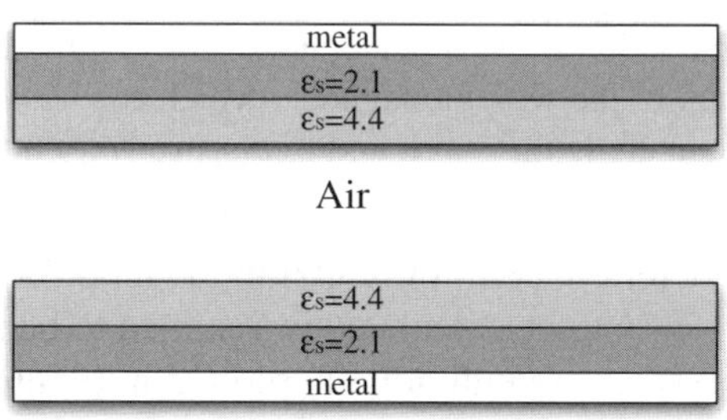

Figure 2.27 Structure configuration in R3

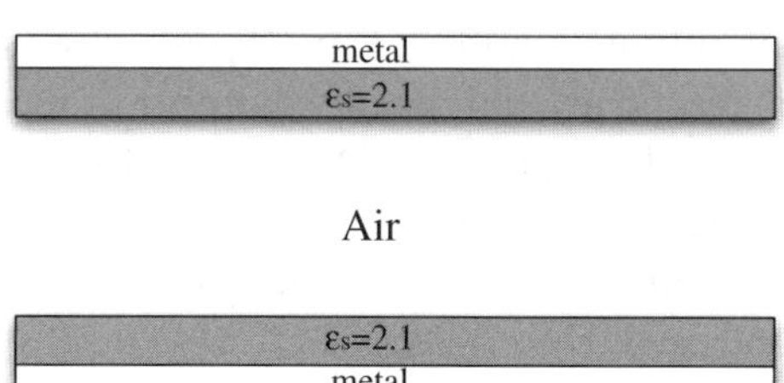

Figure 2.28 Structure configuration in R4

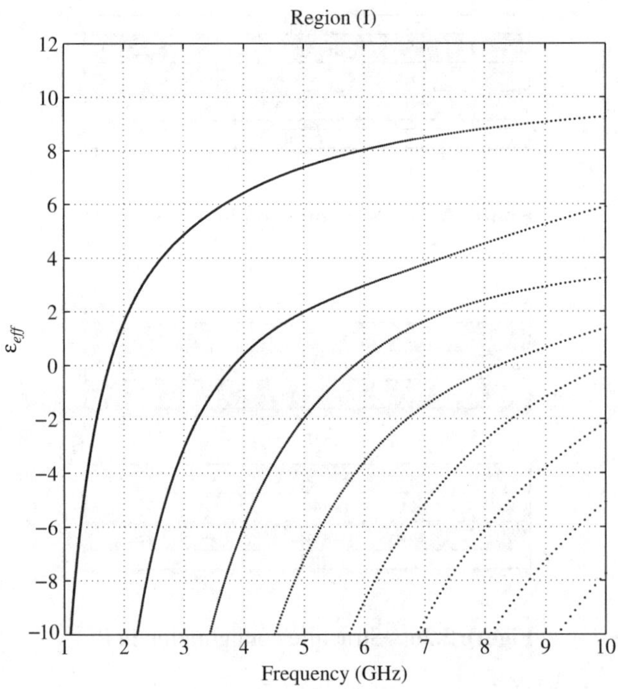

Figure 2.29 Dispersion relation of waveguide modes for Figure 2.25; parameter ε_{eff} is the effective dielectric constant of the waveguide mode, which equals $(k_z/k_o)^2$, where k_z and k_o are the propagation constant along the z-axis and free-space wave number, respectively

in each region are our major concern for realizing the mode conversion through each constituent region, each case to be calculated is assumed to be infinite in extent along the waveguide axis, z-axis, as are individually plotted in Figures 2.29–2.32.

Figure 2.29 shows the dispersion relation of the waveguide modes supported in the first region (R1). Because this area contains the metal object, only half of the structure needs to be considered. Parameter ε_{eff} is defined as the effective dielectric constant of the waveguide mode with the propagation constant defined as $k_z = k_o\sqrt{\varepsilon_{\text{eff}}}$. As is well known, there are an infinite number of modes in a closed waveguide; however, for a certain operational frequency, only a finite number of modes can propagate; the other modes are below cutoff (evanescent). The cutoff frequency of each mode can be easily read from the intersection point between the dispersion curve and $\varepsilon_{\text{eff}} = 0$. From this figure, we know that only the fundamental mode exists before the second mode is excited at around 3.9 GHz.

Figure 2.30 shows the dispersion curves in the second region (R2) with three dielectric layers attached to the sidewall of the waveguide. Owing to the symmetry of the structure with respect to the waveguide axis, the electric field can be decomposed into symmetric and anti-symmetric fields, which correspond to OCB and SCB modes,

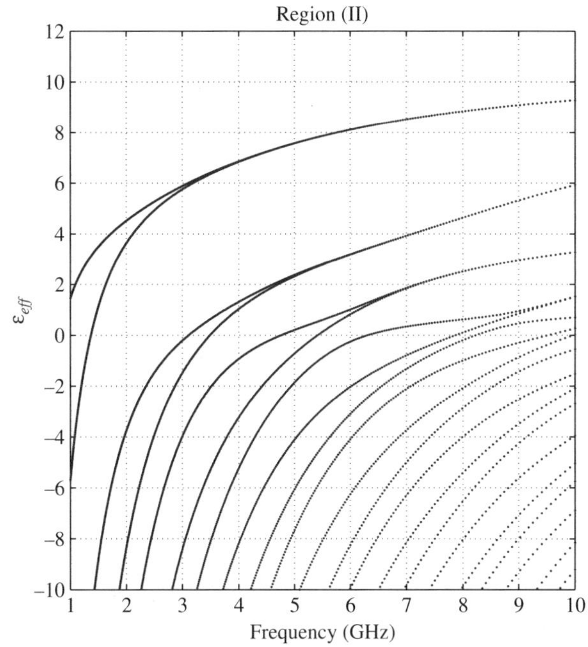

Figure 2.30 Dispersion relation of waveguide modes for Figure 2.26

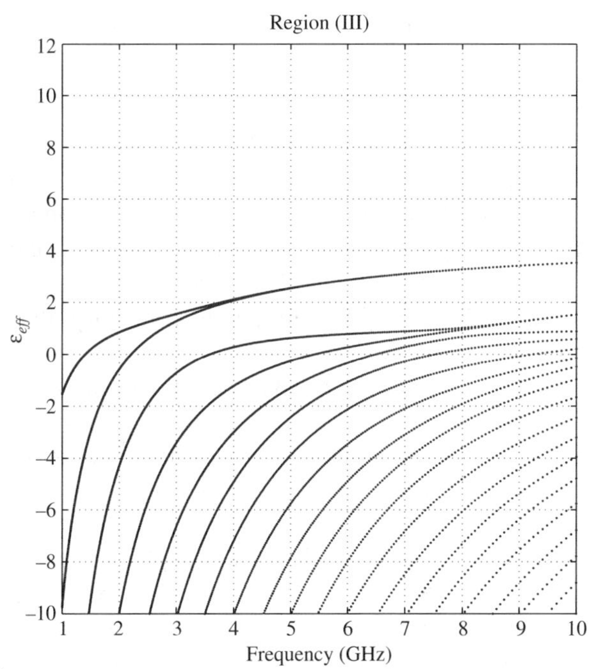

Figure 2.31 Dispersion relation of waveguide modes for Figure 2.27

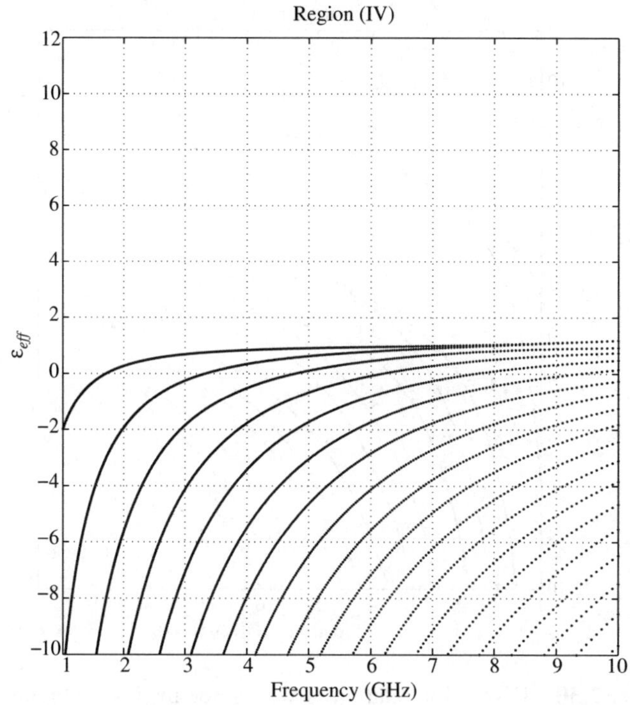

Figure 2.32 Dispersion relation of waveguide modes for Figure 2.28

respectively. Therefore, distinct from Figure 2.29 without symmetry, in this plot the OCS and SCB modes appear at the same time. It is obvious to see that the same order of the OCB and SCB modes converge to each other; for instance, the fundamental modes coincide at a frequency around 3.0 GHz having an effective dielectric constant near 6.0. With $\varepsilon_{\text{eff}} \geq 6$, both the OCB and SCB modes in the air region decay to zero at the termination plane (PEC or PMC). One may conjecture that the fundamental mode dispersion curves coincide for $\varepsilon_{\text{eff}} \geq 6$.

In addition, the dispersion relations of waveguide modes in the environments of R3 and R4 are also plotted in Figures 2.31 and 2.32, respectively. From these figures, we know that the region filled with high dielectric-constant medium, such as R1 and R2, can support more propagating modes than the others filled with low dielectric-constant media.

In addition to the mode dispersion diagram, we plot the mode function of the fundamental mode at 3 GHz in R1 in Figure 2.33. Owing to the boundary condition, the tangential electric field (or mode function) must vanish at the two ends. Significantly, the strongest field strength of the fundamental mode occurs near the sidewall of the waveguide rather than around the center region.

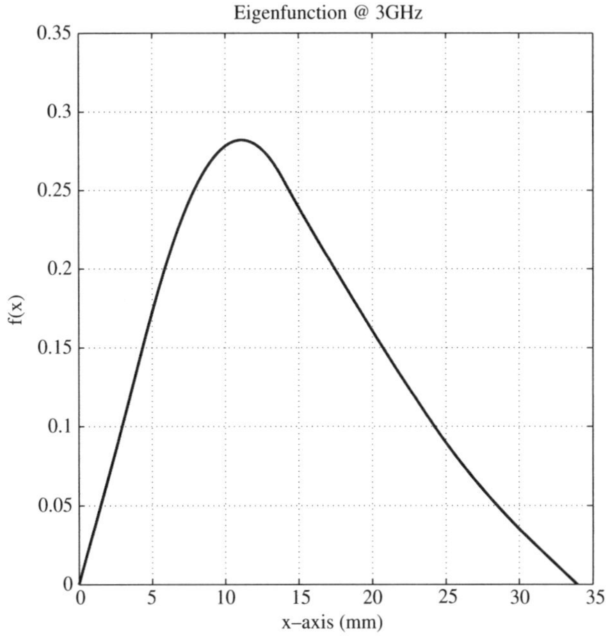

Figure 2.33 Eigenfunction of the fundamental mode at 3 GHz; structure configuration shown in Figure 2.25. The waveguide side wall position is at $x = 0$ mm

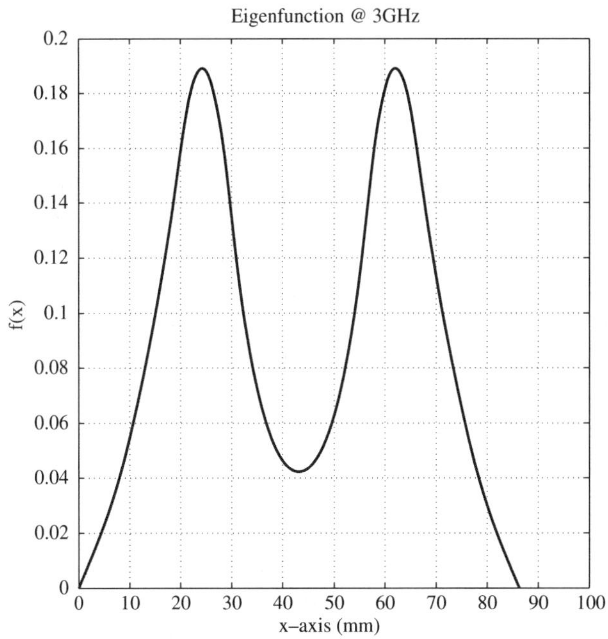

Figure 2.34 Eigenfunction of the fundamental mode at 3 GHz; structure configuration shown in Figure 2.26

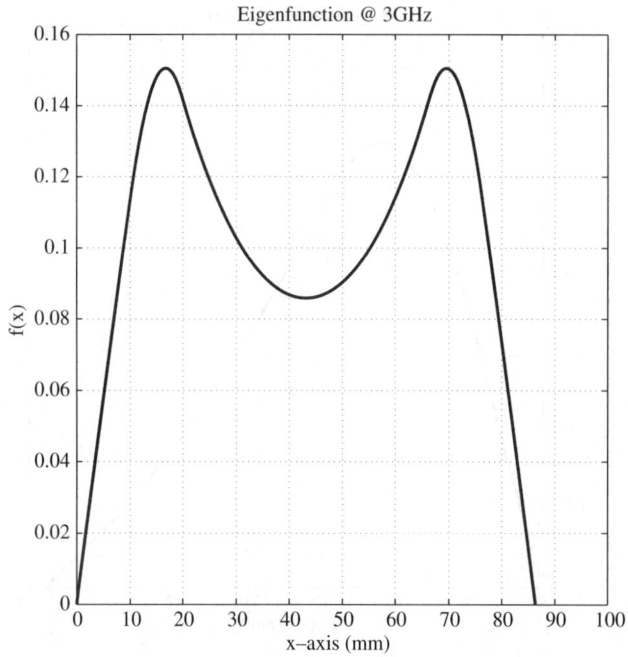

Figure 2.35 Eigenfunction of the fundamental mode at 3 GHz; structure configuration shown in Figure 2.27

Figure 2.34 demonstrates the eigenfunction of the fundamental OCB mode at 3 GHz in R2. At 3 GHz the effective dielectric constant is near 6.0, which means that the transverse electric-field component survives in the dielectric medium with $\varepsilon_s = 10.5$; outside this region it decays exponentially, as shown in the figure. Additionally, the electric-field strength in the central region decays dramatically. As a consequence, if such a mode is excited, the field is concentrated in the cloaking structure instead of the metal block at the center of the waveguide.

Figure 2.35 demonstrates the eigenfunction of the fundamental mode at 3 GHz in R3. Since the effective dielectric constant is less than that of the two dielectric mediums but greater than unity, the strongest field strength concentrates in the dielectric mediums and decays in the air region. Furthermore, in the presence of the boundary condition, the field has to decay to zero at $x = 0$, so that the field profile shall have the shape in this figure.

In Figure 2.36, the effective dielectric constant at 3 GHz is smaller than unity. Thus, the electric field survives in both the dielectric medium with $\varepsilon_s = 2.1$ and air. However, in the presence of the PEC boundary condition, the field inside the dielectric decays to zero at the boundary. The field profile is almost similar to that of the fundamental TE mode, which has the maximum field strength in the central region, in a ppwg.

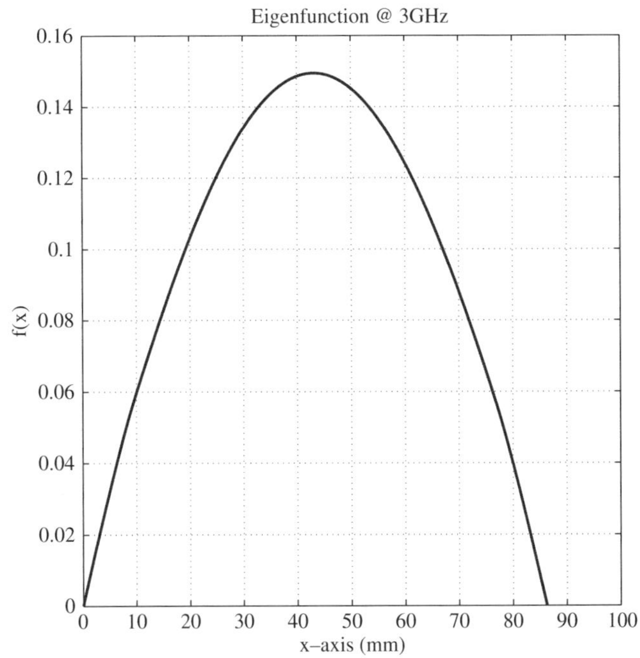

Figure 2.36 Eigenfunction of the fundamental mode at 3 GHz; structure configuration shown in Figure 2.28

Although a scattering analysis is not carried out, we may imagine why such a cloaking structure consisting of composite dielectrics works. Once the fundamental TE mode is excited around 3 GHz, only the OCB fundamental mode is excited in each constituent region. The incident TE_1 mode with maximum electric-field strength in the central region, as shown in Figure 2.36, gradually rearranges its field distribution to one with its maximum located around the region away from the center, as shown in Figure 2.33. Through the process of mode conversion, the incident wave does not hit the metal object, bypassing it by changing its propagation path.

Appendix: Dyadic Definition and Properties

A dyadic is a tensor written in a special notation, formed by juxtaposing pairs of vectors. Each component of a dyadic is a dyad. A dyad is the juxtaposition of a pair of basis vectors and a scalar coefficient. For example:

$$\underline{A} = A_x \underline{x}_o + A_y \underline{y}_o + A_z \underline{z}_o \tag{2A.1}$$

$$\underline{B} = B_x \underline{x}_o + B_y \underline{y}_o + B_z \underline{z}_o \tag{2A.2}$$

$$\underline{AB} = A_x B_x \underline{x}_o \underline{x}_o + A_x B_y \underline{x}_o \underline{y}_o + A_x B_z \underline{x}_o \underline{z}_o + A_y B_x \underline{y}_o \underline{x}_o + A_y B_y \underline{y}_o \underline{y}_o$$
$$+ A_y B_z \underline{y}_o \underline{z}_o + A_z B_x \underline{z}_o \underline{x}_o + A_z B_y \underline{z}_o \underline{y}_o + A_z B_z \underline{z}_o \underline{z}_o \quad \text{(2A.3)}$$

The above dyadic can also be represented as a 3 × 3 matrix:

$$\underline{AB} = \begin{pmatrix} A_x B_x & A_x B_y & A_x B_z \\ A_y B_x & A_y B_y & A_y B_z \\ A_z B_x & A_z B_y & A_z B_z \end{pmatrix} \quad \text{(2A.4)}$$

The 3D unit dyadic is defined as

$$\overline{\overline{I}} = \underline{x}_o \underline{x}_o + \underline{y}_o \underline{y}_o + \underline{z}_o \underline{z}_o \quad \text{(2A.5)}$$

You may also consider such a unit dyadic as a 3 × 3 matrix with ones on the main diagonal and zeros, which is termed the identity matrix and is given below:

$$\overline{\overline{I}} = \begin{pmatrix} 1 & 0 & 0 \\ 0 & 1 & 0 \\ 0 & 0 & 1 \end{pmatrix} \quad \text{(2A.6)}$$

Additionally, the 3D dyadic of A is

$$\overline{\overline{A}} = \underline{AA} = \sum_{i=1}^{3} \sum_{j=1}^{3} A_i A_j \underline{ij} \quad \text{(2A.7)}$$

where index i (j) = 1, 2, and 3 represent x, y, and z, respectively.

Some important properties of the dyadic are listed below for reference:

$$\overline{\overline{I}} \cdot \underline{C} = C_x \underline{x}_o + C_y \underline{y}_o + C_z \underline{z}_o \quad \text{(2A.8)}$$

$$\overline{\overline{I}} \cdot \overline{\overline{I}} = \overline{\overline{I}} \quad \text{(2A.9)}$$

$$\overline{\overline{A}} \cdot \overline{\overline{A}} = \underline{AA} \cdot \underline{AA} = A^2 \underline{AA} = A^2 \overline{\overline{A}} \quad \text{(2A.10)}$$

References

Engheta, N. and Ziolkowski, R. W. (1964) *Metamaterials: Physics and Engineering Explorations*, John Wiley & Sons.

Felsen, L. B. and Marcuvitz, N. (1973) *Radiation and Scattering of Waves*, Prentice-Hall, Englewood Cliffs, NJ.

Hunsperger, R. G. (1985) *Integrated Optics: Theory and Technology*, 2nd edn, Springer-Verlag.

Hwang, R.-B. and Chin, C.-Y. (2011) Broadband cloaking using composite dielectrics. *AIP Advances* **1**, 012112. DOI: 10.1063/1.3562891.
Marcuse, D. (1974) *Theory of Dielectric Optical Waveguides*, Academic Press, New York, NY.
Pozar, D. M. (2005) *Microwave Engineering*, 3rd edn, John Wiley & Sons.
Smith, D. R., Pendry, J. B., and Wiltshire, M. C. K. (2004) Metamaterials and negative refractive index. *Science* **6**, 788–792.
Tamir, T. (ed.) (1979) *Integrated Optics: Topics in Applied Physics*, Springer, Berlin.

Further Reading

Born, M. and Wolf, E. (1964) *Principles of Optics*, Macmillan, New York, NY.

3

One-Dimensional Periodic Medium

The objective of this chapter is to establish a comprehensive and in-depth understanding of the wave propagation in a 1D periodic medium. The Bloch–Floquet theorem will be first introduced for setting up a framework for the fundamental mathematical analysis. Furthermore, a first-order approximation of a dielectric medium with sinusoidal variation of its relative dielectric constant will be employed as an example for illustrating the phase and dispersion relation of a wave supported in the periodic medium. The simple graphic method will also be introduced for predicting the physical phenomenon, such as contraflow interactions between space harmonics occurring in a 1D periodic medium. Additionally, a complex periodic medium consisting of multiple dielectric layers in a unit cell (period) will also be studied using two approaches: the Fourier-method and the modal-transmission-line method. In the former approach, the electric and magnetic fields are expressed in terms of a Fourier series expansion. In the latter, the rigorous modal solutions are directly determined by solving the transmission-line network subject to periodic boundary conditions.

3.1 Bloch–Floquet Theorem

If there is a periodicity along the x-axis, the wave function satisfies the following Bloch–Floquet condition:

$$\psi(x + d_x) = \mathrm{e}^{-\mathrm{j}k_x d_x}\psi(x) \tag{3.1}$$

This equation states that the field solutions, in an infinite 1D periodic environment, at two positions separated by a distance of one period d_x differ by a phase delay. The phase angle is equal to the product of the propagation constant and the period along

Periodic Structures: Mode-Matching Approach and Applications in Electromagnetic Engineering, First Edition.
Ruey-Bing (Raybeam) Hwang.

the periodicity direction, such as the x-axis in Equation (3.1). Furthermore, we may prove that the wave function can be expressed as

$$\psi(x) = \mathrm{e}^{-\mathrm{j}k_x x} p(x) \tag{3.2}$$

where $p(x)$ is a periodic function along the x-axis with period d_x, which satisfies the condition $p(x + d_x) = p(x)$. Consequently, the function in Equation (3.2) means that the wave solution in a 1D periodic medium can be expressed as the product of a plane wave envelope function and a periodic function. The function in Equation (3.2) is the so-called Floquet solution. Furthermore, if the periodic function is expressed in terms of a Fourier series expansion, Equation (3.2) can be rewritten as

$$\psi(x) = \mathrm{e}^{-\mathrm{j}k_x x} \sum_{n=-\infty}^{+\infty} p_n \exp\left(-\mathrm{j}n\frac{2\pi x}{d_x}\right) \tag{3.3}$$

3.2 Eigenwave in a 1D Holographic Grating

In this section we will first formulate the electric and magnetic fields in a periodic medium with relative dielectric function

$$\varepsilon(x) = \varepsilon_{\mathrm{g}}\left(1 + 2\delta\cos\frac{2\pi x}{d_x}\right) \tag{3.4}$$

where ε_{g} and δ are the average dielectric constant and the modulation index of the periodic dielectric function, and d_x is the period along the x-axis. Here, the structure on the y–z plane is uniform; thus, the waves can be individually decomposed into those of TE^x ($E_x = 0$) and TM^x ($H_x = 0$).

We first deal with the TE^x-polarized wave. Let us assume that the electric-field vector is along the y-axis and independent of y ($k_y = 0$), which can be written as follows based on the method of separation of variables:

$$E_y(x, z) = \phi(x)\,\mathrm{e}^{-\mathrm{j}k_z z} \tag{3.5}$$

where $\phi(x)$ is the x-dependent function and k_z is the propagation constant along the z-axis. By substituting Equation (3.5) into the Maxwell equation

$$\nabla \times \underline{E} = -\mathrm{j}\omega\mu\underline{H} \tag{3.6}$$

we may determine the magnetic field components:

$$H_x(x, z) = -\frac{k_z}{\omega\mu_{\mathrm{o}}}\phi(x)\,\mathrm{e}^{-\mathrm{j}k_z z} \tag{3.7}$$

$$H_z(x, z) = -\frac{1}{\mathrm{j}\omega\mu_{\mathrm{o}}}\frac{\mathrm{d}\phi(x)}{\mathrm{d}x}\,\mathrm{e}^{-\mathrm{j}k_z z} \tag{3.8}$$

By substitution of Equations (3.5), (3.7), and (3.8) into the Maxwell equation in Equation (3.9) we may obtain the second-order differential equation in Equation (3.10):

$$\nabla \times \underline{H} = \mathrm{j}\omega\varepsilon_0\varepsilon(x)\underline{E} \tag{3.9}$$

$$\frac{\mathrm{d}^2}{\mathrm{d}x^2}\phi(x) + k_{\mathrm{o}}^2\varepsilon(x)\phi(x) = k_z^2\phi(x) \tag{3.10}$$

Equation (3.10) is a typical Sturm–Liouville differential equation. Parameter k_z is the eigenvalue to be determined. Since the function $\phi(x)$ relates to the electric field in the 1D periodic medium, by incorporating the Bloch–Floquet theory $\phi(x)$ must satisfy the relation in Equation (3.1), and can be written as

$$\phi(x) = \sum_{n=-\infty}^{+\infty} v_n\,\mathrm{e}^{-\mathrm{j}k_{x,n}x} \tag{3.11}$$

Substituting Equation (3.11) into Equation (3.10), we may obtain the difference equation given below:

$$qv_{n-1} + D_n v_n + qv_{n+1} = 0 \tag{3.12}$$

$$D_n = k_{\mathrm{o}}^2\varepsilon_{\mathrm{g}} - k_{x,n}^2 - k_z^2 \tag{3.13}$$

$$q = \delta k_{\mathrm{o}}^2\varepsilon_{\mathrm{g}} \tag{3.14}$$

Equation (3.12) is the so-called three-term recurrence relation that can be solved by the continued fraction method. Alternatively, we may first convert it into a typical eigenvalue problem and solve it by numerical computation. From Equation (3.12), by collecting the index n counting from positive to negative infinities, we may obtain an infinite set of simultaneous equations with an infinite number of unknowns, and it can be expressed in terms of the following matrix equation:

$$\begin{pmatrix} \ddots & q & 0 & 0 & 0 & 0 & 0 \\ q & D_2 & q & 0 & 0 & 0 & 0 \\ 0 & q & D_1 & q & 0 & 0 & 0 \\ 0 & 0 & q & D_0 & q & 0 & 0 \\ 0 & 0 & 0 & q & D_{-1} & q & 0 \\ 0 & 0 & 0 & 0 & q & D_{-2} & q \\ 0 & 0 & 0 & 0 & 0 & q & \ddots \end{pmatrix} \begin{pmatrix} \vdots \\ v_2 \\ v_1 \\ v_0 \\ v_{-1} \\ v_{-2} \\ \vdots \end{pmatrix} = 0 \tag{3.15}$$

In Equation (3.15), the main- and off-diagonal entries are filled with the D_n in Equation (3.13) and q in Equation (3.14), respectively. Otherwise, all remaining entries are filled with zero.

The existence of a nontrivial solution to Equation (3.15) is that the determinant must vanish. This may determine the relationship (phase relation) among the variables k_o, k_x, and k_z. Owing to finite computational resources and power, the infinite matrix in Equation (3.15) has to be truncated into a finite one for determining the phase relation of the waves supported in the 1D periodic medium. By inspecting the matrix in Equation (3.15), for given k_o and k_z, we have to resort to a complex-root searching procedure for determining k_x. More effort is needed to find the accurate solutions. Contrarily, if the k_o and k_x are given to find k_z, such a problem can be converted into an eigenvalue problem; that is, what we will do is to solve the eigenvalue of the following matrix:

$$\begin{pmatrix} \ddots & q & 0 & 0 & 0 & 0 & 0 \\ q & k_o^2\varepsilon_g - k_{x,2}^2 & q & 0 & 0 & 0 & 0 \\ 0 & q & k_o^2\varepsilon_g - k_{x,1}^2 & q & 0 & 0 & 0 \\ 0 & 0 & q & k_o^2\varepsilon_g - k_{x,0}^2 & q & 0 & 0 \\ 0 & 0 & 0 & q & k_o^2\varepsilon_g - k_{x,-1}^2 & q & 0 \\ 0 & 0 & 0 & 0 & q & k_o^2\varepsilon_g - k_{x,-2}^2 & q \\ 0 & 0 & 0 & 0 & 0 & q & \ddots \end{pmatrix} \tag{3.16}$$

The eigenvalue of this matrix, k_z^2, can be readily determined by invoking the commonly used eigenvalue solver without any difficulty.

3.2.1 Two Space-Harmonic Approximation

In the following example, for ease of illustration we truncate the infinite matrix in Equation (3.15) into a 2×2 matrix by merely considering the two space harmonics $n = 0$ and $n = -1$. In doing so, the phase relation can be written as

$$(k_o^2\varepsilon_g - k_x^2 - k_z^2)\left[k_o^2\varepsilon_g - \left(k_x \pm \frac{2\pi}{d_x}\right)^2 - k_z^2\right] = (\delta k_o^2\varepsilon_g)^2 \tag{3.17}$$

The two terms on the left-hand side of Equation (3.17) stand for the three circles with centers located at $k_x = 0$ and $k_x = \pm 2\pi/d_x$. They are the asymptotic circles. Here, we normalize the propagation constants along the x- and z-axes to obtain the origins of the three circles located at integer numbers, as shown in Figure 3.1. Thus, the radius of the circle is $\sqrt{\varepsilon_g}d_x/\lambda$.

In general, for the case with small modulation index δ_x, the curves obtained by solving Equation (3.16) will approach the asymptotic circles except for the region in the vicinity of intersection points between the circles.

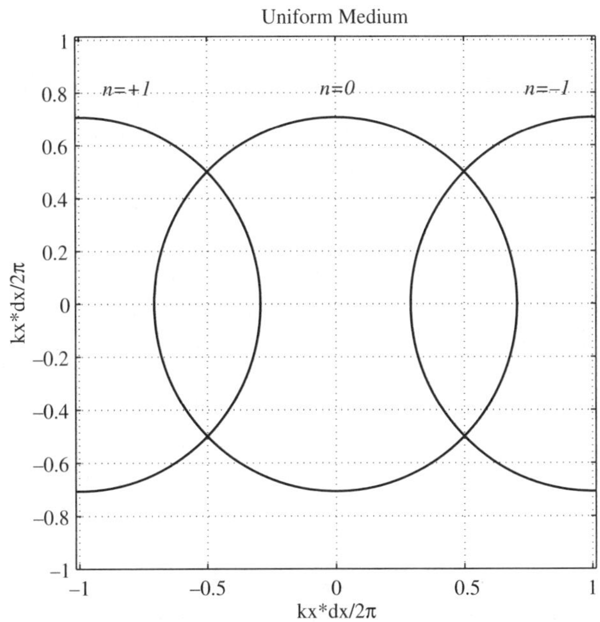

Figure 3.1 Phase relation of the unperturbed periodic medium with vanishing modulation index

In the following example, we will demonstrate a case with the propagation constant along the x-axis is given; therefore, the k_z can be determined by directing solving the eigenvalue of the matrix shown in Equation (3.16).

Example 3.1

Consider a 1D periodic medium with the relative dielectric function given in Equation (3.4). The average dielectric constant, period, and the operational wavelength are designated as $\varepsilon_g = 2$, $d_x = 1$, and $\lambda = 2$, respectively. We assumed that there are no field and structure variations along the y-axis, so that the problem can be regarded as a 2D problem. If the electric field polarization is along the y-axis (TE^x wave), calculate the real and imaginary parts of $k_z d_x/2\pi$ for a given real number of $k_x d_x/2\pi$ (phase relation).

Solution

Figure 3.2 shows the phase relation (real part of k_z) of the case with modulation index $\delta_x = 0.05$. In this case, the exact dispersion curve closes to those of the asymptotic circles since the modulation index is small enough. However, in the vicinity of the intersection points of the two asymptotic circles, the partial enlarged view in Figure 3.3 clearly shows the difference between the exact and unperturbed cases. Apparently, an opening of the gap in the vicinity of the intersection points takes place. This means that there is no real root within this region.

Additionally, we gradually increase the modulation index to see the variation in the phase relation. The examples demonstrating the phase relation with the modulation index, $\delta_x = 0.1, 0.13, 0.16, 0.17$, and 0.18 are respectively shown in Figures 3.4–3.8. It is obvious that the gap width is increasing in accordance with the increase in the modulation index. The gaps enable the presence of the oval-shaped real-root locus. Moreover, an increase in the modulation decreases the dimension of the oval. Specifically, the oval disappears in Figure 3.8 for $\delta = 0.18$; the real part of the propagation constant k_z is varying slowly against the change in k_x.

In Figures 3.2–3.8, only the real part of k_z is shown. In fact, the eigenvalues in the matrix shown in Equation (3.16) generally contain real and complex roots. The real root of k_z means that the wave is propagating in the periodic medium, while the complex root stands for the non-propagating (evanescent) waves. We redraw the phase relation in Figure 3.4 by plotting the imaginary part of the complex roots in Figure 3.9; the dashed line represents the unperturbed phase relation. Here, owing to the lossless nature of the dielectric medium, the complex roots actually are a pure imaginary number. Outside the circles, the pure imaginary roots represent the below-cutoff phenomenon.

In addition to the stop-band being exhibited in the phase-relation diagram of a 1D periodic medium, we may also discover a unique wave-propagation characteristic that is not present in the uniform dielectric medium. Returning to the example shown in Figure 3.8, the solid line represents the phase relation of the 1D periodic medium with modulation index $\delta = 0.18$, while the dashed line is that of the zero modulation index (uniform dielectric medium with relative dielectric constant ε_g). From electromagnetic wave theory, the phase relation allows us to know the direction of energy flow. The direction normal to the phase-relation contour represents the direction of group velocity (energy flow). As shown in Figure 3.10, for example, for a given k_x the direction of energy flow in the uniform dielectric medium is along the wave vector, which is denoted by the arrow on the circle. On the other hand, for the same propagation constant along the x-axis the direction of energy flow is along the direction of arrow that is perpendicular to the phase-relation contour. Additionally, the slowly varying contour indicates that the energy flow tends to be along the z-axis for an arbitrary k_x in such a 1D periodic medium.

In Example 3.1 we calculated the phase relation of the periodic medium. As was demonstrated in Figures 3.4–3.8, for a fixed k_x we may have an infinite number of k_z. Certainly, only a small portion of them are real numbers, and the others are

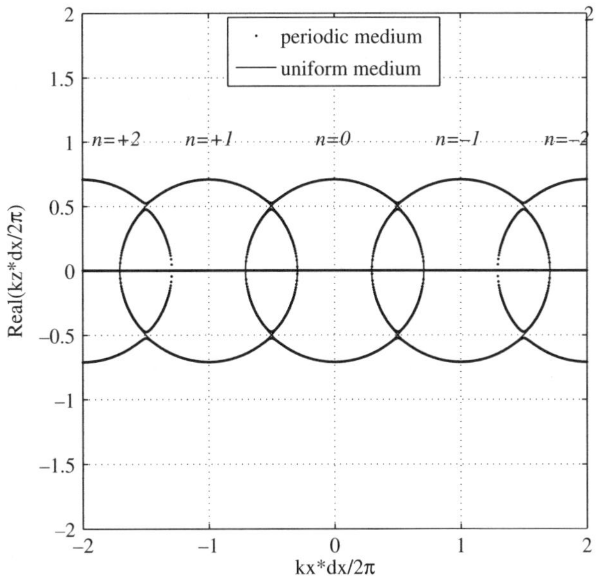

Figure 3.2 Phase relation of the 1D periodic medium with modulation index $\delta_x = 0.05$

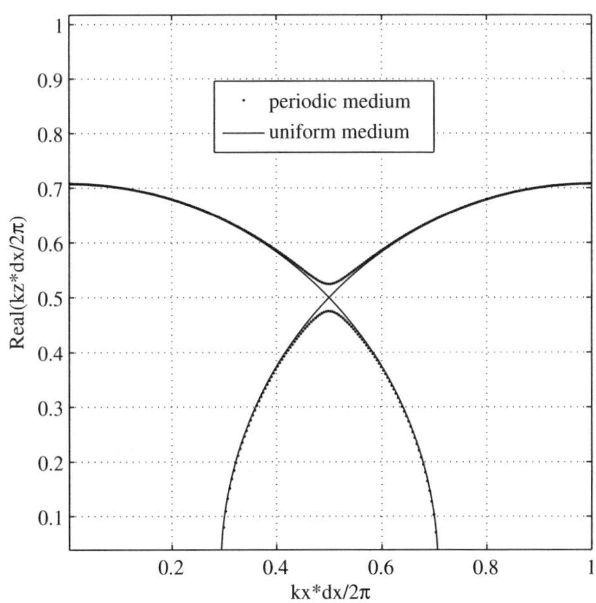

Figure 3.3 Partial enlarged view in the vicinity of intersection point in Figure 3.2

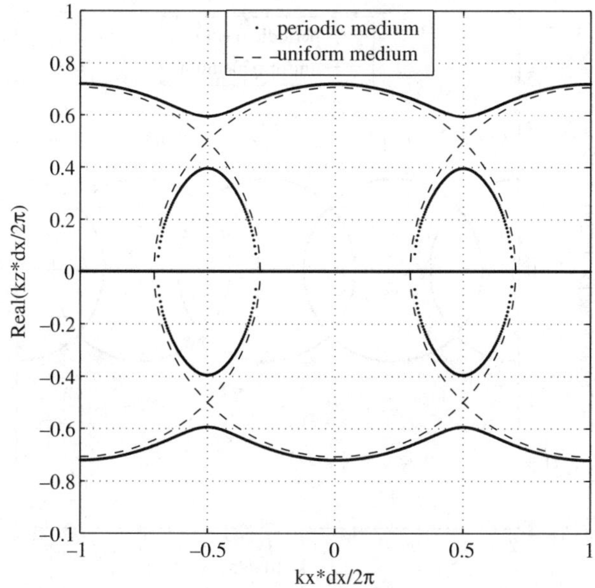

Figure 3.4 Phase relation of the same 1D periodic medium in Figure 3.2 but with modulation index $\delta_x = 0.1$

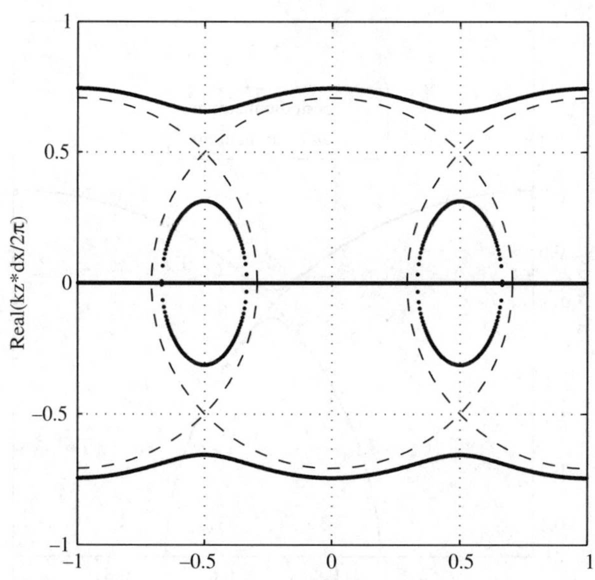

Figure 3.5 Phase relation of the same 1D periodic medium in Figure 3.2 but with modulation index $\delta_x = 0.13$

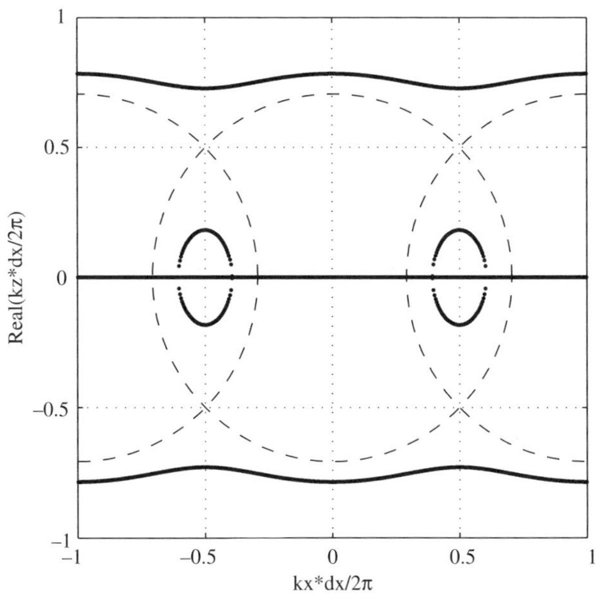

Figure 3.6 Phase relation of the same 1D periodic medium in Figure 3.2 but with modulation index $\delta_x = 0.16$

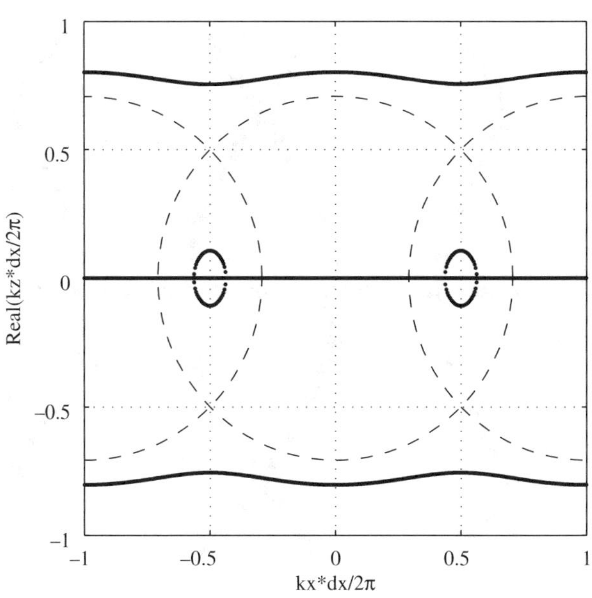

Figure 3.7 Phase relation of the same 1D periodic medium in Figure 3.2 but with modulation index $\delta_x = 0.17$

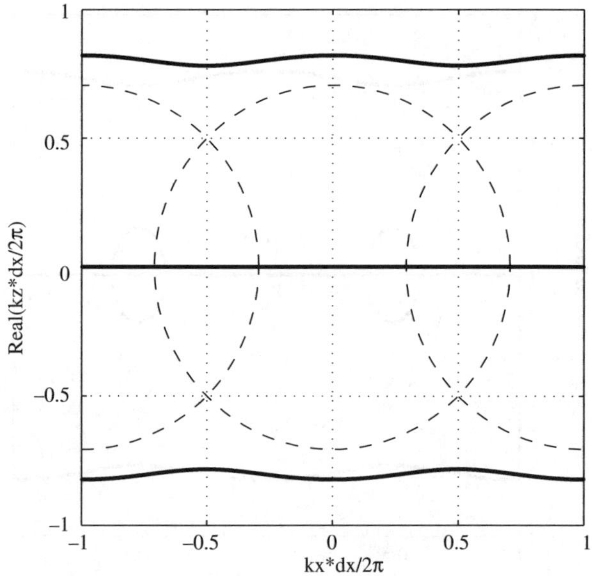

Figure 3.8 Phase relation of the same 1D periodic medium in Figure 3.2 but with modulation index $\delta_x = 0.18$

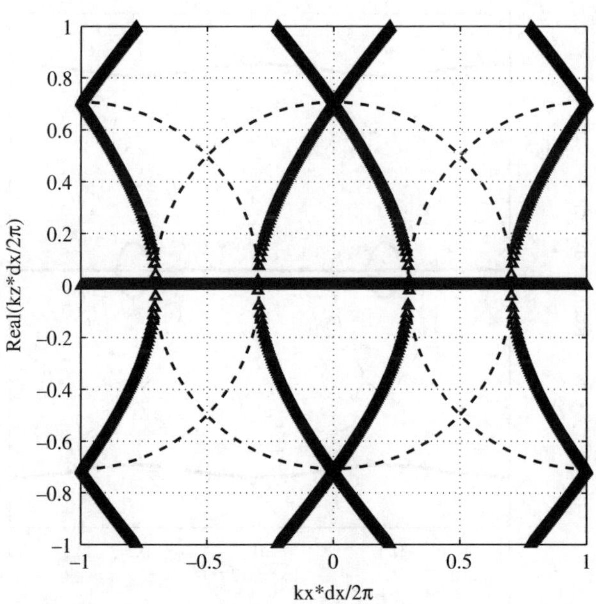

Figure 3.9 Imaginary part of the phase relation in Figure 3.4

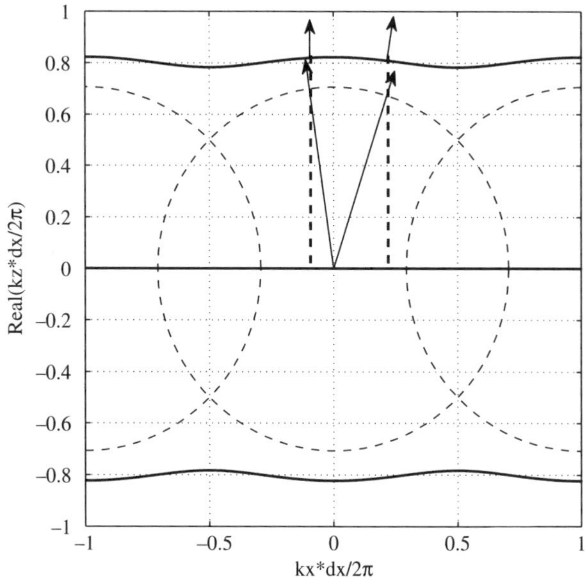

Figure 3.10 Direction of energy flow in a 1D periodic medium with the structure parameters given in Figure 3.8

complex numbers. However, the amplitude of each wave remains to be determined. In Section 3.2.2 we will formulate the electric and magnetic fields in the 1D periodic medium described previously. Furthermore, the discontinuity between the uniform medium and such a 1D periodic medium will be considered by investigating the scattering properties of a plane wave incident into this semi-infinite 1D periodic medium. The excitation of the modes, for which we have demonstrated their phase relation, can be inspected to understand their unique propagation characteristics in the medium.

Returning to Equation (3.16), we may calculate the eigenvalue k_z^2 and its associated eigenvector by invoking the eigenvalue solver. Notably, each eigenvalue has its own eigenvector, and the eigenfunction $\phi(x)$ as well. Thus, the x-dependent function in Equation (3.18) corresponding to each eigenvalue $k_{z,m}^2$ can be rewritten as

$$\phi_m(x) = \sum_{n=-\infty}^{+\infty} v_n^{(m)} \, \mathrm{e}^{-\mathrm{j}k_{x,n}x} \tag{3.18}$$

The electric field component in Equation (3.5) can be rewritten as

$$E_y(x, z) = \sum_{m=-\infty}^{+\infty} (f_m \, \mathrm{e}^{-\mathrm{j}k_{z,m}z} + g_m \, \mathrm{e}^{+\mathrm{j}k_{z,m}z})\phi_m(x) \tag{3.19}$$

In Equation (3.19), the variables f and g represent the forward- and backward-propagating waves along the z-axis, respectively, which will be determined by the electromagnetic boundary condition:

$$E_y(x, z) = \sum_{n=-\infty}^{+\infty} \mathrm{e}^{-\mathrm{j}k_{x,n}x} \left[\sum_{n=-\infty}^{+\infty} v_n^{(m)} \left(f_m \,\mathrm{e}^{-\mathrm{j}k_{z,m}z} + g_m \,\mathrm{e}^{+\mathrm{j}k_{z,m}z} \right) \right] \tag{3.20}$$

Additionally, by substituting Equation (3.20) into Equations (3.6), we may obtain the magnetic field components:

$$H_x(x, z) = -\sum_{n=-\infty}^{+\infty} \mathrm{e}^{-\mathrm{j}k_{x,n}x} \left[\sum_{n=-\infty}^{+\infty} v_n^{(m)} \frac{k_{z,m}}{\omega\mu_{\mathrm{o}}} \left(f_m \,\mathrm{e}^{-\mathrm{j}k_{z,m}z} + g_m \,\mathrm{e}^{+\mathrm{j}k_{z,m}z} \right) \right] \tag{3.21}$$

$$H_z(x, z) = \sum_{n=-\infty}^{+\infty} \frac{k_{x,n}}{\omega\mu_{\mathrm{o}}} \mathrm{e}^{-\mathrm{j}k_{x,n}x} \left[\sum_{n=-\infty}^{+\infty} v_n^{(m)} \left(f_m \,\mathrm{e}^{-\mathrm{j}k_{z,m}z} + g_m \,\mathrm{e}^{+\mathrm{j}k_{z,m}z} \right) \right] \tag{3.22}$$

3.2.2 Single Interface between a Semi-infinite Uniform and a 1D Periodic Medium

In Section 3.2 we formulated the general field solutions in a 1D periodic medium and discussed the wave-propagation characteristics (phase relation) of those eigenwaves; however, the amplitude of each space harmonic is undetermined, since the source-free condition is taken into account. In this section we will proceed to deal with the excitation problem. The problem to be studied is the semi-infinite 1D periodic medium incident by a plane wave. By invoking the electromagnetic boundary condition, the amplitude of each space harmonic can be obtained. Furthermore, the scattering power including the reflection one in uniform medium and the transmission one in 1D periodic medium can be determined. The detail mathematical procedures are given below.

As shown in Figure 3.11, a plane wave is obliquely incident on a semi-infinite periodic medium with the periodic permittivity characterized by $\varepsilon(x)\varepsilon_{\mathrm{o}}$. The incident

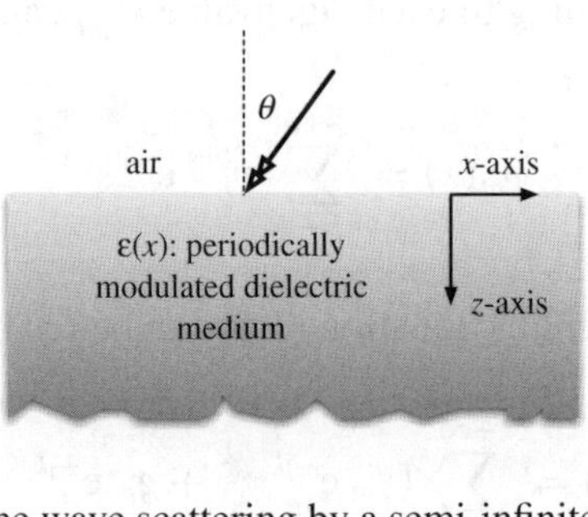

Figure 3.11 Plane wave scattering by a semi-infinite periodic medium

angle counted from z-axis is denoted by θ. The interface between the 1D periodic medium and the air region is located at $z = 0$.

First of all, we assumed that the electric field of the incident plane wave is along the y-axis (TE^x wave). In the presence of the 1D periodic medium, in the air region the tangential electric- and magnetic-field components along the interface can be written as

$$E_y(x, z) = \sum_{n=-\infty}^{+\infty} \mathrm{e}^{-\mathrm{j}k_{x,n}x}(a_n\,\mathrm{e}^{-\mathrm{j}k_{z,n}z} + b_n\,\mathrm{e}^{+\mathrm{j}k_{z,n}z}) \tag{3.23}$$

$$H_x(x, z) = -\sum_{n=-\infty}^{+\infty} \mathrm{e}^{-\mathrm{j}k_{x,n}x}\frac{k_{z,n}}{\omega\mu_o}\left(a_n\,\mathrm{e}^{-\mathrm{j}k_{z,n}z} - b_n\,\mathrm{e}^{+\mathrm{j}k_{z,n}z}\right) \tag{3.24}$$

In Equations (3.23) and (3.24), the variables a_n and b_n are the amplitudes of incident and reflected waves along the z-axis, which will be determined by the electromagnetic boundary conditions. Notably, the propagation constants along the x- and z-axes satisfy the phase relation of a uniform medium, which is given below:

$$k_{z,n} = \sqrt{k_o^2\varepsilon_a - \left(k_x + n\frac{2\pi}{d_x}\right)^2} \tag{3.25}$$

Once the uniform plane wave is incident from the air region shown in Figure 3.11, the waves excited in the second (1D periodic) medium only propagate along the $+z$-axis since no termination is present in the second medium. The tangential electric field in Equation (3.20) and the magnetic field in Equation (3.21) are rewritten as follows:

$$E_y(x, z) = \sum_{n=-\infty}^{+\infty} \mathrm{e}^{-\mathrm{j}k_{x,n}x}\left(\sum_{n=-\infty}^{+\infty} v_n^{(m)} f_m\,\mathrm{e}^{-\mathrm{j}k_{z,m}z}\right) \tag{3.26}$$

$$H_x(x, z) = -\sum_{n=-\infty}^{+\infty} \mathrm{e}^{-\mathrm{j}k_{x,n}x}\left(\sum_{n=-\infty}^{+\infty} v_n^{(m)}\frac{k_{z,m}}{\omega\mu_o} f_m\,\mathrm{e}^{-\mathrm{j}k_{z,m}z}\right) \tag{3.27}$$

Owing to the electromagnetic boundary condition, the tangential electric and magnetic field components must be continuous at the interface ($z = 0$) between two dielectric media. Thus, we can have the following two equations:

$$\sum_{n=-\infty}^{+\infty} \mathrm{e}^{-\mathrm{j}k_{x,n}x}(a_n + b_n) = \sum_{n=-\infty}^{+\infty} \mathrm{e}^{-\mathrm{j}k_{x,n}x}\left(\sum_{n=-\infty}^{+\infty} v_n^{(m)} f_m\right) \tag{3.28}$$

$$-\sum_{n=-\infty}^{+\infty} \mathrm{e}^{-\mathrm{j}k_{x,n}x}\frac{k_{z,n}}{\omega\mu_o}(a_n - b_n) = -\sum_{n=-\infty}^{+\infty} \mathrm{e}^{-\mathrm{j}k_{x,n}x}\left(\sum_{n=-\infty}^{+\infty} v_n^{(m)}\frac{k_{z,m}}{\omega\mu_o} f_m\right) \tag{3.29}$$

Since the Fourier amplitudes corresponding to the same harmonic must be equal to each other at both sides, the above two equation can be rewritten as:

$$a_n + b_n = \sum_{n=-\infty}^{+\infty} v_n^{(m)} f_m \tag{3.30}$$

$$\frac{k_{z,n}}{\omega\mu_o}(a_n - b_n) = \sum_{n=-\infty}^{+\infty} v_n^{(m)} \frac{k_{z,m}}{\omega\mu_o} f_m \tag{3.31}$$

For each index n, Equations (3.30) and (3.31) correspond to a system of simultaneous linear equations with an infinite number of variables. By collecting the index n, the above two equations can be written in matrix form as follows:

$$\underline{a} + \underline{b} = \mathbb{V}\underline{f} \tag{3.32}$$

$$\mathbb{Y}_{\mathrm{a}}(\underline{a} - \underline{b}) = \mathbb{V}\mathbb{Y}_{\mathrm{g}}\underline{f} \tag{3.33}$$

where $\underline{a}$, $\underline{b}$, and $\underline{f}$ are vectors containing the amplitude of each space harmonic. Additionally, parameter $\mathbb{V}$ is a full matrix whose columns are the corresponding eigenvectors. Parameters $\mathbb{Y}_{\mathrm{a}}$ and $\mathbb{Y}_{\mathrm{g}}$ are the diagonal matrices with their admittance of nth space harmonic locating at the nth entry. Through appropriate mathematical procedures, the reflected-wave vector $\underline{b}$ and transmitted-wave vector $\underline{f}$ can be expressed in terms of the incident vector $\underline{a}$:

$$\underline{b} = \mathbb{R}\underline{a} \tag{3.34}$$

$$\underline{f} = \mathbb{V}^{-1}(\mathbb{I} + \mathbb{R})\underline{a} \tag{3.35}$$

$$\mathbb{R} = (\mathbb{Y}_{\mathrm{a}} + \mathbb{V}\mathbb{Y}_{\mathrm{g}}\mathbb{V}^{-1})^{-1}(\mathbb{Y}_{\mathrm{a}} - \mathbb{V}\mathbb{Y}_{\mathrm{g}}\mathbb{V}^{-1}) \tag{3.36}$$

where $\mathbb{R}$ is the reflection matrix defining the relationship between the incident and reflected space harmonics.

Example 3.2

In this example we consider the same 1D periodic medium as Example 3.1, but with the modulation index $\delta = 0.18$. The 1D periodic medium becomes a semi-infinite one attached to the air, as shown in Figure 3.11. We will pay attention to the scattering characteristics and the wave-propagation property in the 1D periodic medium. As usual, the field and structure have no variation along the y-axis; the electric field is assumed to be along the y-axis.

Solution

Based on the mathematical formulation described previously, I have written a computer program to calculate the reflectance and transmittance, which are respectively defined as the ratios of reflected power and of the transmitted power to the incident power. Before the elaborate numerical calculations, we first check the numerical convergence for the reflected and transmitted power against the number of space harmonics employed in the numerical computation. From Figure 3.12, it is apparent that the numerical convergence is excellent; the curves have no obvious variation once the total number of space harmonics is greater than 20. We have also carried out numerous convergence tests by changing the structure and incident parameters and obtained numerically converged results. Now, we are in a good position to study the wave propagation properties in the structure under consideration.

In the following numerical examples, we will change the incident angle to observe the electric- and magnetic-field distribution in the semi-infinite 1D periodic medium. The first example is a plane wave normally incident on the semi-infinite 1D periodic medium. The real parts of the electric- and magnetic-field contour map in the 1D periodic medium are drawn in Figures 3.13, 3.14, and 3.15 for $E_y(x, z)$, $H_x(x, z)$, and $H_z(x, z)$, respectively. Because of there being no termination in the 1D periodic medium, the wave excited at the interface is propagating along the z-axis. Additionally, we plot the intensity distribution of Poynting vector shown in Figure 3.16. As is well known, the Poynting vector indicates the direction of energy flow. The energy propagation direction is indeed generally along the z-axis; however, if we look at the details, we may find that the Poynting vector is bending forward along the z-axis.

Furthermore, we change the incident angle to 60° to observe the electromagnetic-field distribution in the 1D periodic medium. Figures 3.17–3.19 demonstrate the electric- and magnetic-field components, and the intensity distribution of Poynting vector is shown in Figure 3.20. In comparison with Figures 3.13–3.15, the symmetry or anti-symmetry in the fields is no longer present owing to the oblique incidence. Significantly, the energy flow seems to be in the sinusoidal path along each of the unit cells of the 1D periodic medium.

Without loss of generality, the cases with different incident angle are also investigated to understand their propagation characteristics of the Poynting vector. Figures 3.21 and 3.22 present the intensity distributions of the Poynting vector in the 1D periodic medium for the plane-wave incident at angles of 30° and 89°, respectively. As usual, the Poynting vector propagates in a sinusoidal path within the unit cell even for the near-grazing-angle incidence ($\theta = 89°$). For ease of reference, we plot the Poynting vector distribution in the semi-infinite dielectric medium with relative dielectric constant $\varepsilon_s = 2$, which can be regarded as the 1D periodic medium with zero modulation index. Apparently, without the periodic variation in the dielectric medium, the transmitted wave propagates obliquely and the transmitted angle follows Snell's law, as shown in Figure 3.23.

It is astonishing to note that in such a 1D periodic medium the energy propagation is along a sinusoidal path in spite of the angle of incidence, which does not occur in a commonly used dielectric medium. Now what we have to investigate is the underlying physics of this strikingly distinct phenomenon in such a 1D periodic medium. As was mentioned previously, the phase relation in the 1D periodic medium can reveal to us the direction of energy flow (group velocity). Here, we calculated the phase-relation diagram of the 1D periodic medium shown in Figure 3.24 with the parameters indicated in the caption. Considering the plane-wave excitation, the horizontal axis (k_x/k_o) equals $\sin\theta$. For a different incident angle, we may calculate $\sin\theta$ for the z-direction normalized phase constant k_z/k_o. From this figure, we can see that the phase relation exhibits a straight line. This means that the group velocity direction is always along the z-axis in spite of the angle of incidence. That is the reason why all the Poynting vector distributions shown in the previous examples with different incident angles are all along the z-axis even for the case with low grazing angle (89°).

Regarding the sinusoidal path of energy propagation in each unit cell, this is due to the graded-index distribution of the relative dielectric medium. We show the relative dielectric function in a unit cell (period) in Figure 3.25; the profile of the refractive index is similar to that of a graded-index fiber that has a refractive index decreasing with increasing radial distance from the fiber axis. Here, the fiber axis is at the positions $x = n$, where n is an integer ranging from negative to positive infinities.

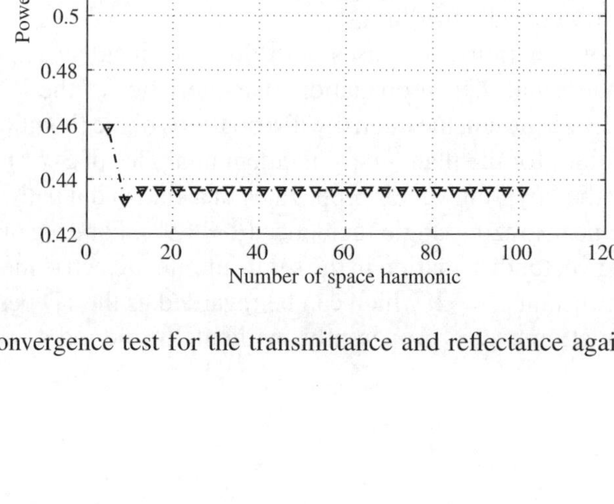

Figure 3.12 Convergence test for the transmittance and reflectance against the number of space harmonics

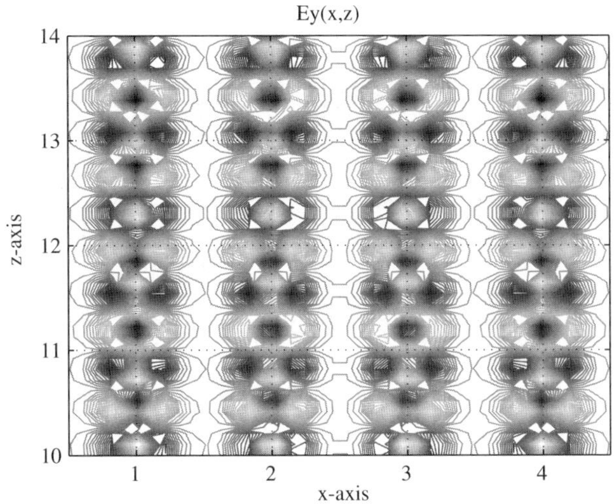

Figure 3.13 Real part of $E_y(x, z)$ in the 1D periodic medium; normal incidence

3.3 Eigenwave in 1D Dielectric Gratings: Modal Transmission-Line Approach

In Section 3.2, the dispersion characteristics of the eigenwave in a dielectric medium having sinusoidal variation in its relative dielectric constant was demonstrated, as well as the scattering properties of a plane wave impinging on a semi-infinite periodic

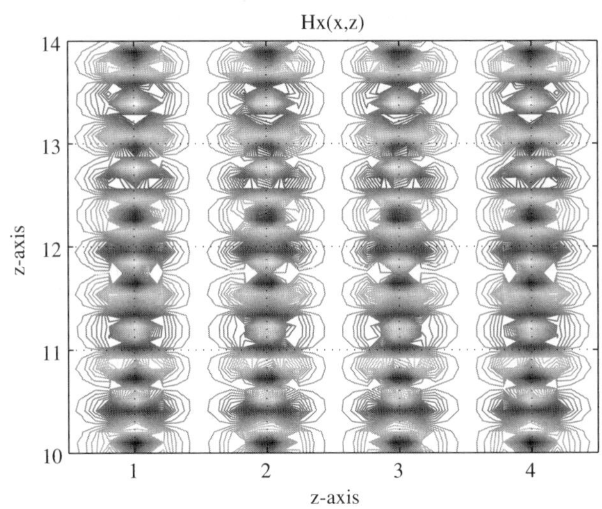

Figure 3.14 Real part of $H_x(x, z)$ in the 1D periodic medium; normal incidence

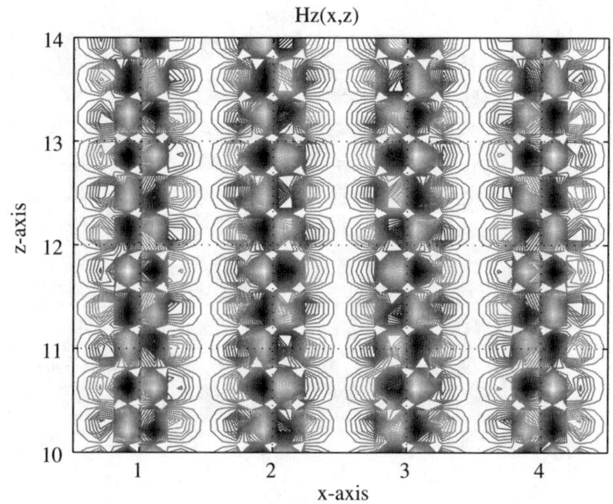

Figure 3.15 Real part of $H_z(x, z)$ in the 1D periodic medium; normal incidence

medium. In fact, the sinusoidal dielectric function employed in Section 3.2 can be regarded as the first-order approximation of a generalized periodic dielectric function. For example, a piecewise continuously periodic dielectric function is expressed as an infinite sum of sine and cosine functions (Fourier series expansion). Consequently, the mathematical procedure for calculating the eigenwaves in the sinusoidal dielectric

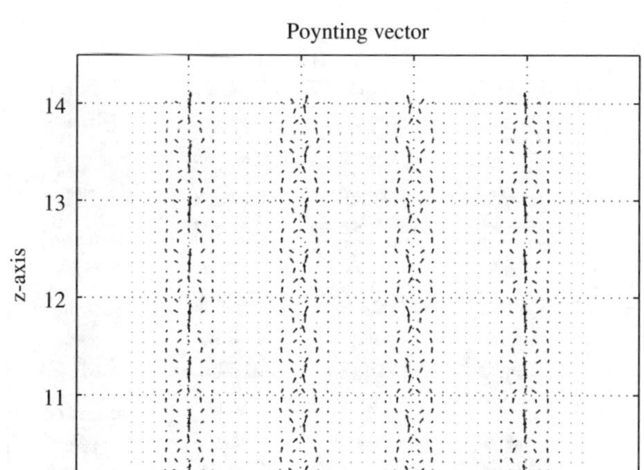

Figure 3.16 Intensity distribution of Poynting vector in the 1D periodic medium; normal incidence

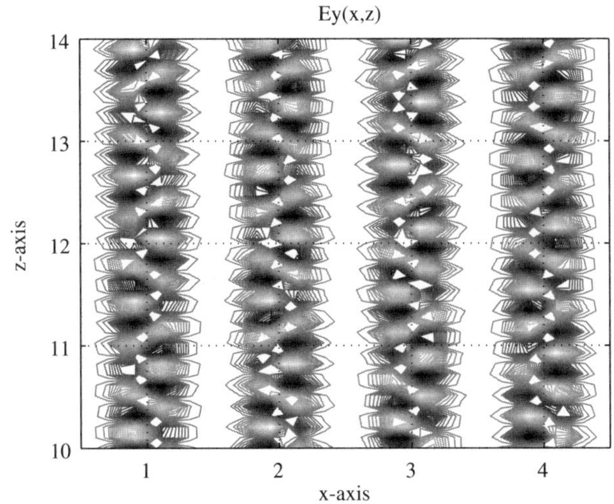

Figure 3.17 Real part of $E_y(x, z)$ in the 1D periodic medium; $\theta = 60°$

function is relatively simple compared with that of the piecewise continuously dielectric function. In this section, we will treat the eigenwave in a 1D periodic medium consisting of multiple uniform dielectric layers.

Distinct from the method described in Section 3.2, we will first calculate the modal solutions in the unit cell (period) by invoking the periodic boundary condition, which

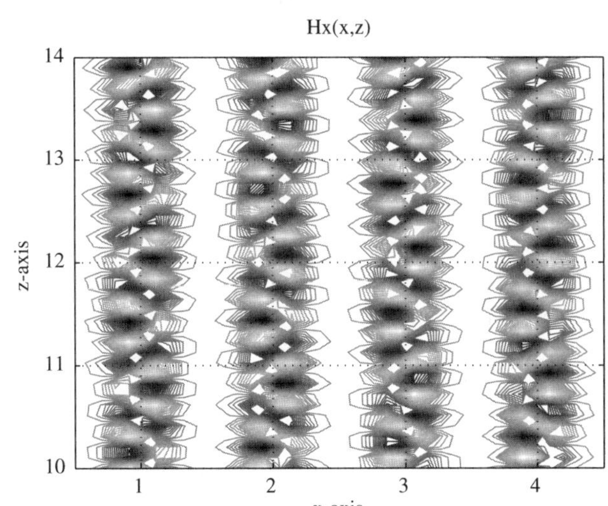

Figure 3.18 Real part of $H_x(x, z)$ in the 1D periodic medium; $\theta = 60°$

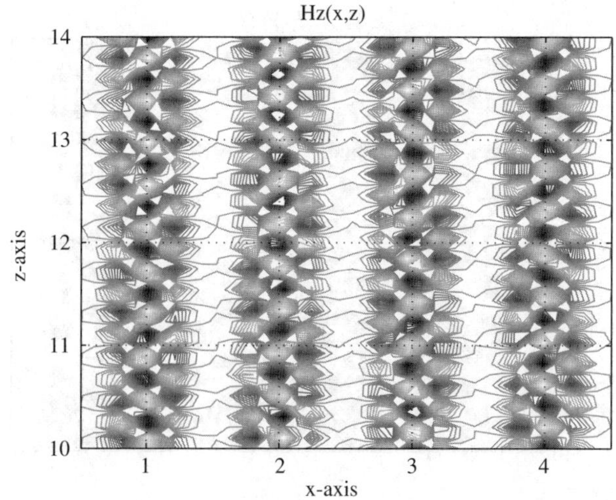

Figure 3.19 Real part of $H_z(x, z)$ in the 1D periodic medium; $\theta = 60^\circ$

is very similar to solving the dispersion relation in a wave-guiding environment. The modal solutions are exact and can be taken as a complete set to expand the general fields excited in the 1D periodic medium. Specifically, some numerical examples are demonstrated to understand the physical pictures of the wave process involved in the 1D periodic medium. Furthermore, we will also study the scattering characteristics

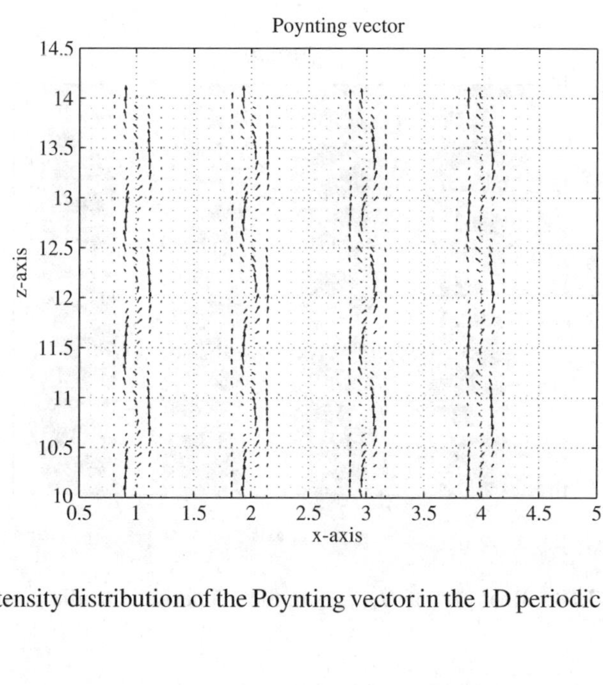

Figure 3.20 Intensity distribution of the Poynting vector in the 1D periodic medium; $\theta = 60^\circ$

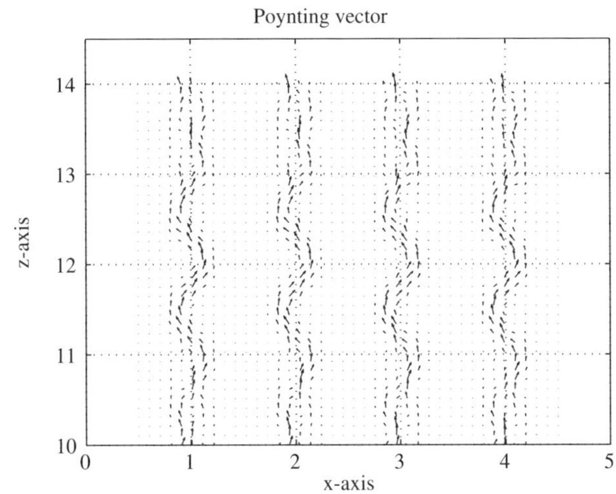

Figure 3.21 Intensity distribution of the Poynting vector in the 1D periodic medium; $\theta = 30^\circ$

of a plane wave by a 1D periodic structure having a finite thickness, as well as the guiding properties of a wave supported by the 1D periodic waveguide.

As shown in Figure 3.26, we consider a 1D periodic structure composed of multiple uniform dielectric layers in the period denoted a. The structure is assumed to be uniform and infinite in extent along the y- and z-axes. The permeability is μ_o for

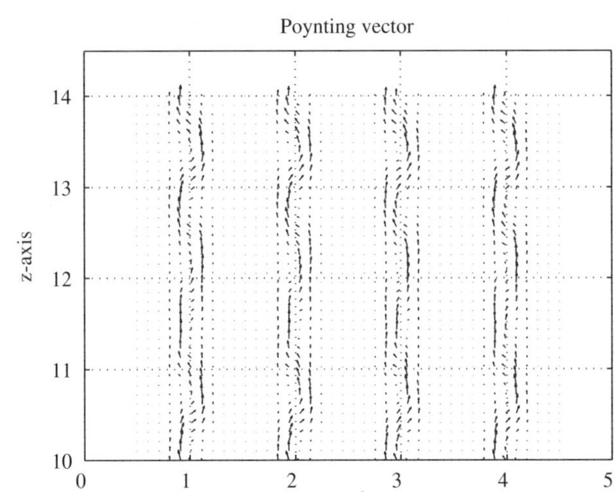

Figure 3.22 Intensity distribution of the Poynting vector in the 1D periodic medium; $\theta = 89^\circ$

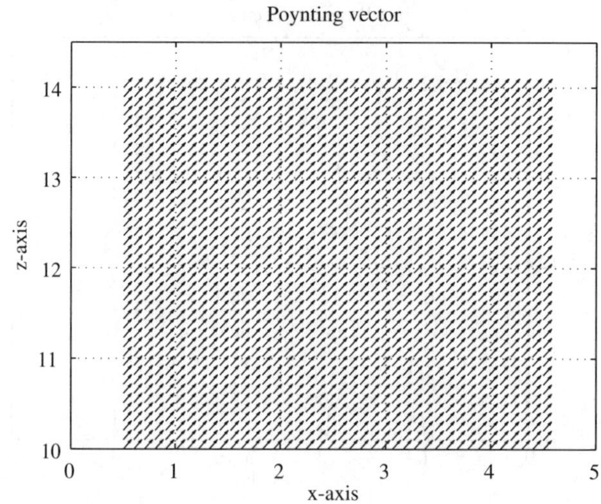

Figure 3.23 Intensity distribution of the Poynting vector in the uniform dielectric medium with relative dielectric constant $\varepsilon_s = 2.0$; $\theta = 60°$

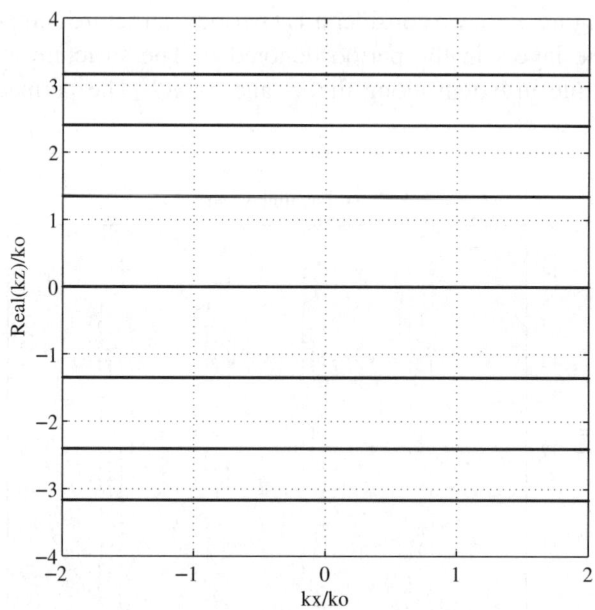

Figure 3.24 Phase-relation diagram of the 1D periodic medium; $\varepsilon_s = 2, \delta_x = 0.18, d_x = 1.0$, $\lambda = 2.0$

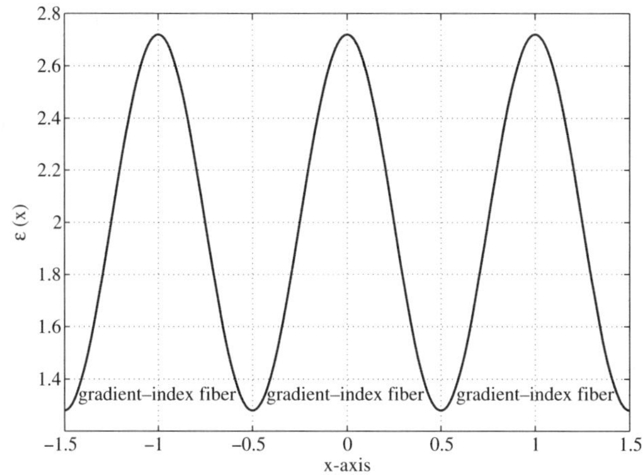

Figure 3.25 Function of the relative dielectric constant of the 1D periodic medium

each dielectric layer. The permittivity of each layer is denoted ε_i, where i is the layer number and is attached in the figure.

$$\varepsilon(x + a) = \varepsilon(x) \tag{3.37}$$

$$\mu(x + a) = \mu(x) \tag{3.38}$$

It is noted that the two functions in Equations (3.37) and (3.38) may be continuous or piecewise continuously periodic functions. Parameter a is the period. Since the structure under consideration is a layered medium and is infinite in extent along the y–z plane, the field solutions can be treated individually as TE^x (the electric-field

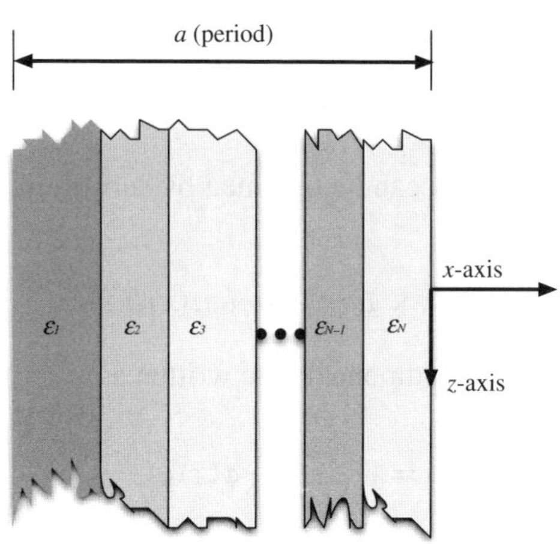

Figure 3.26 A 1D periodic medium made up of multiple uniform dielectric layers

component along the x-axis vanishes) and TM^x modes (the magnetic-field component along the x-axis is zero). Notably, if the 1D periodic medium has finite thickness along the z-axis and the out-of-plane incident condition with non-vanished k_y, the individual TE or TM mode can no longer satisfy the electromagnetic boundary condition at the interface between the uniform and 1D periodic media. Therefore, the general field solutions must be expressed in terms of the superposition of TE and TM modes simultaneously.

3.3.1 In-Plane Incidence: $k_y = 0$

In the following, we first consider that the field has no variation along the y-axis ($k_y = 0$).

3.3.1.1 TE^x Mode

Because the x-directional electric field component vanishes, the electric-field vector can be written as

$$\underline{E}(\underline{r}) = \nabla \times \underline{F} \tag{3.39}$$

$$\underline{F} = \underline{x}_o \psi(x, z) \tag{3.40}$$

From the principle of separation of variables, the scalar potential $\psi(x, z)$ is written as

$$\psi(x, y, z) = \phi(x)\,\mathrm{e}^{-\mathrm{j}k_z z} \tag{3.41}$$

where k_z is the propagation constant along the z-axis. Substituting Equations (3.40) and (3.41) into Equation (3.39), the electric-field component can be determined as follows:

$$E_y = \frac{\partial \psi}{\partial z} = -\mathrm{j}k_z \phi(x)\,\mathrm{e}^{-\mathrm{j}k_z z} \tag{3.42}$$

The magnetic field components can be obtained by substituting Equation (3.42) into the Maxwell equation:

$$\nabla \times \underline{E} = -\mathrm{j}\omega\mu_o\mu(x)\underline{H} \tag{3.43}$$

Therefore, the magnetic field components are written as

$$H_x = \mathrm{j}\frac{k_z^2}{\omega\mu_o\mu(x)}\phi(x)\,\mathrm{e}^{-\mathrm{j}k_z z} \tag{3.44}$$

$$H_z = \frac{k_z}{\omega\mu_o\mu(x)}\frac{\mathrm{d}\phi(x)}{\mathrm{d}x}\,\mathrm{e}^{-\mathrm{j}k_z z} \tag{3.45}$$

So far, all the electric and magnetic fields are expressed in terms of the same potential function, $\phi(x)$. Moreover, by substituting Equations (3.42), (3.44), and (3.45) into the Maxwell equation, Equation (3.46), we can then obtain the second-order differential equation in Equation (3.47):

$$\nabla \times \underline{H} = \mathrm{j}\omega\varepsilon_{\mathrm{o}}\varepsilon(x)\underline{E} \tag{3.46}$$

$$\frac{\mathrm{d}}{\mathrm{d}x}\left[\frac{1}{\mu(x)}\frac{\mathrm{d}\phi(x)}{\mathrm{d}x}\right] + k_{\mathrm{o}}^2\varepsilon(x)\phi(x) = k_z^2\frac{1}{\mu(x)}\phi(x) \tag{3.47}$$

Equation (3.47) is the Sturm–Liouville differential equation, where k_z^2 is the eigenvalue to be determined. By invoking the periodic boundary condition, we may solve the eigenvalue and its associated eigenfunction $\phi(x)$.

3.3.1.2 TM^x Mode

Regarding the TM^x mode, since the mathematical procedures are similar to those for TE we only list several important equations for reference, while the detailed mathematical procedures are given as homework:

$$\underline{H}(\underline{r}) = \nabla \times \underline{A} \tag{3.48}$$

$$\underline{A} = \underline{x}_{\mathrm{o}}\psi(x, z) \tag{3.49}$$

$$\psi(x, z) = \phi(x)\,\mathrm{e}^{-\mathrm{j}k_z z} \tag{3.50}$$

$$H_y = \frac{\partial\psi}{\partial z} \tag{3.51}$$

$$\frac{\mathrm{d}}{\mathrm{d}x}\left[\frac{1}{\varepsilon(x)}\frac{\mathrm{d}\phi(x)}{\mathrm{d}x}\right] + k_{\mathrm{o}}^2\mu(x)\phi(x) = k_z^2\frac{1}{\varepsilon(x)}\phi(x) \tag{3.52}$$

Similarly, we get a governing equation in Equation (3.52). Additionally, the governing equation, boundary condition, and the electric- and magnetic-field components for the TE^x and TM^x modes are given in Table 3.1 for ease of reference.

3.3.2 *Out-of-Plane Incidence: $k_y \neq 0$*

3.3.2.1 TE^x Mode

Since the electric-field component along the x-axis vanishes, we assume that the electric field is the curl of a vector potential along the x-axis, which is written as follows:

$$\underline{E}(\underline{r}) = \nabla \times \underline{F} \tag{3.53}$$

$$\underline{F} = \underline{x}_{\mathrm{o}}\psi(x, y, z) \tag{3.54}$$

Table 3.1 Governing equation, boundary condition, and the electric- and magnetic-field components in a 1D periodic medium

TE^x	TM^x
Electric-field components	
$E_y = -\phi(x)\,\mathrm{e}^{-\mathrm{j}k_z z}$	$E_x = \frac{k_z}{\omega\mu_0\varepsilon(x)}\phi(x)\,\mathrm{e}^{-\mathrm{j}k_z z}$
	$E_z = \frac{-\mathrm{j}}{\omega\mu_0\varepsilon(x)}\frac{\mathrm{d}\phi(x)}{\mathrm{d}x}\,\mathrm{e}^{-\mathrm{j}k_z z}$
Magnetic-field components	
$H_x = \frac{k_z}{\omega\mu_0\mu(x)}\phi(x)\,\mathrm{e}^{-\mathrm{j}k_z z}$	$H_y = \phi(x)\,\mathrm{e}^{-\mathrm{j}k_z z}$
$H_z = \frac{-\mathrm{j}}{\omega\mu_0\mu(x)}\frac{\mathrm{d}\phi(x)}{\mathrm{d}x}\,\mathrm{e}^{-\mathrm{j}k_z z}$	
Governing equation	
$\frac{\mathrm{d}}{\mathrm{d}x}\left[p(x)\frac{d\phi(x)}{dx}\right] + q(x)\phi(x) = k_z^2 r(x)\phi(x)$	
$p(x) = r(x) = \frac{1}{\mu(x)}$	$p(x) = r(x) = \frac{1}{\varepsilon(x)}$
$q(x) = k_0^2\varepsilon(x)$	$q(x) = k_0^2\mu(x)$
Boundary condition	
$q(x) = k_0^2\varepsilon(x)$	$q(x) = k_0^2\mu(x)$

Based on the principle of separation of variables, the scalar potential ψ can be expressed as

$$\psi(x, y, z) = \phi(x)\,\mathrm{e}^{-\mathrm{j}k_y y}\,\mathrm{e}^{\mp\mathrm{j}k_z z} \tag{3.55}$$

In fact, ψ is the 3D plane-wave solution. Here, we reserve the flexibility to consider that the wave may propagate along the positive and negative z-axis for a 1D periodic structure with finite thickness along the z-axis. Contrarily, the structure along the y-axis is infinite in extent, so that we may assume that the wave is propagating along the $+y$-axis. By substituting Equations (3.54) and (3.55) into Equation (3.53), we may determine the electric-field components along the y- and z-axes as

$$E_y = \frac{\partial\psi}{\partial z} = \mp\mathrm{j}k_z\phi(x)\,\mathrm{e}^{-\mathrm{j}k_y y}\,\mathrm{e}^{\mp\mathrm{j}k_z z} \tag{3.56}$$

$$E_z = -\frac{\partial\psi}{\partial y} = \mathrm{j}k_y\phi(x)\,\mathrm{e}^{-\mathrm{j}k_y y}\,\mathrm{e}^{\mp\mathrm{j}k_z z} \tag{3.57}$$

Substituting the electric field components in Equations (3.56) and (3.57) into Equation (3.6), we may obtain the magnetic-field components given as

$$\left(\underline{x}_{\text{o}}\frac{\partial}{\partial x}+\underline{y}_{\text{o}}\frac{\partial}{\partial y}+\underline{z}_{\text{o}}\frac{\partial}{\partial z}\right)\times\left(\underline{y}_{\text{o}}\frac{\partial\psi}{\partial z}-\underline{z}_{\text{o}}\frac{\partial\psi}{\partial y}\right)=-\text{j}\omega\mu_{\text{o}}\underline{H} \tag{3.58}$$

$$\frac{\partial}{\partial x}\frac{\partial\psi}{\partial z}=-\text{j}\omega\mu_{\text{o}}H_z \tag{3.59}$$

$$\frac{\partial}{\partial x}\frac{\partial\psi}{\partial y}=-\text{j}\omega\mu_{\text{o}}H_y \tag{3.60}$$

$$-\frac{\partial}{\partial y}\frac{\partial\psi}{\partial y}-\frac{\partial}{\partial z}\frac{\partial\psi}{\partial z}=-\text{j}\omega\mu_{\text{o}}H_x \tag{3.61}$$

By substituting the scalar potential function Equation (3.55) into the above equations, we may determine the magnetic field components explicitly as

$$H_z=\pm\frac{k_z}{\omega\mu_{\text{o}}}\frac{\text{d}\phi(x)}{\text{d}x}\,\text{e}^{-\text{j}k_y y}\,\text{e}^{\mp\text{j}k_z z} \tag{3.62}$$

$$H_y=\frac{k_y}{\omega\mu_{\text{o}}}\frac{\text{d}\phi(x)}{\text{d}x}\,\text{e}^{-\text{j}k_y y}\,\text{e}^{\mp\text{j}k_z z} \tag{3.63}$$

$$H_x=\text{j}\frac{k_{\text{t}}^2}{\omega\mu_{\text{o}}}\phi(x)\,\text{e}^{-\text{j}k_y y}\,\text{e}^{\mp\text{j}k_z z} \tag{3.64}$$

$$k_{\text{t}}^2=k_y^2+k_z^2 \tag{3.65}$$

Again, substituting the electric and magnetic fields expressed in terms of the scalar potential $\psi(x, y, z)$ into Ampere's equation with Maxwell's modification, we may obtain the following second-order differential:

$$\frac{\text{d}^2\phi(x)}{\text{d}x^2}+k_{\text{o}}^2\varepsilon(x)\phi(x)=k_{\text{t}}^2\phi(x) \tag{3.66}$$

Equation (3.66) typically is the Sturm–Liouville differential equation, where k_{t}^2 is the eigenvalue to be solved. By incorporating the proper boundary conditions into the above differential equation, we may determine the eigenfunction $\phi(x)$ and its associated eigenvalue. For ease of reference, we list all the electric- and magnetic-field components as follows:

$$E_y=\mp\text{j}k_z\phi(x)\,\text{e}^{-\text{j}k_y y}\,\text{e}^{\mp\text{j}k_z z} \tag{3.67}$$

$$E_z=\text{j}k_y\phi(x)\,\text{e}^{-\text{j}k_y y}\,\text{e}^{\mp\text{j}k_z z} \tag{3.68}$$

$$H_x=\text{j}\frac{k_{\text{t}}^2}{\omega\mu_{\text{o}}}\phi(x)\,\text{e}^{-\text{j}k_y y}\,\text{e}^{\mp\text{j}k_z z} \tag{3.69}$$

$$H_y = \frac{k_y}{\omega\mu_o}\frac{\mathrm{d}\phi(x)}{\mathrm{d}x}\,\mathrm{e}^{-\mathrm{j}k_y y}\,\mathrm{e}^{\mp\mathrm{j}k_z z} \tag{3.70}$$

$$H_z = \pm\frac{k_z}{\omega\mu_o}\frac{\mathrm{d}\phi(x)}{\mathrm{d}x}\,\mathrm{e}^{-\mathrm{j}k_y y}\,\mathrm{e}^{\mp\mathrm{j}k_z z} \tag{3.71}$$

If we divide each of the above equations by $\mp\mathrm{j}k_\mathrm{t}$, they can be rewritten as follows:

$$E_y = \frac{k_z}{k_\mathrm{t}}\phi\,(x)\,\mathrm{e}^{-\mathrm{j}k_y y}\,\mathrm{e}^{\mp\mathrm{j}k_z z} \tag{3.72}$$

$$E_z = \mp\frac{k_y}{k_\mathrm{t}}\phi(x)\,\mathrm{e}^{-\mathrm{j}k_y y}\,\mathrm{e}^{\mp\mathrm{j}k_z z} \tag{3.73}$$

$$H_x = \mp\frac{k_\mathrm{t}}{\omega\mu_o}\phi(x)\,\mathrm{e}^{-\mathrm{j}k_y y}\,\mathrm{e}^{\mp\mathrm{j}k_z z} \tag{3.74}$$

$$H_y = \pm\mathrm{j}\frac{k_y}{k_\mathrm{t}\omega\mu_o}\frac{\mathrm{d}\phi(x)}{\mathrm{d}x}\,\mathrm{e}^{-\mathrm{j}k_y y}\,\mathrm{e}^{\mp\mathrm{j}k_z z} \tag{3.75}$$

$$H_z = \mathrm{j}\frac{k_z}{k_\mathrm{t}\omega\mu_o}\frac{\mathrm{d}\phi(x)}{\mathrm{d}x}\,\mathrm{e}^{-\mathrm{j}k_y y}\,\mathrm{e}^{\mp\mathrm{j}k_z z} \tag{3.76}$$

Notably, if $k_y = 0$ is considered, then the above equations will return to those in Table 3.1.

Returning to Equations (3.72)–(3.76), for given k_x and k_y, we may determine the eigenvalues k_z and their associated eigenfunction $\phi(x)$. In fact, an infinite number of eigenvalues exist. If a discontinuity (or interface) between the periodic medium and uniform medium is present, which is shown in Figure 3.11, the field solutions should take into account the contribution from all the eigenmodes to ensure the continuity of the tangential electric- and magnetic-fields over the boundary. The tangential components of the electric and magnetic fields (here, we assume that the interface is perpendicular to the z-axis; therefore, the tangential directions are along the x- and y-axes) can be written as:

$$E_y = \mathrm{e}^{-\mathrm{j}k_y y}\sum_m \frac{k_{z,m}}{k_{\mathrm{t},m}}\phi_m(x)\left(v_m^+\,\mathrm{e}^{-\mathrm{j}k_{z,m}z} + v_m^-\,\mathrm{e}^{+\mathrm{j}k_{z,m}z}\right) \tag{3.77}$$

$$H_x = -\frac{\mathrm{e}^{-\mathrm{j}k_y y}}{\omega\mu_o}\sum_m k_{\mathrm{t},m}\phi_m(x)(v_m^+\,\mathrm{e}^{-\mathrm{j}k_{z,m}z} - v_m^-\,\mathrm{e}^{+\mathrm{j}k_{z,m}z}) \tag{3.78}$$

$$H_y = \frac{\mathrm{j}\,\mathrm{e}^{-\mathrm{j}k_y y}}{\omega\mu_o}\sum_m \frac{k_y}{k_{\mathrm{t},m}}\frac{\mathrm{d}\phi_m(x)}{\mathrm{d}x}(v_m^+\,\mathrm{e}^{-\mathrm{j}k_{z,m}z} - v_m^-\,\mathrm{e}^{+\mathrm{j}k_{z,m}z}) \tag{3.79}$$

where m represents the index of the mth eigenmode having eigenvalue $k_{z,m}$ and eigenfunction $\phi_m(x)$. Parameters v_m^+ and v_m^- individually are the mode amplitudes corresponding to the forward- and backward-propagating waves, which will be determined by the excitation condition.

So far, the rigorous electric and magnetic fields in a 1D periodic medium have been determined by solving the Sturm–Liouville differential equation together with the periodic boundary condition. The eigenfunction has a closed-form solution for some specific cases; for example, a stratified structure. However, for a general dielectric function $\varepsilon(x)$, no closed-form solution exists. We may approximate it by

$$\phi_m(x) = \mathrm{e}^{-\mathrm{j}k_x x} \sum_n q_n^{(m)} \, \mathrm{e}^{-\mathrm{j}\frac{n2\pi}{a}x} \tag{3.80}$$

In Equation (3.80), the periodic function is expressed in terms of the sum of Fourier series, where $q_n^{(m)}$ is the expansion coefficient of the nth harmonic of the mth eigenmode. Moreover, we may replace the modal function in Equations (3.77)–(3.79) by Equation (3.80), and they can be rewritten as follows:

$$E_y = \mathrm{e}^{-\mathrm{j}k_y y} \sum_n \mathrm{e}^{-\mathrm{j}k_{x,n}x} \sum_m q_n^{(m)} (v_m^+ \, \mathrm{e}^{-\mathrm{j}k_{z,m}z} + v_m^- \, \mathrm{e}^{+\mathrm{j}k_{z,m}z}) \tag{3.81}$$

$$H_x = -\mathrm{e}^{-\mathrm{j}k_y y} \sum_n \mathrm{e}^{-\mathrm{j}k_{x,n}x} \sum_m q_n^{(m)} \frac{k_{\mathrm{t},m}}{\omega\mu_{\mathrm{o}}} (v_m^+ \, \mathrm{e}^{-\mathrm{j}k_{z,m}z} - v_m^- \, \mathrm{e}^{+\mathrm{j}k_{z,m}z}) \tag{3.82}$$

$$H_y = k_y \, \mathrm{e}^{-\mathrm{j}k_y y} \sum_n k_{x,n} \, \mathrm{e}^{-\mathrm{j}k_{x,n}x} \sum_m \frac{q_n^{(m)}}{\omega\mu_{\mathrm{o}}k_{\mathrm{t},m}} (v_m^+ \, \mathrm{e}^{-\mathrm{j}k_{z,m}z} - v_m^- \, \mathrm{e}^{+\mathrm{j}k_{z,m}z}) \tag{3.83}$$

3.3.2.2 TM^x Mode

Similar procedures can be applied to obtain the governing equation for solving the eigenvalue and eigenfunction of the eigenmode supported in the 1D periodic medium. Here, I will not repeat the process and just list the results as follows for reference:

$$\underline{H}(\underline{r}) = \nabla \times \underline{A} \tag{3.84}$$

$$\underline{A} = \underline{x}_{\mathrm{o}} \psi(x, y, z) \tag{3.85}$$

$$\psi(x, y, z) = \phi(x) \, \mathrm{e}^{-\mathrm{j}k_y y} \mathrm{e}^{\mp\mathrm{j}k_z z} \tag{3.86}$$

The governing equation is as follows:

$$\frac{\mathrm{d}}{\mathrm{d}x}\left[\frac{1}{\varepsilon(x)}\frac{\mathrm{d}\phi(x)}{\mathrm{d}x}\right] + k_{\mathrm{o}}^2\phi(x) = \frac{k_{\mathrm{t}}^2}{\varepsilon(x)}\phi(x) \tag{3.87}$$

$$H_y = \mp\mathrm{j}k_z\phi(x) \, \mathrm{e}^{-\mathrm{j}k_y y} \, \mathrm{e}^{\mp\mathrm{j}k_z z} \tag{3.88}$$

$$H_z = \mathrm{j}k_y\phi(x) \, \mathrm{e}^{-\mathrm{j}k_y y} \, \mathrm{e}^{\mp\mathrm{j}k_z z} \tag{3.89}$$

$$E_x = \frac{k_{\mathrm{t}}^2}{\mathrm{j}\omega\varepsilon_{\mathrm{o}}} \frac{\phi(x)}{\varepsilon(x)} \, \mathrm{e}^{-\mathrm{j}k_y y} \, \mathrm{e}^{\mp\mathrm{j}k_z z} \tag{3.90}$$

$$E_y = \frac{-k_y}{\omega\varepsilon_{\mathrm{o}}} \frac{1}{\varepsilon(x)} \frac{\mathrm{d}\phi(x)}{\mathrm{d}x} \, \mathrm{e}^{-\mathrm{j}k_y y} \, \mathrm{e}^{\mp\mathrm{j}k_z z} \tag{3.91}$$

$$E_z = \frac{\mp k_z}{\omega\varepsilon_o} \frac{1}{\varepsilon(x)} \frac{\mathrm{d}\phi(x)}{\mathrm{d}x} \mathrm{e}^{-\mathrm{j}k_y y} \mathrm{e}^{\mp \mathrm{j}k_z z} \tag{3.92}$$

$$\frac{\phi_m(x)}{\varepsilon(x)} = \sum_n q_n^{(m)} \mathrm{e}^{-\mathrm{j}k_{x,n}x} \tag{3.93}$$

$$\phi_m(x) = \sum_n p_n^{(m)} \mathrm{e}^{-\mathrm{j}k_{x,n}x} \tag{3.94}$$

$$\frac{1}{\varepsilon(x)} \frac{\mathrm{d}\phi_m(x)}{\mathrm{d}x} = \sum_n t_n^{(m)} \mathrm{e}^{-\mathrm{j}k_{x,n}x} \tag{3.95}$$

$$E_x = \frac{\phi(x)}{\varepsilon(x)} \mathrm{e}^{-\mathrm{j}k_y y} \mathrm{e}^{\mp \mathrm{j}k_z z} \tag{3.96}$$

$$H_y = \pm \frac{\omega\varepsilon_o}{k_\mathrm{t}^2/k_z} \phi(x) \mathrm{e}^{-\mathrm{j}k_y y} \mathrm{e}^{\mp \mathrm{j}k_z z} \tag{3.97}$$

$$E_y = \frac{-\mathrm{j}k_y}{k_\mathrm{t}^2} \frac{1}{\varepsilon(x)} \frac{\mathrm{d}\phi(x)}{\mathrm{d}x} \mathrm{e}^{-\mathrm{j}k_y y} \mathrm{e}^{\mp \mathrm{j}k_z z} \tag{3.98}$$

$$E_x = \mathrm{e}^{-\mathrm{j}k_y y} \sum_n \mathrm{e}^{-\mathrm{j}k_{x,n}x} \sum_m q_n^{(m)} (v_m^+ \mathrm{e}^{-\mathrm{j}k_{z,m}z} + v_m^- \mathrm{e}^{+\mathrm{j}k_{z,m}z}) \tag{3.99}$$

$$E_y = \mathrm{e}^{-\mathrm{j}k_y y} \sum_n \mathrm{e}^{-\mathrm{j}k_{x,n}x} \sum_m \frac{-\mathrm{j}k_y}{k_{\mathrm{t},m}^2} t_n^{(m)} (v_m^+ \mathrm{e}^{-\mathrm{j}k_{z,m}z} + v_m^- \mathrm{e}^{+\mathrm{j}k_{z,m}z}) \tag{3.100}$$

$$H_y = \mathrm{e}^{-\mathrm{j}k_y y} \sum_n \mathrm{e}^{-\mathrm{j}k_{x,n}x} \sum_m \frac{\omega\varepsilon_o k_{z,m}}{k_{\mathrm{t},m}^2} p_n^{(m)} (v_m^+ \mathrm{e}^{-\mathrm{j}k_{z,m}z} - v_m^- \mathrm{e}^{+\mathrm{j}k_{z,m}z}) \tag{3.101}$$

3.3.3 Eigenwave in a Two-Tone Periodic Medium

In Section 3.3.2 we demonstrated that the field solutions in a 1D periodic medium can be determined by solving the boundary-value problem. Nevertheless, the method to resolve the boundary-value problem has not yet been introduced. In this section, a two-tone periodic medium (i.e., a unit cell consisting of two dielectric slabs) will be taken as an example to calculate the eigenwave solution. The mathematical procedures were developed by Peng *et al.* (1975) for calculating the guiding characteristics of a grating-assisted waveguide. In this example, we begin with the TM^x mode and this time the TE^x is left as an exercise for the reader.

From Equations (3.92) and (3.97), the electric and magnetic fields on the transverse (y–z) plane can be rewritten as

$$E_z(x, y, z) = \frac{1}{\mathrm{j}\omega\varepsilon_o\varepsilon_i} \frac{\mathrm{d}\phi_i(x)}{\mathrm{d}x} \mathrm{e}^{-\mathrm{j}k_y y} \mathrm{e}^{\mp \mathrm{j}k_z z} \tag{3.102}$$

$$H_y(x, y, z) = \phi_i(x) \mathrm{e}^{-\mathrm{j}k_y y} \mathrm{e}^{\mp \mathrm{j}k_z z} \tag{3.103}$$

where i represents the first or second region.

In each uniform layer, the potential function $\phi(x)$ can be easily obtained by solving the second-order differential equation with constant coefficients in Equation (3.87), which can be expressed as the sum of forward- and backward-propagating waves along the positive and negative x-axis. If $\phi(x)$ is designated as current function, then $I(x)$, the x-dependent function in Equation (3.102), can be regarded as the voltage function. The voltage and current functions satisfy the transmission-line equation given below:

$$\frac{\mathrm{d}I_i(x)}{\mathrm{d}x} = -\mathrm{j}k_x^{(i)} Y_i V_i(x) \tag{3.104}$$

$$Z_i = \frac{1}{Y_i} = \frac{k_x^{(i)}}{\omega \varepsilon_0 \varepsilon_i} \tag{3.105}$$

Thus, E_z and H_y can be expressed in terms of the voltage and current waves, as was indicated in Chapter 2:

$$E_z = -V_i(x)\,\mathrm{e}^{-\mathrm{j}k_y y}\,\mathrm{e}^{\mp \mathrm{j}k_z z} \tag{3.106}$$

$$H_y = I_i(x)\,\mathrm{e}^{-\mathrm{j}k_y y}\,\mathrm{e}^{\mp \mathrm{j}k_z z} \tag{3.107}$$

Additionally, we have the other set of tangential electric and magnetic fields, E_y and H_z, which are as follows:

$$E_y = \frac{-1}{\mathrm{j}\omega \varepsilon_0 \varepsilon_i} \frac{\mathrm{d}\phi_i(x)}{\mathrm{d}x}\,\mathrm{e}^{-\mathrm{j}k_y y}\,\mathrm{e}^{\mp \mathrm{j}k_z z} \tag{3.108}$$

$$H_z = \phi_i(x)\,\mathrm{e}^{-\mathrm{j}k_y y}\,\mathrm{e}^{\mp \mathrm{j}k_z z} \tag{3.109}$$

Again, if the potential function $\phi_i(x)$ is set as the current function $I_i(x)$, then the x-dependent function in Equation (3.108) can be considered as the voltage function; they satisfy the same transmission-line equation in Equations (3.104) and (3.105).

Owing to the phase-match condition, the propagation constants along the y- and z-axes, k_y and k_z must be continuous at the interface between two dielectric mediums. The tangential electromagnetic fields are represented by the voltage and current functions. Thus, the electromagnetic fields in the unit cell can be formulated by considering the cascade of two transmission-line sections. We apply the transfer (or ABCD) matrix method demonstrated in Chapter 2 for determining the input–output relation of the voltage and current between the input and output terminals. Subject to the boundary condition (i.e., continuity of the tangential electric and magnetic fields at the interface), we may obtain the dispersion relation given as follows:

$$\cos k_x^{(1)} d_1 \cos k_x^{(2)} d_2 - Q \sin k_x^{(1)} d_1 \sin k_x^{(2)} d_2 = \cos k_x a \tag{3.110}$$

$$k_x^{(1)} = \sqrt{k_0^2 \varepsilon_1 \mu_1 - k_t^2} \tag{3.111}$$

$$k_x^{(2)} = \sqrt{k_0^2 \varepsilon_2 \mu_2 - k_t^2} \tag{3.112}$$

$$k_{\mathrm{t}}^2 = k_y^2 + k_z^2 \tag{3.113}$$

$$Q = \frac{1}{2}\left(\frac{Z_1}{Z_2} + \frac{Z_2}{Z_1}\right) \tag{3.114}$$

$$Z_i = \begin{cases} \frac{\omega\mu_i}{k_x^{(i)}}; & \text{TE} \\ \frac{\omega\varepsilon_0\varepsilon_i}{k_x^{(i)}}; & \text{TM} \end{cases} \tag{3.115}$$

where k_x is the propagation constant along the x-axis. In Equation (3.74), we may solve the relation between k_x and k_z for a given k_y and frequency of operation. Such a problem can be classified into two categories: one is k_z given to find k_x, and the other is k_x given to find k_z. Regarding the first category, this is the single mode problem; i.e., the propagation constant k_x can be directly obtained by evaluating Equation (3.74). The wave is exponentially decaying along the propagating direction when k_x is a complex number, which is the so-called the stop-band behavior. Otherwise, k_x is a real number, which is operated in the pass-band region.

On the other hand, for the second category we have to resolve the nonlinear equation for determining the unknown k_z. There are an infinite number of solutions satisfying such a nonlinear equation, so that it is so-called the multi-mode problem.

3.3.4 Sturm–Liouville Differential Equation with Periodic Boundary Condition

In Section 3.3.3 we solved the eigenwaves in a two-tone periodic medium using the transfer matrix method incorporating the periodic boundary condition. Although the formulation is rigorous, it has difficulty in searching complex roots. In this section we will introduce the eigenfunction expansion method for solving the boundary-value problem. Significantly, the Sturm–Liouville differential equation having the coefficient of arbitrary periodic functions will be treated.

In numerical analysis, the shooting method is a way to solve a boundary-value problem by reducing it to the solution of an initial-value problem. However, for practical applications, the 1D periodic medium is made by stacking the uniform dielectric slabs, so the dielectric function of the medium is a piecewise continuous function. The shooting method fails under this situation.

The Sturm–Liouville differential equation is

$$\frac{\mathrm{d}}{\mathrm{d}x}\left[p(x)\frac{\mathrm{d}}{\mathrm{d}x}\psi(x)\right] + q(x)\psi(x) = \lambda r(x)\psi(x) \tag{3.116}$$

From the Bloch–Floquet theory, the field solution in a 1D periodic medium can be expressed as the product of the envelope of the plane-wave solution by a periodic function. Here, the periodic function is expanded by the series sum of complex exponentials:

$$\psi(x) = \sum_n f_n \phi_n(x) \tag{3.117}$$

$$\phi_n(x) = \frac{1}{\sqrt{a}} \mathrm{e}^{-\mathrm{j}k_{x,n}x} \tag{3.118}$$

$$k_{x,n} = k_x + n\frac{2\pi}{a} \tag{3.119}$$

where $\phi_n(x)$ forms a complete set satisfying the following orthogonal property:

$$\left\langle \phi_m^\dagger(x) \,\middle|\, \phi_n(x) \right\rangle = \delta_{mn} \tag{3.120}$$

where δ is the Kronecker delta.

Multiplying by $\phi_m^\dagger(x)$ on both sides of Equation (3.116) and taking the overlap integral over one period, we may obtain the equation given below.

$$\left\langle \phi_m^\dagger(x) \,\middle|\, \frac{\mathrm{d}}{\mathrm{d}x}\left[p(x)\frac{\mathrm{d}}{\mathrm{d}x}\psi(x)\right] + q(x)\psi(x) \right\rangle = \lambda \left\langle \phi_m^\dagger(x) \,\middle|\, r(x)\psi(x) \right\rangle \tag{3.121}$$

Using the principle of integration by parts, the first inner product in the left-hand side of Equation (3.121) can be written as

$$\left\langle \phi_m^\dagger(x) \,\middle|\, \frac{\mathrm{d}}{\mathrm{d}x}\left[p(x)\frac{\mathrm{d}}{\mathrm{d}x}\psi(x)\right] \right\rangle = \phi_m^\dagger(x)p(x)\frac{\mathrm{d}}{\mathrm{d}x}\psi(x)\bigg|_0^a - \int_0^a p(x)\frac{\mathrm{d}}{\mathrm{d}x}\psi(x)\,\mathrm{d}\phi_m^\dagger(x) \tag{3.122}$$

Moreover, the first term on the right-hand side of Equation (3.122) can be further reduced to the following:

$$\phi_m^\dagger(x)p(x)\frac{\mathrm{d}}{\mathrm{d}x}\psi(x)\bigg|_0^a = \mathrm{j}\frac{p(0)-p(a)}{a}\sum_n f_n k_{x,n} \tag{3.123}$$

The second term in Equation (3.122) can be transformed into the results in Equations (3.124) and (3.125):

$$-\int_0^a p(x)\frac{\mathrm{d}}{\mathrm{d}x}\psi(x)\,\mathrm{d}\phi_m^\dagger(x) = \sum_n f_n p_{mn} \tag{3.124}$$

$$p_{mn} = -\frac{k_{x,m}k_{x,n}}{a}\int_0^a p(x)\,\mathrm{e}^{\mathrm{j}2\pi\frac{m-n}{a}x}\,\mathrm{d}x \tag{3.125}$$

After some mathematical procedures, the overlap integral in the second term in the left-hand side of Equation (3.121) is written as

$$\left\langle \phi_m^\dagger(x) \,\middle|\, q(x)\psi(x) \right\rangle = \sum_n f_n q_{mn} \tag{3.126}$$

$$q_{mn} = \frac{1}{a}\int_0^a \mathrm{e}^{\mathrm{j}2\pi\frac{m-n}{a}x} q(x)\,\mathrm{d}x \tag{3.127}$$

The overlap integral in the right-hand side of Equation (3.121) can be written as

$$\left\langle \phi_m^\dagger(x) \,\middle|\, r(x)\psi(x) \right\rangle = \sum_n f_n r_{mn} \tag{3.128}$$

$$r_{mn} = \frac{1}{a}\int_0^a \mathrm{e}^{\mathrm{j}2\pi\frac{m-n}{a}x} r(x)\,\mathrm{d}x \tag{3.129}$$

Finally, Equation (3.121) can be further reduced to

$$\sum_n f_n(p_{mn} + q_{mn}) = \lambda \sum_n f_n r_{mn} \tag{3.130}$$

By collecting m, we may obtain an infinite set of simultaneous equations containing an infinite number of unknowns, which can be expressed in terms of the following matrix-vector equation:

$$\mathbb{A}\underline{f} = \lambda \mathbb{B}\underline{f} \tag{3.131}$$

$$\mathbb{A} = \{p_{mn} + q_{mn}\}_{m=-\infty,\ldots,+\infty,\, n=-\infty,\ldots,+\infty} \tag{3.132}$$

$$\mathbb{B} = \{r_{mn}\}_{m=-\infty,\ldots,+\infty,\, n=-\infty,\ldots,+\infty} \tag{3.133}$$

$$\underline{f} = \{f_m\}_{m=-\infty,\ldots,+\infty} \tag{3.134}$$

Equation (3.131) is a typical generalized eigenvalue problem. As far as an infinite number of space harmonics is concerned, the mathematical analysis is rigorous. However, owing to limited computational resources and power, we have to truncate matrices $\mathbb{A}$ and $\mathbb{B}$ into finite ones, which will cause some errors compared with the exact ones.

3.4 Eigenwave in a 1D Metallic Periodic Medium

In Sections 3.2 and 3.3 we derived the mathematical formulations for determining the eigenwave in a 1D periodic medium consisting of dielectric material. Actually, the above-mentioned mathematical procedures still work for a metallic 1D periodic medium, since a real metal can be regarded as a dielectric medium with the relative dielectric constant characterized as $\varepsilon_\mathrm{m} = 1 - \mathrm{j}60\sigma\lambda$, where σ and λ are respectivel the conductivity and operational wavelength. However, for microwaves and millimeter-waves the electrical property of a metal is pretty much similar to a perfect electrical conductor (σ approaches infinity). Under this situation, the electric field in the metal vanishes, while the electric field merely survives within the region enclosed by conductors. The wave-propagation characteristic is determined by that of the waveguiding environment, which can be independently treated without considering the periodic nature.

In this section, we will consider the scattering characteristics of a single interface between a semi-infinite 1D metallic periodic medium and the air region. The metal in the periodic medium is assumed to be a PEC; the electromagnetic fields in the periodic medium will be the typical parallel-plate waveguide (ppwg) fields. The input–output relation of the electric and magnetic fields at the interface will be presented using the generalized scattering matrix representation. With the generalized scattering matrix, we are able to deal with the eigenwave in a 2D metallic periodic medium to be introduced in Chapter 4.

3.4.1 Generalized Scattering Matrix at the Interface between a 1D Metallic Periodic Medium and Uniform Medium

As shown in Figure 3.27, a 1D metallic periodic structure of infinite thickness is connected to a semi-infinite uniform medium with relative dielectric constant ϵ_r. Since the metal in the 1D periodic medium is assumed to be a perfect conductor, the electric field inside the conductor vanishes. The electrical field exists only in the region between adjacent conductors. Therefore, the tangential electric and magnetic fields inside the ppwg are expressed in terms of the superposition of the waveguide modes, given below:

$$E_z(x, y) = \sum_{n=1}^{\infty} V_{p,n}(y)\phi_n(x) \tag{3.135}$$

$$H_x(x, y) = \sum_{n=1}^{\infty} I_{p,n}(y)\phi_n(x) \tag{3.136}$$

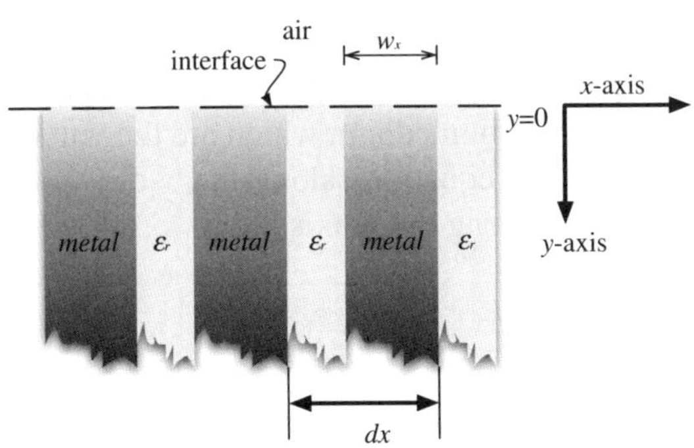

Figure 3.27 A semi-infinite metallic periodic medium

$$\phi_n(x) = \sqrt{\frac{2}{d_x - w_x}} \sin \frac{n\pi x}{d_x - w_x}, n = 1, 2, 3, ... \tag{3.137}$$

Parameters d_x and w_x are respectively the period along the x-axis and the width of the metallic layer. The functions $V_{p,n}(y)$ and $I_{p,n}(y)$ represent the voltage and current waves inside the ppwg for each mode with the mode index n. They satisfy the transmission-line equations given below:

$$\frac{dV_{p,n}(y)}{dy} = -jk_{y,n}^{(p)} Z_n^{(p)} I_{p,n}(y) \tag{3.138}$$

$$\frac{dI_{p,n}(y)}{dy} = -jk_{y,n}^{(p)} Y_n^{(p)} V_{p,n}(y) \tag{3.139}$$

$$k_{y,n}^{(p)} = \sqrt{k_o^2 \varepsilon_r - [n\pi/(d_x - w_x)]^2} \tag{3.140}$$

$$Z_n^{(p)} = \omega\mu_o / k_{y,n}^{(p)} \tag{3.141}$$

On the other hand, the electric and magnetic fields in the uniform dielectric medium in the presence of periodic perturbation can be expressed as the superposition of plane waves (space harmonics):

$$E_z(x, y) = \sum_{n=-\infty}^{n=+\infty} V_{u,n}(y)\varphi_n(x) \tag{3.142}$$

$$H_x(x, y) = \sum_{n=-\infty}^{n=+\infty} I_{u,n}(y)\varphi_n(x) \tag{3.143}$$

$$\varphi_n(x) = \exp(-jk_{x,n}x)/\sqrt{d_x}; \quad n = 0, \pm 1, \pm 2, \ldots \tag{3.144}$$

$$k_{x,n} = k_{x,o} + n2\pi/d_x \tag{3.145}$$

where the functions $V_{u,n}(y)$ and $I_{u,n}(y)$ are the expansion coefficients of the electric and magnetic fields in the uniform medium, which are the voltage and current waves satisfying the transmission-line equations along the y-direction. Their propagation constant and wave impedance along the y-axis are

$$k_{y,n} = \sqrt{k_o^2 \varepsilon_r - k_{x,n}^2} \tag{3.146}$$

$$Z_{u,n} = \omega\mu_o / k_{y,n} \tag{3.147}$$

Based on the electromagnetic boundary conditions, the tangential electric and magnetic fields must be continuous across the interface between the 1D metallic periodic medium and uniform medium, at $y = 0$, shown in Figure 3.27. This yields the following equations:

$$\sum_{n=-\infty}^{\infty} V_{u,n}(0)\varphi_n(x) = \begin{cases} \sum\limits_{n=1}^{\infty} V_{p,n}(0)\phi_n(x); & \text{for } x \in \text{slot} \\ 0; & \text{for } x \in \text{metal} \end{cases} \tag{3.148}$$

$$\sum_{n=-\infty}^{\infty} I_{u,n}(0)\varphi_n(x) = \sum_{n=1}^{\infty} I_{p,n}(0)\phi_n(x); \quad \text{for } x \in \text{slot} \tag{3.149}$$

After taking the overlap integral on both sides by the modal function in a ppwg and Floquet's mode in uniform medium, we can obtain the relationship for the voltage and current wave vectors in each region, at the interface of a step discontinuity ($y = 0$):

$$\underline{V}_{\text{p}}(0) = \mathbf{R}\underline{V}_{\text{u}}(0) \tag{3.150}$$

$$\underline{I}_{\text{p}}(0) = \mathbf{R}\underline{I}_{\text{u}}(0) \tag{3.151}$$

The subscripts "p" and "u" denote the voltage (current) wave in the parallel-plate region and uniform medium, respectively. Parameter $\mathbf{R}$ is a matrix with its elements representing the coupling between the modes inside parallel-plate region and uniform medium, with the mnth entry represented as

$$r_{mn} = \langle \phi_m(x) \mid \varphi_n(x) \rangle \tag{3.152}$$

The symbol "H" designates the operation of transpose and complex conjugation for the matrix (Hermitian matrix). The voltage and current waves at the discontinuity can be expressed as the superposition of incident and reflected waves, a and b, in each region, respectively. Thus, Equations (3.150) and (3.151) can be rewritten as

$$\underline{b}_{\text{p}}(0) + \underline{a}_{\text{p}}(0) = \mathbf{R}[\underline{a}_{\text{u}}(0) + \underline{b}_{\text{u}}(0)] \tag{3.153}$$

$$\mathbf{Y}_{\text{p}}[\underline{b}_{\text{p}}(0) - \underline{a}_{\text{p}}(0)] = \mathbf{RY}[\underline{a}_{\text{u}}(0) - \underline{b}_{\text{u}}(0)] \tag{3.154}$$

Diagonal matrices $\mathbf{Y}_{\text{u}}$ and $\mathbf{Y}_{\text{p}}$ are the admittance matrices in the uniform and parallel-plate regions, respectively. After reformulating Equations (3.153) and (3.154), we obtain, at the discontinuity, the following generalized scattering matrix (a scattering matrix including the propagating as well as nonpropagating waves (Hall *et al.* 1988)):

$$\begin{pmatrix} \underline{b}_{\text{u}}(0) \\ \underline{b}_{\text{p}}(0) \end{pmatrix} = \mathbf{S} \begin{pmatrix} \underline{a}_{\text{u}}(0) \\ \underline{a}_{\text{p}}(0) \end{pmatrix} \tag{3.155}$$

$$\mathbf{S} = \begin{pmatrix} -\mathbf{R} & \mathbf{I} \\ \mathbf{RY}_{\text{u}} & \mathbf{Y}_{\text{p}} \end{pmatrix}^{-1} \begin{pmatrix} \mathbf{R} & -\mathbf{I} \\ \mathbf{RY}_{\text{u}} & \mathbf{Y}_{\text{p}} \end{pmatrix} \tag{3.156}$$

3.5 Hybrid-Mode Analysis of a 1D Dielectric Grating: Fourier-Modal Approach

In the Section 3.4 the modal transmission-line approach was introduced to determine the *rigorous* solution of eigenwaves supported in the 1D periodic medium. Although the field solutions in the 1D periodic medium are exact, they have to be expanded in terms of the Floquet solution, once the 1D periodic medium is terminated by a uniform medium. This is because the electric and magnetic fields in a uniform medium are expressed as the product of a plane-wave envelope function and a periodic function, which is decomposed into the sum of a set of complex exponentials.

In this section we will introduce the Fourier-modal approach for formulating the eigenwaves (source-free solutions) in a 1D periodic medium consisting of a stack of uniform dielectric layers or a dielectric medium with sinusoidal variation on the relative dielectric constant. Consequently, in this section, we directly expand the electric and magnetic fields in the periodic medium using the complex exponentials (Floquet solutions). Through the mathematical procedures to be demonstrated, the propagation constants of the eigenwaves and their associated eigenfunctions will be achieved by solving a typical eigenvalue problem. Such a systematic approach is convenient for solving the guiding and scattering characteristics of a 1D periodic structure with finite thickness.

The TE- and TM-polarized electromagnetic fields in a 1D periodic medium of infinite extent can be independently treated, as mentioned in Section 3.3; however, for a 1D periodic medium connecting to a uniform dielectric medium and obliquely impinged by a plane wave (k_y exists), both of the two orthogonal polarization waves must be taken into account for matching the electromagnetic boundary condition. The formulation regarding the general electric and magnetic fields in a 1D periodic medium based on the transformation between the space- and eigen-coordinate systems was rigorously derived by Peng (1989).

First of all, we assume that the periodicity of the 1D periodic medium is along the x-axis, shown in Figure 3.26. The structure is infinite in extent along the x- and y-axes, while it has finite thickness along the z-axis. Considering the oblique incident condition, the propagation constant along the y-axis, k_y, exists and is determined by the incident condition. From the modal transmission-line approach described previously, the electric- and magnetic-field solutions in a 1D periodic medium can be written as

$$\phi(x, y, z) = \mathrm{e}^{-\mathrm{j}k_y y} \sum_m \mathrm{e}^{-\mathrm{j}k_z^{(m)} z} \varphi_m(x) \tag{3.157}$$

where $k_z^{(m)}$ is the mth eigenvalue and $\varphi_m(x)$ is its associated eigenfunction. From the Bloch–Floquet theory, the x-dependent function $\varphi_m(x)$ can be expressed as a periodic function $p(x)$ multiplied by $\mathrm{e}^{-\mathrm{j}k_{x,o}x}$. Moreover, the periodic function can be expressed in terms of the Fourier series expansion with complex exponentials as its basis as follows:

$$\varphi_m(x) = \mathrm{e}^{-\mathrm{j}k_{x,o}x}\, p(x) \tag{3.158}$$

$$p(x) = p(x+a) = \sum_{n=-\infty}^{n=+\infty} p_n^{(m)}\, \mathrm{e}^{-\mathrm{j}\frac{n2\pi x}{a}} \tag{3.159}$$

$$k_{x,n} = k_x + n\frac{2\pi}{a} \tag{3.160}$$

Substituting Equations (3.158) and (3.159) into Equation (3.157), the general field solution in Equation (3.157) can be rewritten as

$$\phi(x, y, z) = \sum_{n=-\infty}^{n=+\infty} \mathrm{e}^{-\mathrm{j}k_y y}\, \mathrm{e}^{-\mathrm{j}k_{x,n}x} \left[\sum_m p_n^{(m)}\, \mathrm{e}^{-\mathrm{j}k_z^{(m)}z}\right] \tag{3.161}$$

The z-dependent function inside the square brackets in Equation (3.161) is designated as a new function, $v_n(z)$. Therefore, Equation (3.161) can be rewritten as

$$\phi(x, y, z) = \sum_{n=-\infty}^{n=+\infty} \mathrm{e}^{-\mathrm{j}\underline{k}_{t,n}\cdot\underline{\rho}}\, v_n(z) \tag{3.162}$$

$$\underline{k}_{t,n} = \underline{x}_o k_{x,n} + \underline{y}_o k_y \tag{3.163}$$

$$\underline{\rho} = \underline{x}_o x + \underline{y}_o y \tag{3.164}$$

Owing to the general field solution demonstrated previously, the electric- and magnetic-field solutions of each component in a rectangular coordinate system can be written as

$$E_i(x, y, z) = \sum_n \mathcal{E}_{i,n}(z) \exp(-\mathrm{j}\underline{k}_{t,n}\cdot\underline{\rho}) \tag{3.165}$$

$$H_i(x, y, z) = \sum_n \mathcal{H}_{i,n}(z) \exp(-\mathrm{j}\underline{k}_{t,n}\cdot\underline{\rho}) \tag{3.166}$$

where i represents x, y or z. Variables $\mathcal{E}_{i,n}(z)$ and $\mathcal{H}_{i,n}(z)$ are the z-dependent functions of the electric- and magnetic-field components, respectively. Furthermore, the vector tangential electric and magnetic fields are

$$\underline{E}_t(\underline{r}) = \sum_n \underline{\mathcal{E}}_{t,n}(z) \exp(-\mathrm{j}k_{t,n}\cdot\underline{\rho}) \tag{3.167}$$

$$\underline{H}_t(\underline{r}) = \sum_n \underline{\mathcal{H}}_{t,n}(z) \exp(-\mathrm{j}k_{t,n}\cdot\underline{\rho}) \tag{3.168}$$

By decomposing the electric and magnetic fields into the superposition of tangential (x–y plane) and longitudinal (z-axis) components, the Maxwell equation in Equation (3.169) can be rewritten in terms of Equations (3.170) and (3.171), respectively representing the longitudinal and transverse components:

$$\nabla \times \underline{E} = -\mathrm{j}\omega\mu_{\mathrm{o}}\underline{H} \tag{3.169}$$

$$\nabla_{\mathrm{t}} \times \underline{E}_{\mathrm{t}} = -\mathrm{j}\omega\mu_{\mathrm{o}}\underline{H}_{z} \tag{3.170}$$

$$\nabla_{\mathrm{t}} \times \underline{E}_{z} + \frac{\partial}{\partial z}\underline{z}_{\mathrm{o}} \times \underline{E}_{\mathrm{t}} = -\mathrm{j}\omega\mu_{\mathrm{o}}\underline{H}_{\mathrm{t}} \tag{3.171}$$

Substituting Equations (3.166) and (3.167) into Equation (3.170), we obtain

$$\sum_{n} \underline{k}_{\mathrm{t},n} \times \underline{\mathcal{E}}_{\mathrm{t},n}(z)\,\mathrm{e}^{-\mathrm{j}\underline{k}_{\mathrm{t},n}\cdot\underline{\rho}} = \omega\mu_{\mathrm{o}} \sum_{n} \underline{\mathcal{H}}_{z,n}(z)\,\mathrm{e}^{-\mathrm{j}\underline{k}_{\mathrm{t},n}\cdot\underline{\rho}} \tag{3.172}$$

By equating the same Fourier amplitude corresponding to the same harmonic on both sides of Equation (3.172), we have the following relationship:

$$\underline{k}_{\mathrm{t},n} \times \underline{\mathcal{E}}_{\mathrm{t},n}(z) = \omega\mu_{\mathrm{o}}\underline{\mathcal{H}}_{z,n}(z) \tag{3.173}$$

Furthermore, Equation (3.173) can be rewritten as

$$-k_{y}\mathcal{E}_{x,n}(z) + k_{x,n}\mathcal{E}_{y,n}(z) = \omega\mu_{\mathrm{o}}\mathcal{H}_{z,n}(z) \tag{3.174}$$

Collecting each of the harmonics of the electric- and magnetic-field components and putting into vectors, the above equation can be rewritten in a matrix-vector form:

$$\underline{\mathcal{H}}_{z}(z) = -\frac{k_{y}}{\omega\mu_{\mathrm{o}}}\underline{\mathcal{E}}_{x}(z) + \frac{\mathbb{K}_{x}}{\omega\mu_{\mathrm{o}}}\underline{\mathcal{E}}_{y}(z) \tag{3.175}$$

$$\underline{\mathcal{E}}_{x}(z) = \{\mathcal{E}_{x,n}\}, \quad n = +\infty, \ldots, -\infty \tag{3.176}$$

$$\underline{\mathcal{E}}_{y}(z) = \{\mathcal{E}_{y,n}\}, \quad n = +\infty, \ldots, -\infty \tag{3.177}$$

$$\underline{\mathcal{H}}_{z}(z) = \{\mathcal{H}_{z,n}\}, \quad n = +\infty, \ldots, -\infty \tag{3.178}$$

Parameter $\mathbb{K}_x$ is a diagonal matrix with its diagonal element filled with $k_{x,n}$, where n ranges from positive to negative infinities.

By substituting Equations (3.165), (3.167) and (3.168) into Equation (3.171), we obtain

$$\begin{aligned} &\sum_{n} -\mathrm{j}\underline{k}_{\mathrm{t},n} \times \underline{\mathcal{E}}_{z,n}(z)\,\mathrm{e}^{-\mathrm{j}\underline{k}_{\mathrm{t},n}\cdot\underline{\rho}} \\ &\quad + \sum_{n} \frac{\partial}{\partial z}\underline{z}_{\mathrm{o}} \times \underline{\mathcal{E}}_{\mathrm{t},n}(z)\,\mathrm{e}^{-\mathrm{j}\underline{k}_{\mathrm{t},n}\cdot\underline{\rho}} \\ &\quad = -\mathrm{j}\omega\mu_{\mathrm{o}} \sum_{n} \underline{\mathcal{H}}_{\mathrm{t},n}(z)\,\mathrm{e}^{-\mathrm{j}\underline{k}_{\mathrm{t},n}\cdot\underline{\rho}} \end{aligned} \tag{3.179}$$

By equating the same Fourier harmonics on both sides of Equation (3.179), we obtain the relationship between the electric and magnetic field component:

$$-\mathrm{j}\underline{k}_{\mathrm{t},n} \times \underline{\mathcal{E}}_{z,n}(z) + \frac{\partial}{\partial z}\underline{z}_{\mathrm{o}} \times \underline{\mathcal{E}}_{\mathrm{t},n}(z) = -\mathrm{j}\omega\mu_{\mathrm{o}}\underline{\mathcal{H}}_{\mathrm{t},n}(z) \tag{3.180}$$

Following the same procedure described previously, we could rewrite the above equation in terms of the following two matrix-vector equations:

$$-\mathrm{j}k_y\underline{\mathcal{E}}_z(z) - \frac{\mathrm{d}}{\mathrm{d}z}\underline{\mathcal{E}}_y(z) = -\mathrm{j}\omega\mu_{\mathrm{o}}\underline{\mathcal{H}}_x(z) \tag{3.181}$$

$$\mathrm{j}\mathbb{K}_x\underline{\mathcal{E}}_z(z) + \frac{\mathrm{d}}{\mathrm{d}z}\underline{\mathcal{E}}_x(z) = -\mathrm{j}\omega\mu_{\mathrm{o}}\underline{\mathcal{H}}_y(z) \tag{3.182}$$

By decomposing the curl operator and the field components in Equation (3.183) into the transverse and longitudinal ones, we obtain the two equations given in Equations (3.184) and (3.185):

$$\nabla \times \underline{H} = \mathrm{j}\omega\varepsilon_{\mathrm{o}}\varepsilon(x)\underline{E} \tag{3.183}$$

$$\nabla_{\mathrm{t}} \times \underline{H}_{\mathrm{t}} = \mathrm{j}\omega\varepsilon_{\mathrm{o}}\varepsilon(x)\underline{E}_z \tag{3.184}$$

$$\nabla_{\mathrm{t}} \times \underline{H}_z + \frac{\partial}{\partial z}\underline{z}_{\mathrm{o}} \times \underline{H}_{\mathrm{t}} = \mathrm{j}\omega\varepsilon_{\mathrm{o}}\varepsilon(x)\underline{E}_{\mathrm{t}} \tag{3.185}$$

Substituting Equations (3.165) and (3.168) into Equation (3.184), we obtain

$$\sum_n \underline{k}_{\mathrm{t},n} \times \underline{\mathcal{H}}_{\mathrm{t},n}(z)\,\mathrm{e}^{-\mathrm{j}\underline{k}_{\mathrm{t},n}\cdot\underline{\rho}} = -\omega\varepsilon_{\mathrm{o}}\varepsilon(x)\sum_n \underline{\mathcal{E}}_{z,n}(z)\,\mathrm{e}^{-\mathrm{j}\underline{k}_{\mathrm{t},n}\cdot\underline{\rho}} \tag{3.186}$$

Since $\varepsilon(x)$ is the relative dielectric constant, it cannot be zero. We could divide Equation (3.186) on both sides by $\varepsilon(x)$, and the above equation can be rewritten as

$$-\sum_n \underline{k}_{\mathrm{t},n} \times \underline{\mathcal{H}}_{\mathrm{t},n}(z)\frac{1}{\varepsilon(x)}\,\mathrm{e}^{-\mathrm{j}\underline{k}_{\mathrm{t},n}\cdot\underline{\rho}} = \omega\varepsilon_{\mathrm{o}}\sum_n \underline{\mathcal{E}}_{z,n}(z)\,\mathrm{e}^{-\mathrm{j}\underline{k}_{\mathrm{t},n}\cdot\underline{\rho}} \tag{3.187}$$

By multiplying both sides of Equation (3.187) with $\exp(\mathrm{j}\underline{k}_{t,m}\cdot\underline{\rho})$ and integrating over a period, we obtain

$$\mathcal{E}_{z,m}(z) = -\sum_n \frac{\eta_{mn}}{\omega\varepsilon_{\mathrm{o}}}\underline{z}_{\mathrm{o}} \times \underline{k}_{\mathrm{t},n}\cdot\underline{\mathcal{H}}_{\mathrm{t},n}(z) \tag{3.188}$$

$$\langle \mathrm{e}^{\mathrm{j}\underline{k}_{\mathrm{t},m}\cdot\underline{\rho}} | \mathrm{e}^{-\mathrm{j}\underline{k}_{\mathrm{t},n}\cdot\underline{\rho}}\rangle = d_x\delta_{mn} \tag{3.189}$$

$$\left\langle \mathrm{e}^{\mathrm{j}\underline{k}_{\mathrm{t},m}\cdot\underline{\rho}} \left| \frac{1}{\varepsilon(x)}\,\mathrm{e}^{-\mathrm{j}\underline{k}_{\mathrm{t},n}\cdot\underline{\rho}} \right.\right\rangle = d_x\eta_{mn} \tag{3.190}$$

Equation (3.187) can be rewritten in terms of the matrix-vector form as follows:

$$\underline{\mathcal{E}}_z(z) = \frac{1}{\omega\varepsilon_{\mathrm{o}}}\left[\|\eta\|k_y\underline{\mathcal{H}}_x(z) - \|\eta\|\mathbb{K}_x\underline{\mathcal{H}}_y(z)\right] \tag{3.191}$$

In the above equation, the longitudinal component of the electric field vector is replaced by the tangential component of the magnetic field vectors.

Substituting Equations (3.166), (3.167) and (3.168) into Equation (3.185), we obtain

$$\sum_n -\mathrm{j}\underline{k}_{\mathrm{t},n} \times \underline{\mathcal{H}}_{z,n}(z)\,\mathrm{e}^{-\mathrm{j}\underline{k}_{\mathrm{t},n}\cdot\underline{\rho}} + \sum_n \frac{\partial}{\partial z}\underline{z}_{\mathrm{o}} \times \underline{\mathcal{H}}_{\mathrm{t},n}(z)\,\mathrm{e}^{-\mathrm{j}\underline{k}_{\mathrm{t},n}\cdot\underline{\rho}} = \mathrm{j}\omega\varepsilon_{\mathrm{o}} \sum_n \underline{\mathcal{E}}_{\mathrm{t},n}(z)\varepsilon(x)\,\mathrm{e}^{-\mathrm{j}\underline{k}_{\mathrm{t},n}\cdot\underline{\rho}} \tag{3.192}$$

By multiplying both sides of Equation (3.192) with $\exp(\mathrm{j}\underline{k}_{\mathrm{t},m}\cdot\underline{\rho})$ and taking the overlap integral over a period, we obtain the following equation:

$$-\mathrm{j}\underline{k}_{\mathrm{t},m} \times \underline{\mathcal{H}}_{z,m}(z) + \frac{\partial}{\partial z}\underline{z}_{\mathrm{o}} \times \underline{\mathcal{H}}_{\mathrm{t},m}(z) = \mathrm{j}\omega\varepsilon_{\mathrm{o}} \sum_n \underline{\mathcal{E}}_{\mathrm{t},n}(z)\epsilon_{mn} \tag{3.193}$$

$$\langle \mathrm{e}^{\mathrm{j}\underline{k}_{\mathrm{t},m}\cdot\underline{\rho}} \mid \varepsilon(x)\,\mathrm{e}^{-\mathrm{j}\underline{k}_{\mathrm{t},n}\cdot\underline{\rho}} \rangle = d_x \epsilon_{mn} \tag{3.194}$$

Equation (3.193) can be expressed in matrix-vector form as

$$\mathrm{j}\mathbb{K}_x \underline{\mathcal{H}}_z(z) + \frac{\mathrm{d}}{\mathrm{d}z}\underline{\mathcal{H}}_x(z) = \mathrm{j}\omega\varepsilon_{\mathrm{o}}\|\epsilon\|\underline{\mathcal{E}}_y(z) \tag{3.195}$$

$$-\mathrm{j}k_y \underline{\mathcal{H}}_z(z) - \frac{\mathrm{d}}{\mathrm{d}z}\underline{\mathcal{H}}_y(z) = \mathrm{j}\omega\varepsilon_{\mathrm{o}}\|\epsilon\|\underline{\mathcal{E}}_x(z) \tag{3.196}$$

So far, we have determined the relationship between the electric- and magnetic-field components through the matrix-vector equations obtained previously. For ease of inspecting their relationship, we rewrite them as follows:

$$-\mathrm{j}k_y \underline{\mathcal{E}}_z(z) - \frac{\mathrm{d}}{\mathrm{d}z}\underline{\mathcal{E}}_y(z) = -\mathrm{j}\omega\mu_{\mathrm{o}}\underline{\mathcal{H}}_x(z) \tag{3.197}$$

$$\mathrm{j}\mathbb{K}_x \underline{\mathcal{E}}_z(z) + \frac{\mathrm{d}}{\mathrm{d}z}\underline{\mathcal{E}}_x(z) = -\mathrm{j}\omega\mu_{\mathrm{o}}\underline{\mathcal{H}}_y(z) \tag{3.198}$$

$$\underline{\mathcal{E}}_z(z) = \frac{1}{\omega\varepsilon_{\mathrm{o}}}\left[\|\eta\|\,k_y \underline{\mathcal{H}}_x(z) - \|\eta\|\,\mathbb{K}_x \underline{\mathcal{H}}_y(z)\right] \tag{3.199}$$

In Equation (3.199), the z-component of the electric field is expressed in terms of the tangential magnetic field components. Therefore, we could substitute Equation (3.199) into Equations (3.197) and (3.198) to eliminate the z-component to obtain a vector differential equation given as

$$\frac{\mathrm{d}}{\mathrm{d}z}\begin{bmatrix} -\underline{\mathcal{E}}_y(z) \\ \underline{\mathcal{E}}_x(z) \end{bmatrix} = -\mathrm{j}\omega\mu_{\mathrm{o}}\mathbb{Z}\begin{bmatrix} \underline{\mathcal{H}}_x(z) \\ \underline{\mathcal{H}}_y(z) \end{bmatrix} \tag{3.200}$$

$$\mathbb{Z} = \begin{bmatrix} \mathbb{I} - \dfrac{k_y\|\eta\|k_y}{k_{\mathrm{o}}^2} & \dfrac{k_y\|\eta\|\mathbb{K}_x}{k_{\mathrm{o}}^2} \\ \dfrac{\mathbb{K}_x\|\eta\|k_y}{k_{\mathrm{o}}^2} & \mathbb{I} - \dfrac{\mathbb{K}_x\|\eta\|\mathbb{K}_x}{k_{\mathrm{o}}^2} \end{bmatrix} \tag{3.201}$$

By employing a similar procedure mentioned previously, we may determine the differential equation of tangential magnetic fields by eliminating the z-component in Equations (3.202) and (3.203) via Equation (3.204):

$$\mathrm{j}\mathbb{K}_x \underline{\mathcal{H}}_z(z) + \frac{\mathrm{d}}{\mathrm{d}z}\underline{\mathcal{H}}_x(z) = \mathrm{j}\omega\varepsilon_\mathrm{o} \|\epsilon\| \underline{\mathcal{E}}_y(z) \tag{3.202}$$

$$-\mathrm{j}k_y \underline{\mathcal{H}}_z(z) - \frac{\mathrm{d}}{\mathrm{d}z}\underline{\mathcal{H}}_y(z) = \mathrm{j}\omega\varepsilon_\mathrm{o} \|\epsilon\| \underline{\mathcal{E}}_x(z) \tag{3.203}$$

$$\underline{\mathcal{H}}_z(z) = -\frac{k_y}{\omega\mu_\mathrm{o}}\underline{\mathcal{E}}_x(z) + \frac{\mathbb{K}_x}{\omega\mu_\mathrm{o}}\underline{\mathcal{E}}_y(z) \tag{3.204}$$

The vector differential equation is

$$\frac{\partial}{\partial z}\begin{bmatrix} \underline{\mathcal{H}}_x(z) \\ \underline{\mathcal{H}}_y(z) \end{bmatrix} = -\mathrm{j}\omega\varepsilon_\mathrm{o}\mathbb{Y}\begin{bmatrix} -\underline{\mathcal{E}}_y(z) \\ \underline{\mathcal{E}}_x(z) \end{bmatrix} \tag{3.205}$$

$$\mathbb{Y} = \begin{bmatrix} \|\epsilon\| - \frac{\mathbb{K}_x^2}{k_\mathrm{o}^2} & -\frac{\mathbb{K}_x k_y}{k_\mathrm{o}^2} \\ -\frac{k_y \mathbb{K}_x}{k_\mathrm{o}^2} & \|\epsilon\| - \frac{k_y^2}{k_\mathrm{o}^2}\mathbb{I} \end{bmatrix} \tag{3.206}$$

By differentiating Equation (3.200) with respect to z on both sides, we obtain the equation given below:

$$\frac{d^2}{dz^2}\begin{bmatrix} -\underline{\mathcal{E}}_y(z) \\ \underline{\mathcal{E}}_x(z) \end{bmatrix} = -\mathrm{j}\omega\mu_\mathrm{o}\mathbb{Z}\frac{\mathrm{d}}{\mathrm{d}z}\begin{bmatrix} \underline{\mathcal{H}}_x(z) \\ \underline{\mathcal{H}}_y(z) \end{bmatrix} \tag{3.207}$$

Substituting Equation (3.205) into Equation (3.207), we obtain a second-order differential equation.

$$\frac{\mathrm{d}^2}{\mathrm{d}z^2}\begin{bmatrix} -\underline{\mathcal{E}}_y(z) \\ \underline{\mathcal{E}}_x(z) \end{bmatrix} = -\mathbb{A}\begin{bmatrix} -\underline{\mathcal{E}}_y(z) \\ \underline{\mathcal{E}}_x(z) \end{bmatrix} \tag{3.208}$$

$$\mathbb{A} = k_\mathrm{o}^2\mathbb{Z}\mathbb{Y} \tag{3.209}$$

Owing to finite computational resources and power, we have to truncate the vectors and matrix in Equation (3.208) into finite ones. In the numerical examples to be demonstrated in Section 3.6, we will discuss the numerical convergence of the eigenvalue against the number of space harmonics employed.

As is well known in matrix theory, if we truncate the matrix $\mathbb{A}$ into a finite and square one, it can be factorized as

$$\mathbb{A} = \mathbb{Q}\mathbb{D}\mathbb{Q}^{-1} \tag{3.210}$$

where $\mathbb{Q}$ is a square matrix whose ith column is the eigenvector $\underline{q}_i$ of $\mathbb{A}$ and $\mathbb{D}$ is the diagonal matrix whose diagonal elements are the corresponding eigenvalues; i.e., $\lambda_i = d_{ii}$.

Consequently, the general solution to Equation (3.208) can be written as

$$\begin{bmatrix} -\underline{\mathcal{E}}_y(z) \\ \underline{\mathcal{E}}_x(z) \end{bmatrix} = \mathbb{Q}[\exp(-\mathrm{j}\mathbb{K}_z z)\underline{f} + \exp(+\mathrm{j}\mathbb{K}_z z)\underline{g}] \tag{3.211}$$

where $\exp(-\mathrm{j}\mathbb{K}_z z)$ is a diagonal matrix whose ith diagonal element equals $\exp(-\mathrm{j}\sqrt{d_{ii}}z)$. Vectors $\underline{f}$ and $\underline{g}$ are the amplitudes of the forward- and backward-propagating space harmonics to be determined by the excitation condition.

Substituting Equation (3.211) into Equation (3.200), we may derive the tangential magnetic field as follows:

$$\begin{bmatrix} \underline{\mathcal{H}}_x(z) \\ \underline{\mathcal{H}}_y(z) \end{bmatrix} = \mathbb{P}[\exp(-\mathrm{j}\mathbb{K}_z z)\underline{f} - \exp(+\mathrm{j}\mathbb{K}_z z)\underline{g}] \tag{3.212}$$

$$\mathbb{P} = \frac{\mathbb{Z}^{-1}\mathbb{Q}\mathbb{K}_z}{\omega\mu_o} \tag{3.213}$$

3.6 Input–Output Relation of a 1D Periodic Medium of Finite Thickness

With the general electric- and magnetic-field solutions in a 1D periodic medium being determined using the modal transmission-line or modal Fourier approach, now we are able to carry out the calculation of scattering analysis of the input–output relation of a single interface between the 1D periodic medium and uniform dielectric medium. Moreover, the scattering characteristics of a single 1D periodic medium of finite thickness can be further determined. In fact, such a 1D periodic layer is a called a dielectric diffraction grating and is commonly used in the optics community.

To begin the mathematical formulation, first of all we have to recall the general field solutions in a uniform dielectric medium, which are given in Chapter 2. Consider the periodic nature; the tangential components (in the x–y plane) of the electric and magnetic fields in a uniform dielectric medium are expressed as

$$\underline{z}_0 \times \underline{E}_\mathrm{t}(x, y, z) = \sum_n [\underline{a}'_n v'_n(z) + \underline{a}''_n v''_n(z)]\,\mathrm{e}^{-\mathrm{j}\underline{k}_{\mathrm{t},n}\underline{\rho}} \tag{3.214}$$

$$\underline{H}_\mathrm{t}(x, y, z) = \sum_n [\underline{a}'_n i'_n(z) + \underline{a}''_n i''_n(z)]\,\mathrm{e}^{-\mathrm{j}\underline{k}_{\mathrm{t},n}\underline{\rho}} \tag{3.215}$$

Since the tangential electric and magnetic fields in the 1D periodic medium were expressed in terms of vectors on the order of the space harmonic – which were given in Equations (3.211) and (3.212) – to match the electromagnetic boundary condition, those in the uniform dielectric medium have to be expressed in the same form for ease of performing the mode-matching procedure. In doing so, the tangential electric and magnetic fields are written as follows:

$$\begin{bmatrix} -\underline{E}_y(z) \\ \underline{E}_x(z) \end{bmatrix} = \begin{pmatrix} \mathbf{A}_x & -\mathbf{A}_y \\ \mathbf{A}_y & \mathbf{A}_x \end{pmatrix} \begin{bmatrix} \underline{V}'(z) \\ \underline{V}''(z) \end{bmatrix} \tag{3.216}$$

$$\begin{bmatrix} \underline{H}_x(z) \\ \underline{H}_y(z) \end{bmatrix} = \begin{pmatrix} \mathbf{A}_x & -\mathbf{A}_y \\ \mathbf{A}_y & \mathbf{A}_x \end{pmatrix} \begin{bmatrix} \underline{I}'(z) \\ \underline{I}''(z) \end{bmatrix} \tag{3.217}$$

where $\mathbf{A}_x$ and $\mathbf{A}_y$ are diagonal matrices with $k_{x,n}/k_{t,n}$ and $k_y/k_{t,n}$ filling at their nth entry, respectively. The voltage (current) mode amplitude of the nth space harmonic locates at the nth element of the voltage (current) vector. Moreover, the voltage and current wave satisfy the transmission-line equation along the z-axis. Thus, the voltage and current vectors in the previous two equations can be written as

$$\underline{V}(z) = \begin{bmatrix} \underline{V}'(z) \\ \underline{V}''(z) \end{bmatrix} = \exp(-\mathrm{j}\mathbf{k}_z z)\underline{a} + \exp(+\mathrm{j}\mathbf{k}_z z)\underline{b} \tag{3.218}$$

$$\underline{I}(z) = \begin{bmatrix} \underline{I}'(z) \\ \underline{I}''(z) \end{bmatrix} = \mathbf{Y}[\exp(-\mathrm{j}\mathbf{k}_z z)\underline{a} - \exp(+\mathrm{j}\mathbf{k}_z z)\underline{b}] \tag{3.219}$$

$$\mathbf{Y} = \begin{pmatrix} \mathbf{Y}' & \mathbf{0} \\ \mathbf{0} & \mathbf{Y}'' \end{pmatrix} \tag{3.220}$$

$$\mathbf{k}_z = \begin{pmatrix} \mathbf{k}_z' & \mathbf{0} \\ \mathbf{0} & \mathbf{k}_z'' \end{pmatrix} \tag{3.221}$$

where $\mathbf{k}_z'$ and $\mathbf{k}_z''$ are diagonal matrices, which are the same in a uniform dielectric medium. The nth diagonal element is equal to $\sqrt{k_0^2\varepsilon_s - k_t^2}$, where ε_s is the relative dielectric constant. The prime and double prime symbols denote the TE- and TM-polarized plane waves in a uniform medium. Matrix $\mathbf{0}$ is the null matrix. The vectors $\underline{a}$ and $\underline{b}$ are respectively the forward- and backward-propagating waves along the transmission line, each of which contains the TE and TM space-harmonic amplitudes, with the representation given below:

$$\underline{a} = \begin{pmatrix} \underline{a}' \\ \underline{a}'' \end{pmatrix} \tag{3.222}$$

$$\underline{b} = \begin{pmatrix} \underline{b}' \\ \underline{b}'' \end{pmatrix} \tag{3.223}$$

We assume that a 1D periodic dielectric medium has thickness t_g along the z-axis and period d along the x-axis. If such a 1D dielectric grating is terminated by an impedance matrix denoted $\mathbf{Z}_{\text{out}}$, the tangential electric and magnetic fields, which are given in Equations (3.211) and (3.212), satisfy the relation written as follows:

$$\mathbf{Q}[\exp(-\mathbf{j}\mathbf{k}_z^{(g)}t_g)\underline{f} + \exp(+\mathbf{j}\mathbf{k}_z^{(g)}t_g)\underline{g}] = \mathbf{Z}_{\text{out}}\mathbf{P}[\exp(-\mathbf{j}\mathbf{k}_z^{(g)}t_g)\underline{f} - \exp(+\mathbf{j}\mathbf{k}_z^{(g)}t_g)\underline{g}] \tag{3.224}$$

The vectors $\underline{f}$ and $\underline{g}$ are respectively the forward- and backward-propagating waves along the z-axis. To distinguish the propagation constant between the periodic (grating) and uniform mediums, the superscript "g" in the above equation represents that in the grating region. After simple mathematical derivation, we can determine the relationship between the vectors $\underline{f}$ and $\underline{g}$, as given below:

$$\underline{g} = \Gamma\underline{f} \tag{3.225}$$

$$\Gamma = \exp(-\mathbf{j}\mathbf{k}_z^{(g)}t_g)\Gamma_l\exp(-\mathbf{j}\mathbf{k}_z^{(g)}t_g) \tag{3.226}$$

$$\Gamma_l = (\mathbf{Z}_{\text{out}}\mathbf{P} + \mathbf{Q})^{-1}(\mathbf{Z}_{\text{out}}\mathbf{P} - \mathbf{Q}) \tag{3.227}$$

The matrix Γ is called the reflection matrix, which relates each of the forward- and backward-propagating space harmonics. Once the relationship between $\underline{f}$ and $\underline{g}$ is determined, we could substitute it into the tangential electric and magnetic fields in Equations (3.211) and (3.212). Moreover, the input impedance matrix $\mathbf{Z}_{\text{in}}$, which is defined as the relationship between the tangential electric and magnetic field vectors, and the transfer matrix $\mathbf{T}$, for the tangential electric field between the input and output interfaces, can be obtained and are written as follows:

$$\underline{\mathcal{E}}(0) = \mathbf{Z}_{\text{in}}\underline{\mathcal{H}}(0) \tag{3.228}$$

$$\mathbf{Z}_{\text{in}} = \mathbf{Q}(\mathbf{I} + \Gamma)(\mathbf{I} - \Gamma)^{-1}\mathbf{P}^{-1} \tag{3.229}$$

$$\underline{\mathcal{E}}(t_g) = \mathbf{T}(t_g; 0)\underline{\mathcal{E}}(0) \tag{3.230}$$

$$\mathbf{T} = \mathbf{Q}(\mathbf{I} + \Gamma_l)\exp(-\mathbf{j}\mathbf{k}_z^{(g)}t_g)(\mathbf{I} + \Gamma)^{-1}\mathbf{Q}^{-1} \tag{3.231}$$

Problem 3.1

Prove the input impedance matrix and electric-field transfer matrix given in Equations (3.229) and (3.231).

In addition, Equations (3.211) and (3.212) can also be applied in the case of a uniform dielectric layer, which is in an environment having a periodic nature, by replacing $\mathbf{Q}$ with $\mathbf{A}$, and $\mathbf{P}$ with $\mathbf{AY}$. The matrix $\mathbf{A}$ is defined as

$$\mathbf{A} = \begin{pmatrix} \mathbf{A}_x & -\mathbf{A}_y \\ \mathbf{A}_y & \mathbf{A}_x \end{pmatrix} \tag{3.232}$$

Notably, matrix $\mathbf{A}$ is the coordinate transformation matrix which transforms the electric- and magnetic-field quantities from the eigen-coordinate system to the spatial-coordinate system. If the plane wave is incident in the principal plane ($\phi = 0$), the propagation constant in the y-axis, k_y vanishes. The matrix $\mathbf{A}$ returns to an identity matrix.

3.7 Scattering Characteristics of a Grating Consisting of Multiple 1D Periodic Layers

3.7.1 *Building-Block Approach*

Consider a structure consisting of N 1D periodic layers sandwiched by two semi-infinite uniform mediums with relative dielectric constants ε_+ and ε_- in the upper and lower regions, respectively (Figure 3.28). A plane wave with incident angles θ and ϕ from the upper region impinges on the structure obliquely. The propagation constants along the x- and y-axes are $k_x = k_o\sqrt{\varepsilon_+}\sin\theta\cos\phi$ and $k_y = k_o\sqrt{\varepsilon_+}\sin\theta\sin\phi$. We begin with the last (Nth) 1D periodic layer. Since the output region is unterminated, the input impedance matrix looking from the output interface of the Nth layer is the characteristic impedance matrix of the medium, which is given as $\mathbf{Z}_{\text{in}} = \mathbf{A}\mathbf{Z}_-\mathbf{A}^{-1}$, where $\mathbf{Z}_-$ is the inverse of the admittance matrix in Equation (3.220). Here, the input impedance matrix, in fact, is the output impedance matrix of the Nth layer. By substituting this impedance matrix into Equations (3.229) and (3.231), the input impedance matrix looking into the Nth layer as well as the electric-field transfer matrix through

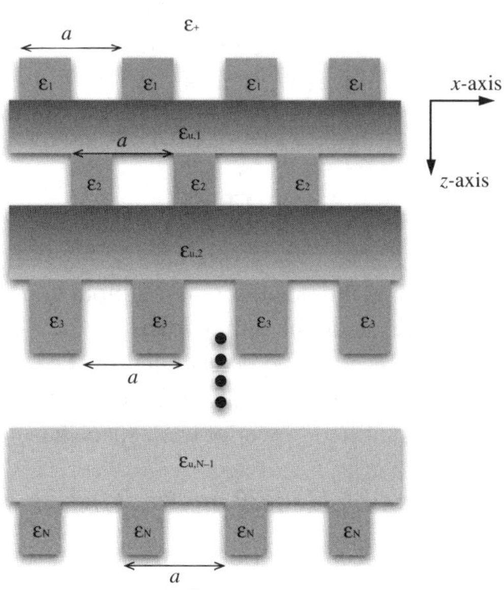

Figure 3.28 A stack of 1D periodic layers

the layer can be obtained. Moreover, the above input impedance matrix can be regarded as the output impedance matrix of the $(N-1)$th layer. Again, the formulas for the input impedance matrix and the electric-field transfer matrix are employed to determine the input impedance matrix looking into the $(N-1)$th layer and the transfer matrix as well.

After recursively employing the procedure described previously, we can finally determine the input impedance matrix looking into the first layer, which is denoted $\mathbf{Z}_{\text{in}}^{(1)}$. At the input interface $z=0$, the tangential electric and magnetic fields are

$$\underline{\mathcal{E}}(0) = \mathbf{A}(\underline{a} + \underline{b}) \tag{3.233}$$

$$\underline{\mathcal{H}}(0) = \mathbf{AY}(\underline{a} - \underline{b}) \tag{3.234}$$

where $\underline{a}$ and $\underline{b}$ are the incident and reflected wave vectors storing the amplitude of each space harmonic. Matrix $\mathbf{Y}$ is a diagonal matrix with its elements representing the admittance of the TE and TM space harmonic. Incorporating the boundary condition (input impedance matrix $\mathbf{Z}_{\text{in}}^{(1)}$) given at $z=0$, the reflection matrix relating the incident and reflected wave vectors is written as

$$\underline{b} = \Gamma \underline{a} \tag{3.235}$$

$$\Gamma = (\mathbf{A}^{-1}\mathbf{Z}_{\text{in}}\mathbf{AY} + \mathbf{I})^{-1}(\mathbf{A}^{-1}\mathbf{Z}_{\text{in}}\mathbf{AY} - \mathbf{I}) \tag{3.236}$$

Notably, Γ is the reflection matrix of the overall structure. Once the incident wave vector $\underline{a}$ is given, the reflected amplitude of each space harmonic can be determined by Equation (3.235). Additionally, the electric-field transfer matrix of the overall structure composed of N 1D periodic layers can be obtained by successively multiplying the transfer matrix of each layer, which is written as

$$\underline{\mathcal{E}}(z_{\text{out}}) = \mathbf{T}\underline{\mathcal{E}}(z_{\text{in}}) \tag{3.237}$$

$$\mathbf{T} = \prod_{i=N}^{1} \mathbf{T}_i \tag{3.238}$$

where z_{in} and z_{out} are denoted as the positions of the input and output interface.

Since the electric field vector at the input interface is $\mathbf{A}(\mathbf{I}+\Gamma)\underline{a}$, the electric-field vector at the output interface can be determined by $\mathbf{TA}(\mathbf{I}+\Gamma)\underline{a}$. Therefore, the forward-propagating (along z-axis) wave amplitude of each space harmonic can be completely obtained. The transmitted power of each space harmonic can be achieved.

3.7.2 Scattering Analysis of 1D Diffraction Gratings

As is well known, the 1D dielectric periodic structure is traditionally used as a diffraction grating for diffracting an incident beam into several beams, as shown in Figure 3.29. Such a structure can serve as a beam splitter or a device for optical interconnects. In this section, several examples concerning the scattering analysis of

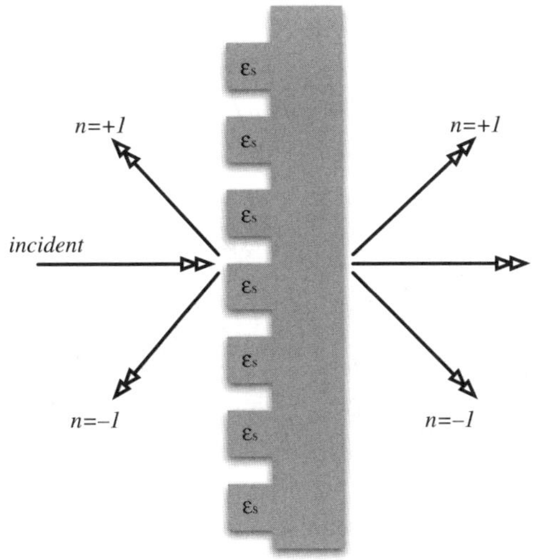

Figure 3.29 A diffraction grating employed as a beam splitter

the 1D dielectric grating will be studied for exploring its scattering properties. I have implemented a computer program based on the hybrid-mode approach described previously to calculate the reflectance and transmittance of propagating space harmonics in an environment consisting of multiple 1D periodic layers having the same period. To verify the accuracy of my approach I use the case (structural parameters) of Sharon *et al.* (1997) as an example. In their paper, they fabricated the grating-waveguide structure by electron-beam lithography and chemical vapor deposition and measured the transmittance and reflectance. Their measurement results will be used to examine the numerical accuracy of my computer codes. As shown in Figure 3.30, the structure is composed of two layers: a grating layer and a dielectric waveguide mounted on the silica substrate. The top one is the grating layer made up of silicon nitride–SiO_2 with refractive index $n = 1.73$; the second one is the waveguide, which is also made of silicon nitride. Below the dielectric waveguide is a high-optical-quality fused silica substrate that had an index of refraction of ~ 1.46.

Figure 3.31 demonstrates the reflection and transmission intensities, which are normalized to the incident power intensity, against the operation wavelength. The TE-polarized plane wave (electric field is along the y-axis) impinges on the structure with incident angle 1°. The numerical results obtained from my code agree very well with measured ones, although not shown here, confirming the numerical accuracy of the computer code.

With the reliable numerical results compared with the literature, we may carry out numerous numerical experiments for demonstrating the scattering characteristics of 1D dielectric grating. The following example is a grating waveguide having the structure configuration shown in Figure 3.32. The 1D grating has period and thickness

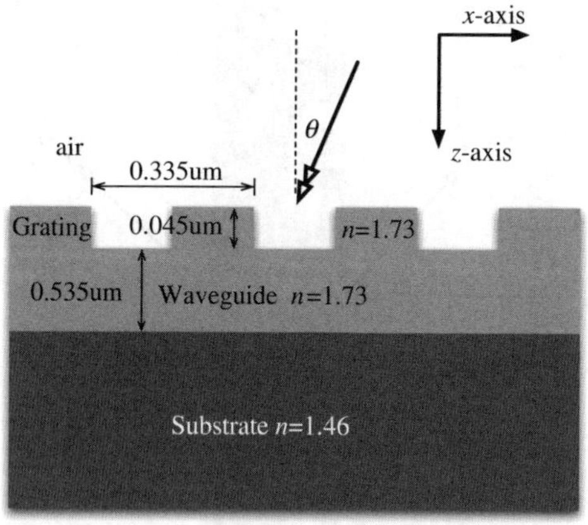

Figure 3.30 A resonance grating-waveguide structure: TE-polarized plane-wave incidence with incident angle 1°

1.0 and 0.25, respectively; the duty cycle is 50%. The thickness of the waveguide is 0.5. The grating and waveguide share the same material with the relative dielectric constant 4.0. Notably, all the dimensions are in arbitrary units. We first carry out the scattering analysis by considering the normal incidence case; that is, the uniform plane wave with TE (or TM) polarization is individually incident on the grating-waveguide structure.

Figures 3.33 and 3.34 demonstrate the reflection and transmission efficiencies (normalized to the incident power) of the fundamental and ±1th-order space harmonics for TE-polarized (E_y mode) plane-wave incidence, while Figures 3.35 and 3.36 are

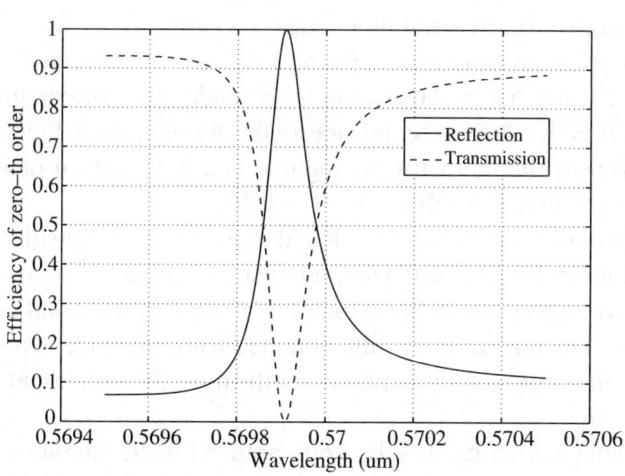

Figure 3.31 Reflectance and transmittance of plane-wave impinging on the grating-waveguide shown in Figure 3.30

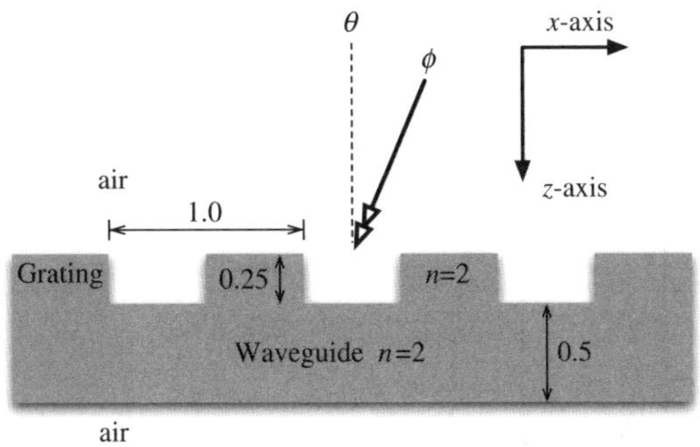

Figure 3.32 A grating-waveguide illuminated by a plane wave with oblique incidence

the results of TM-polarized (H_y mode) plane-wave incidence. Owing to the normal incidence, the $+1$th and -1th space harmonics have the same response; therefore, the dashed and dotted lines coincide with each other in the figures.

As shown in Figures 3.33 and 3.34, this structure can provide us two total reflection windows and one total transmission window for a normally incident TE-polarized plane wave in the wavelength range of operation. On the other hand, there are two total reflection windows for TM-polarized plane wave with normal incidence. Notably, the first higher order mode with space harmonic index $n = \pm 1$ occurs at $\lambda = 1$ (cutoff wavelength); that is, theses two higher order space harmonics become propagating

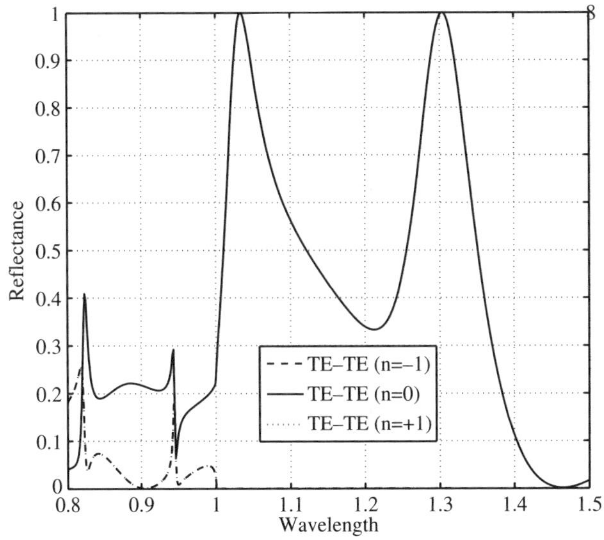

Figure 3.33 Reflectance of fundamental and ±1th-order space harmonics for normal incidence: TE polarization

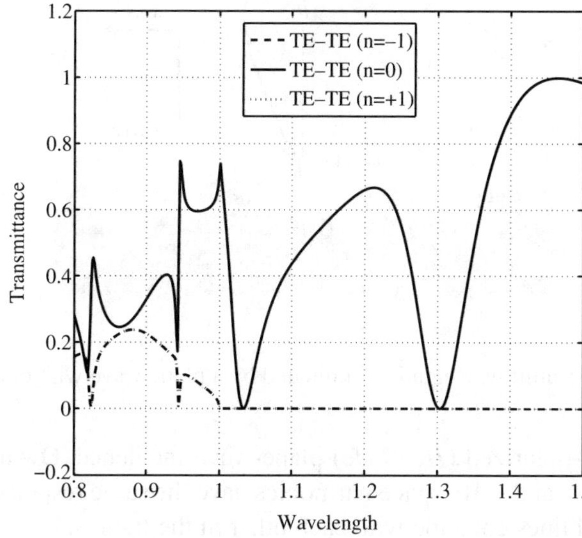

Figure 3.34 Transmittance of fundamental and ±1th-order space harmonics for normal incidence: TE polarization

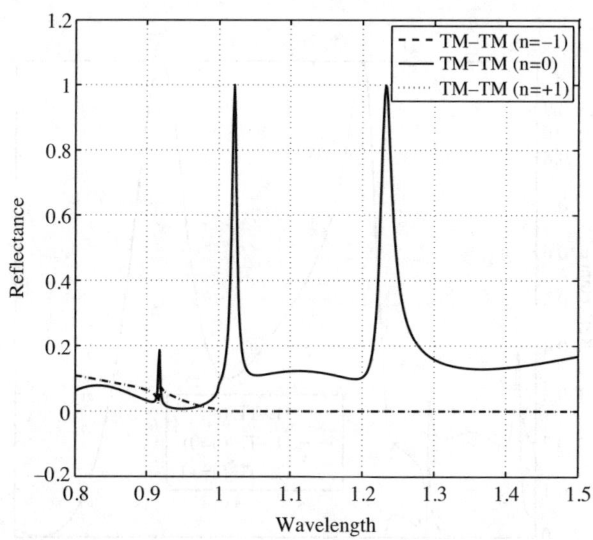

Figure 3.35 Reflectance of fundamental and ±1th-order space harmonics for normal incidence: TM polarization

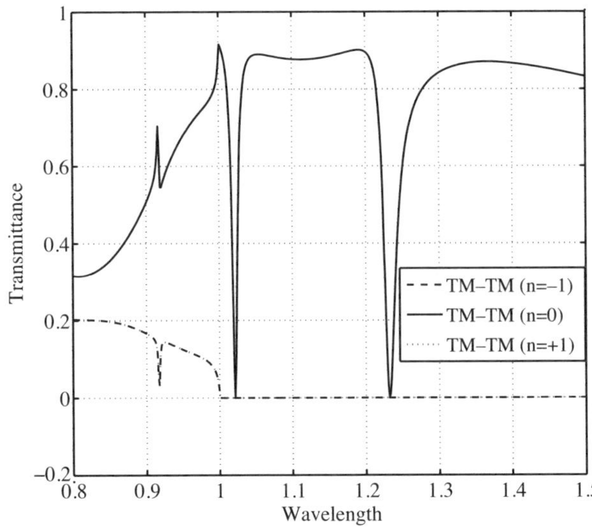

Figure 3.36 Transmittance of fundamental and ±1th-order space harmonics for normal incidence: TM polarization

ones as $\lambda \leq 1.0$. In fact, the cutoff wavelength can be figured out by the phase relation of the space harmonic in the surrounding medium, such as air, which is written as

$$k_{xn}^2 + k_y^2 = k_o^2 \tag{3.239}$$

Equation (3.239) is the phase relation of the nth space harmonic (plane wave), with the phase constants along the x and y-axes, $k_{xn} = k_o\sqrt{\varepsilon_i}\sin\theta\cos\phi$ and $k_y = k_o\sqrt{\varepsilon_i}\sin\theta\sin\phi$. Thus, the cutoff wavelength for the nth space harmonic can be determined as follows:

$$\lambda_c^{(\pm|n|)} = \frac{d_x}{|n|}\sqrt{\varepsilon_j - \varepsilon_i \sin^2\theta \sin^2\phi} \mp \sqrt{\varepsilon_i}\sin\theta\cos\phi \tag{3.240}$$

where ε_j is the relative dielectric constant in the input or output region (uniform medium). For the example depicted previously, the cutoff wavelength of the nth space harmonic is $1/|n|$. Consequently, three propagating modes exist, including the fundamental and $n = \pm 1$th order in the input and output regions, as the operational wavelength is in between 1.0 and 2.0. Moreover, the diffraction angle of the first-order space harmonics ($n = \pm 1$) will be $\theta_{\pm 1} = \sin^{-1}\theta_{inc} \pm \lambda/d_x$, where θ_{inc} is the incident angle in radians. For example, the two diffraction angles will be $\pm 53.13^\circ$ for $n = \pm 1$ as the operational wavelength is 0.8.

In the above example, the dielectric medium is assumed to be lossless; therefore, the incident power is equal to the sum of transmission and reflection power. In the following example, we consider a lossy dielectric medium with loss tangent 0.001; that is, $\varepsilon_r = 4(1 - j0.001)$. In Figure 3.37, this structure exhibits wavelength-selective

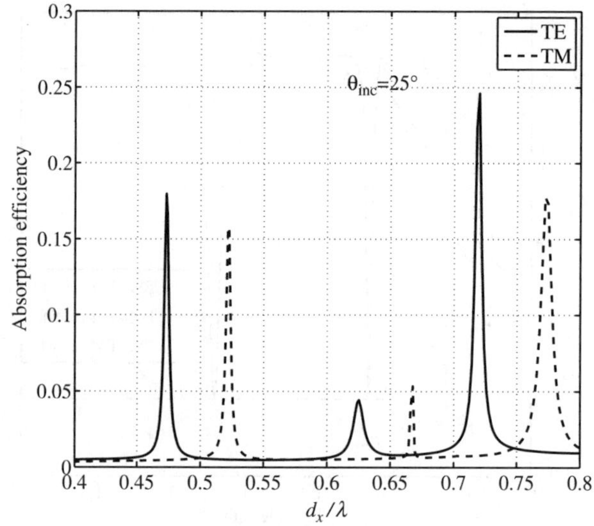

Figure 3.37 Absorption efficiency of the 1D grating shown in Figure 3.32 with incident angle $\theta = 25^{\circ}$

power absorption. Such a distinct phenomenon is called the *Wood's anomaly* in a diffraction grating, and its concise theoretical explanation was developed by Hessel and Oliner (1965). For ease of comparison, we calculate the absorption efficiency of the corresponding structure by replacing the grating layer with a dielectric layer with the same height and relative dielectric constant. Figure 3.38 demonstrates its absorption

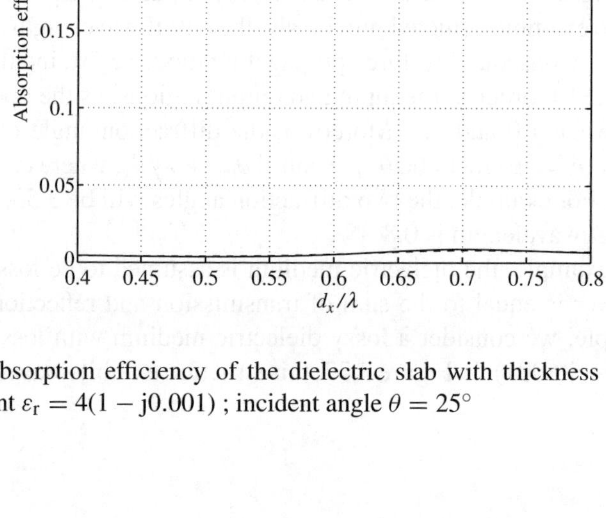

Figure 3.38 Absorption efficiency of the dielectric slab with thickness 0.75 and relative dielectric constant $\varepsilon_r = 4(1 - j0.001)$; incident angle $\theta = 25^{\circ}$

efficiency against the wavelength. It is obvious that the absorption efficiency is not significant even though the new structure has a large surface area of lossy dielectric medium. To explain such an interesting phenomenon, the resonant coupling between the space harmonic of the guided mode and incident plane wave have to be understood first; therefore, we have to stop here and study the guiding characteristics of the grating waveguide.

3.8 Guiding Characteristics of Waveguides Consisting of Multiple 1D Periodic Layers

3.8.1 Transverse Resonance Technique

In Chapter 2 we introduced the *transverse resonance technique* for determining the guiding characteristics of waves supported in a wave-guiding environment consisting of multiple uniform dielectric layers. The same theory can be applied here for solving the phase and dispersion relations of a waveguide (Peng *et al.* 1975; Tamir 1979). In Equation (3.235), when the incident wave vector $\underline{a}$ vanishes, the necessary and sufficient condition for the existence of nontrivial solution ($\underline{b} \neq 0$) is

$$\det(\mathbf{A}^{-1}\mathbf{Z}_{\text{in}}\mathbf{AY} + \mathbf{I}) = 0 \tag{3.241}$$

Since the elements of the matrix in Equation (3.241) are a function of variables k_x, k_y, k_o, and structure parameters, the solution can allow us to know the relationship among k_x, k_y, and k_o for a prescribed structure. Specifically, for a given frequency (or k_o), the relationship between k_x and k_y is called the *phase relation*. On the other hand, the relation between k_x and k_o for a given k_y is called the *dispersion relation*. Notably, the eigensolution to Equation (3.241) is the waveguide mode propagating along the transverse (or x–y) plane. Certainly, the waveguide mode contains two types: one is the surface wave bound to the surface; the other is the leaky wave which is able to radiate its energy into the surrounding medium. The type of wave can be judged by the effective dielectric constant, which is defined by $\varepsilon_{\text{eff}} = (k_t/k_o)^2$.

3.8.2 Dispersion Relation of a 1D Grating Waveguide

A *Brillouin diagram* is the most convenient way to understand the dispersion characteristics of the waveguide mode in a periodic structure. In the following, the procedures for drawing a Brillouin diagram will be illustrated. Considering a 1D grating waveguide surrounded by air, owing to the periodic nature the propagation constant along the periodicity (x-axis, for example) becomes $k_{x,n} = k_x + n2\pi/d_x$, where k_x is the propagation constant along the waveguide axis and n is the index of the space harmonic ranging from negative to positive infinity. For the nth space harmonic, the cutoff condition (the transition state from nonpropagating to propagating in the air region) is written as

$$|k_x + n2\pi/d_x| = k_o \tag{3.242}$$

By multiplying the above equation with $d_x/2\pi$ on both sides, Equation (3.242) represents the two straight-line equations

$$k_x d_x/2\pi + n = \pm d_x/\lambda \tag{3.243}$$

If we plot the two straight-line equations by taking $k_x d_x/2\pi$ and $d_x/\lambda(k_o d_x/2\pi)$ as x- and y-axis in a rectangular coordinate system, they intersect at $k_x d_x/2\pi = -n$. In doing so, we may draw the cutoff condition of each space harmonic on the same plot; see Figure 3.39. The legend attached to each straight line indicates the index of the space harmonic.

Additionally, if the absolute value of the propagation constant $k_{x,n}$ is greater than k_o, the propagation constant along the z-axis, k_z, is a complex number in free space. This means such a space harmonic is nonpropagating in the surrounding medium, which is termed the *slow wave*. Contrarily, if $|k_{x,n}| < k_o$, the real part of k_z exists; the space harmonic is radiating its power into the surrounding medium. Such a mode is called the *leaky wave* or *fast wave*. If we convert the above equations into geometry, the two straight lines define the boundary of the slow- and fast-wave regions. The region in between the two lines supports the fast wave, while the slow wave exists outside that region; see Figure 3.40. The triangular shaded region is the *bound-wave* triangular region. Notably, due to the periodic nature, the dispersion curves exhibit

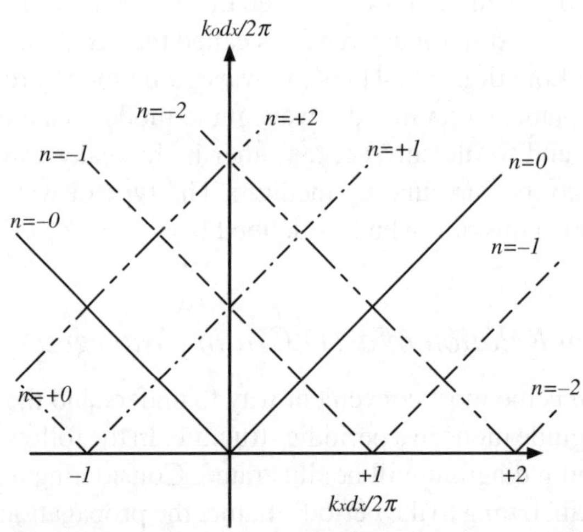

Figure 3.39 Brillouin diagram: the cutoff condition of each space harmonic

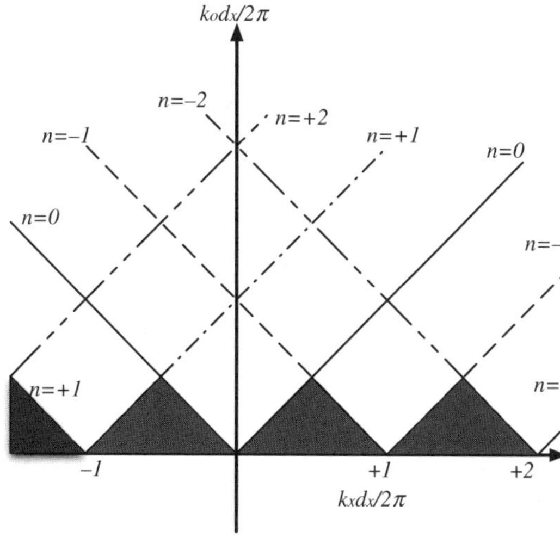

Figure 3.40 Brillouin diagram: slow-wave regions are highlighted in gray

periodic variation along the axis $k_x d_x/2\pi$ with period 1; thus, the dispersion relation within the bound-wave region remains the same.

We again return to the previous example shown in Figure 3.32 for calculating the dispersion relation of waveguide modes supported in the grating-assisted waveguide. Using the transverse resonance technique described at the beginning of this section, we may obtained the dispersion relation, which will be demonstrated in the ensuing paragraphs. Before that we will first provide the background information concerning the dispersion roots. For a lossless grating-waveguide surrounding by free space, inside the bound-wave region the electromagnetic field is concentrated in the dielectric layer with a propagation constant $k_x = k_o n_{\text{eff}}$ greater than that of free space wavenumber k_o, where n_{eff} is the effective refractive index and is greater than unity. Such a waveguide mode is called a *surface wave* or *substrate mode*. Contrarily, outside the bound-wave region, the waveguide mode can no longer be bound to the surface since its real part of the propagation constant ($k_x = \beta_x - j\alpha_x$) is less than k_o; that is, $\beta_x < k_o$. The real part of k_z, β_z, exists, enabling the radiation (power leakage) into the surrounding medium.

With the basic knowledge in mind, we first consider the unperturbed case corresponding to Figure 3.32 by replacing the grating layer by a dielectric layer having the same relative dielectric constant and thickness. The dispersion relation in the bound-wave region is shown in Figures 3.41 and 3.42 for TE- and TM-polarization, respectively. As shown in these two figures, inside the bound-wave region there are two surface modes and their first-order space harmonic ($n = -1$). Evidently, the space harmonic $n = -1$ is a backward wave having the opposite direction of group velocity but the same direction of phase velocity compared with the fundamental mode. Notably, from the coupled-mode theory described previously, the intersection points

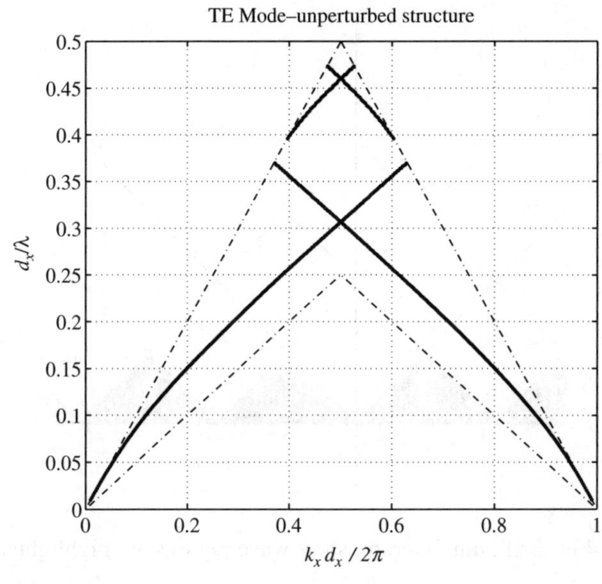

Figure 3.41 Dispersion relation, shown in the format of a Brillouin diagram, of the dielectric layer with thickness 0.75 and relative dielectric constant $\varepsilon_r = 4$; TE polarization

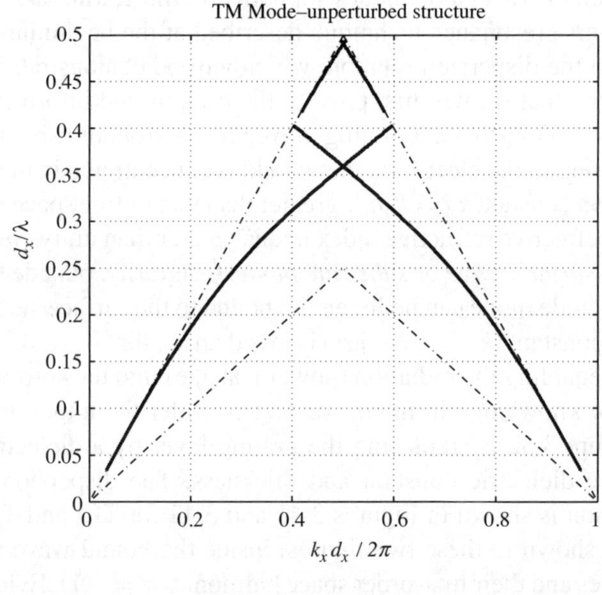

Figure 3.42 Dispersion relation, shown in the format of Brillouin diagram, of the dielectric layer with thickness 0.75 and relative dielectric constant $\varepsilon_r = 4$; TM polarization

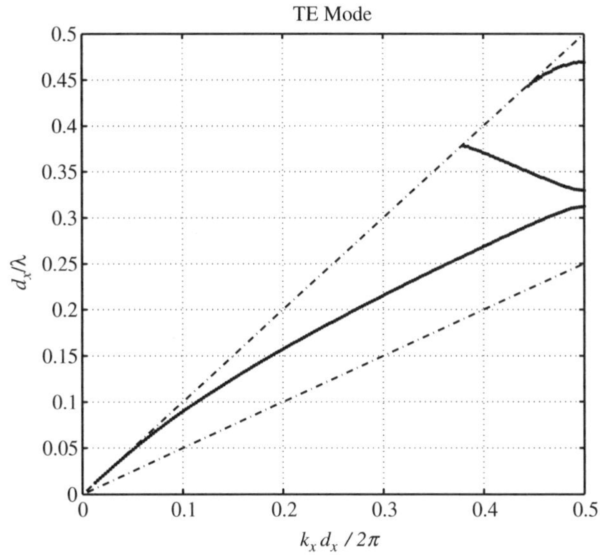

Figure 3.43 Dispersion relation, shown in the format of Brillouin diagram, of the grating waveguide shown in Figure 3.32; TE polarization

between the fundamental mode and space harmonic $n = -1$ represent the phase-match condition, resulting in the contraflow coupling between these two modes, which will become clear later on.

Figures 3.43 and 3.44 show the real roots of the dispersion relation of the grating waveguide, shown in Figure 3.32, in the bound-wave region. Here, we only show half of the first bound-wave triangle, since they are mirror reflections with respect to $k_x d_x/2\pi = 0.5$; the half bound-wave region is termed the *irreducible Brillouin zone* in solid-state physics. In Figure 3.43, the band gap around $d_x/\lambda = 0.32$ corresponds to the contraflow interaction between the fundamental mode and space harmonic $n = -1$. Within the band gap, all the propagation constants ($k_x = \beta_x - j\alpha_x$) are complex numbers; therefore, no real roots are presented. A nonzero attenuation constant α_x means that the fundamental mode experiences attenuation during its propagation. If the number of the period (unit cell) is considerable, the incident fundamental mode will totally transfer its power to the space harmonic $n = -1$ and reflect back. That is the reason why the grating waveguide can be designed as a reflector.

We trace the fundamental mode and its space harmonic in Figures 3.43 and 3.44 for plotting the dispersion relation including the phase and attenuation constants shown in Figures 3.45 and 3.46, respectively. In comparison with the real roots shown in Figure 3.43 or 3.44, the only differences between them exist in the stop-band region. Within the stop-band, the propagation constant is indeed a complex number with the attenuation constant distribution given on the left-hand side of the figure. At the band

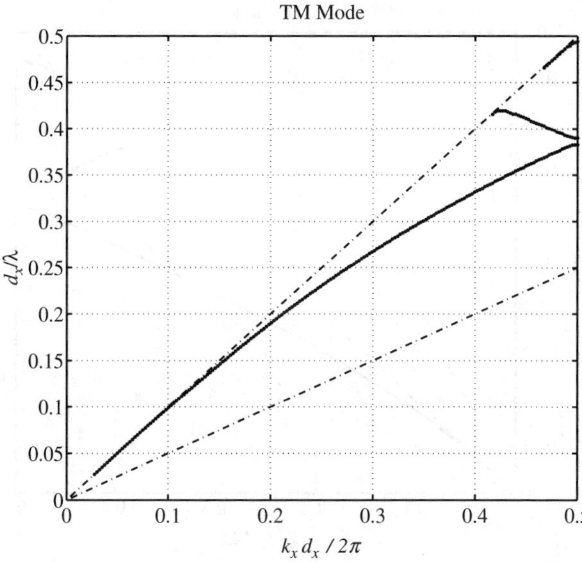

Figure 3.44 Dispersion relation, shown in the format of Brillouin diagram, of the grating waveguide shown in Figure 3.32; TM polarization

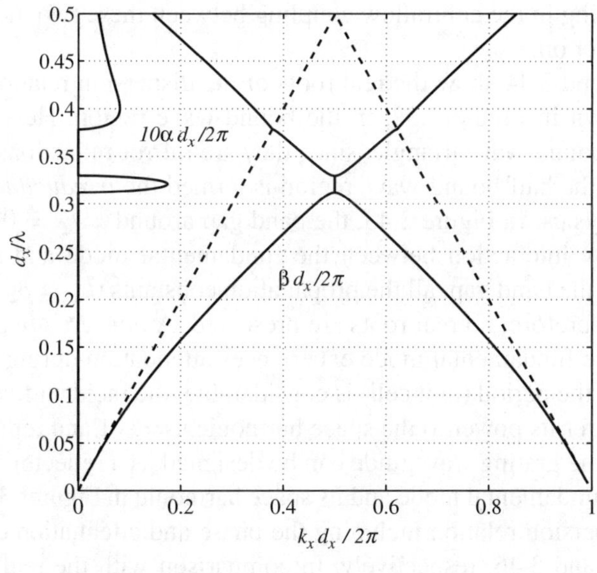

Figure 3.45 Bound and leaky waves, shown in the format of a Brillouin diagram, of the grating waveguide shown in Figure 3.32; TE polarization

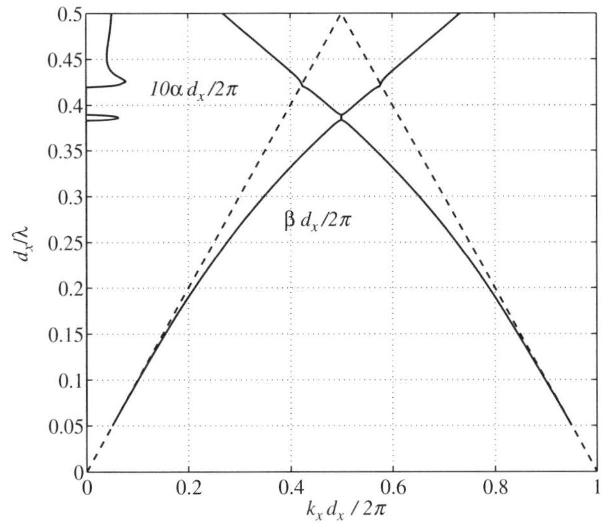

Figure 3.46 Bound and leaky waves, shown in the format of Brillouin diagram, of the grating waveguide shown in Figure 3.32; TM polarization

edges, the attenuation constant is zero; the maximum attenuation constant takes place at the center of the band-gap. The attenuation constant again returns to zero when the normalized frequency (d_x/λ) is away from the stop-band but inside the bound-wave region. Interestingly, when the normalized frequency starts to leave the border of the bound-wave triangle, the attenuation constant increases dramatically from zero. Away from the bound-wave region, the attenuation generally is not zero; it becomes a leaky wave in the fast-wave region. The normalized frequencies at 0.3782 (for TE-polarization) and 0.4194 (for TM-polarization) are termed the onset point of the leaky wave.

Additionally, it is interesting to find that the stop-bands (or band-gaps) in Figures 3.45 and 3.46 do not overlap; this means that within the band-gap of TE polarization the incident TE-polarized wave is totally reflected (as long as the number of periods is large enough) while the TM-polarized incident wave can pass through. For a hybrid mode incident into the grating waveguide, the TE-polarized component will be reflected and the TM-polarized can be received at the output port; this structure can be employed as a waveguide-based polarization filter (Figure 3.47).

Returning to Figure 3.37, the anomalous absorption occurs at several specific wavelengths for the grating waveguide incident by a plane wave with incident angle 25°. The underlying physics has not yet been explained. With a thorough understanding regarding the dispersion relation expressed in terms of the Brillouin diagram, we are now in a good position to explore the physical picture of the wave process involved.

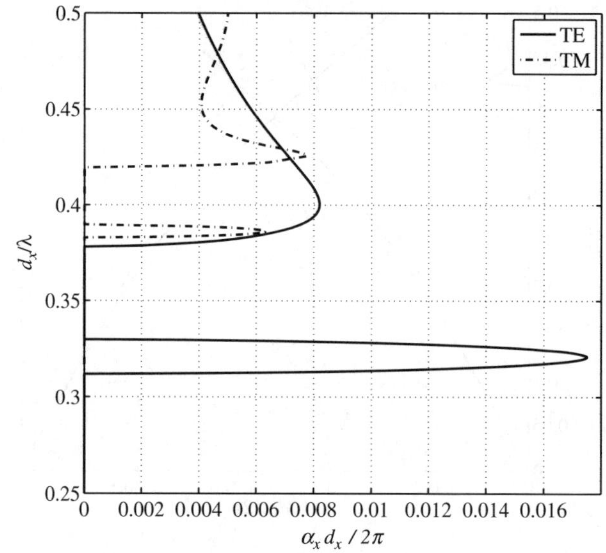

Figure 3.47 Using the different locations of band-gap for designing a polarization filter

As is well known, the plane wave cannot couple its energy into a dielectric slab waveguide since the phase-match condition cannot take place. Furthermore, if the guiding axis of the dielectric slab waveguide is along the x-axis, the propagation constant is $k_x = k_o n_{\text{eff}}$, where $n_{\text{eff}} > 1.0$. For the plane wave obliquely incident, the phase constant along the guiding axis is $k_x = k_o \sin\theta_{\text{inc}} < k_o$, where θ_{inc} is the incident angle. This is the reason that the phase-match condition never takes place. Nevertheless, if the grating layer is imposed on the dielectric slab waveguide, the propagation constant of the fundamental mode becomes $k_{x,n} = k_x + n2\pi/d_x$, where n is the index of space harmonic running from negative to positive infinity. For negative index n, propagation constants of the nth order have the opportunity to match with $k_o \sin\theta_{\text{inc}}$; however, in this situation, the nth space harmonic must be in the fast-wave region with a complex propagation constant $k_{x,n} = \beta_{x,n} - j\alpha_x$. The phase-match condition expressed in terms of a mathematical equation is

$$\beta_{x,n} = k_o \sin\theta_{\text{inc}} \tag{3.244}$$

If we plot the x-component phase constant of the incident plane wave on the Brillouin diagram, it is a straight-line equation given as $d_x/\lambda = \csc\theta_{\text{inc}} k_x d_x/2\pi$, with slope $\csc\theta_{\text{inc}}$. Figures 3.48 and 3.49 are plots of the dispersion relation of the grating-waveguide in the format of a Brillouin diagram together with that of the plane wave. The intersection points, which are denoted in alphabetical order, correspond to the phase-match condition, thus enabling the coupling between the plane wave and the space harmonic $n = -1$ of the fundamental surface mode. Notably, since the space

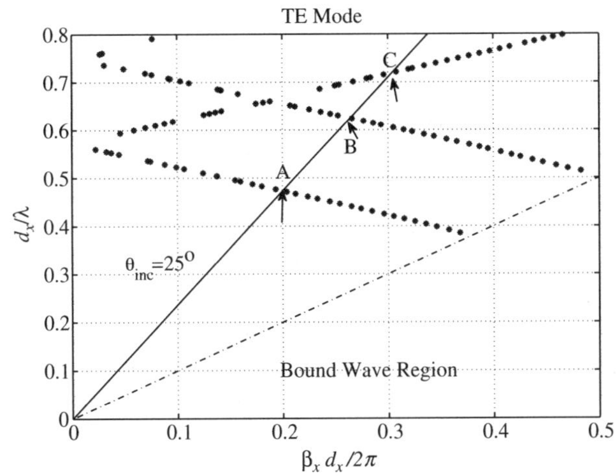

Figure 3.48 Real part of the dispersion root in the fast-wave region: TE-polarization

harmonic is a leaky wave, it will again radiate its energy into the surrounding medium (Bertoni *et al.* 1989). Therefore, for a lossless grating waveguide, the power coupling into the grating wave will be totally radiated into the surrounding medium – becoming the transmitting and reflecting waves – so that no power is absorbed by the structure. Contrarily, if we put a little loss on the structure with loss-tangent 0.001, the resonant coupling power is absorbed by the grating waveguide, as was shown in Figure 3.37. Significantly, the normalized frequencies of anomalous absorption can be predicted by

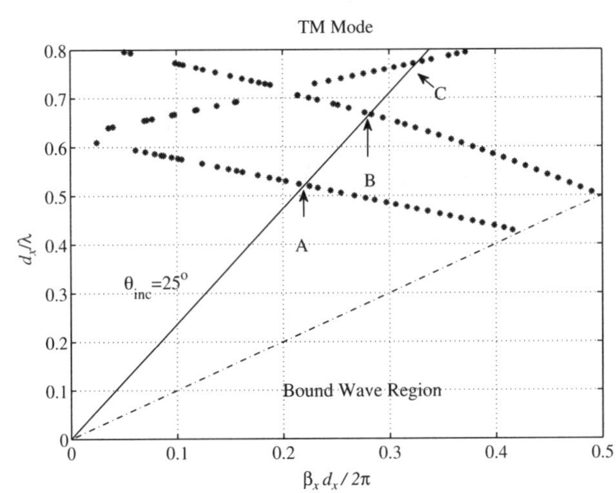

Figure 3.49 Real part of the dispersion root in the fast-wave region: TM-polarization

the intersection points indicated in Figures 3.48 and 3.49 for TE- and TM-polarization, respectively.

The previous examples that we treated are single polarization problems, which is TE- or TM-polarization independently; the y-axis component of the propagation constant, k_y, vanishes and the wave propagation is along the x-axis. Certainly, for an infinite 1D periodic medium, the single polarization wave indeed exists even though k_y is present. However, for the 1D periodic medium with finite thickness as the 1D grating shown in Figure 3.32, the presence of the discontinuities between the dielectric layer and the 1D grating forces us to take into account both polarization modes. In doing so, the electromagnetic boundary condition – that is, the tangential electric and magnetic fields must be continuous at the discontinuity between two dielectric mediums – can be attained at the discontinuity. For the guided mode propagating in the grating waveguide, shown in Figure 3.32, that k_y exists means that the guided wave is propagating like that depicted in Figure 3.50. Figure 3.51 shows the phase relation, which is the relationship between β_x and β_y (where β_y is a real number and k_x, in general, is a complex number) at $d_x/\lambda = 0.4$. In fact, the value of $\beta_x d_x/2\pi$ at $\beta_y = 0$ can be read from Figures 3.45 and 3.46. Recalling the phase relation of a wave in a 1D periodic medium in Section 3.2, the phase relation of the fundamental mode and each higher order space harmonic form circles. The electromagnetic coupling takes place at the intersection points between circles; thus, the stop-band (or band-gap) occurs therein, as does the phase relation of a surface wave in a 1D grating-waveguide structure shown in Figure 3.32. Moreover, the radii of the circles ($k_o n_{\text{eff}}$) are different for TE- and TM-polarized surface waves. Here, the radius of the TE mode is greater than that of the TM mode; see Figure 3.51. The upper stop-band at $\beta_x d_x/2\pi = 0.5$ with the legend TE(0)–TE(−1) is due to the coupling between the fundamental TE mode and its −1th-order space harmonic, while the lower one

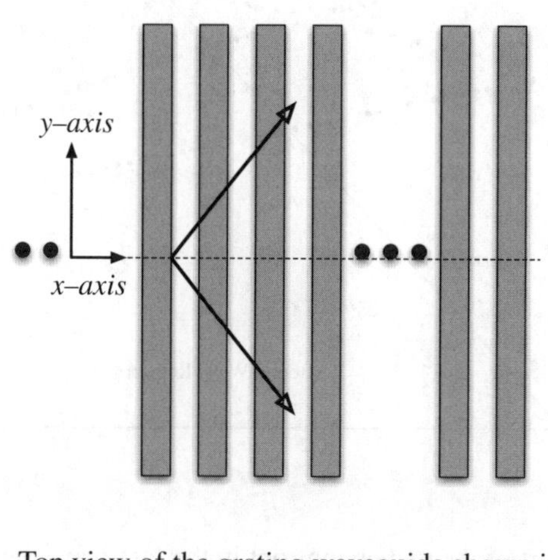

Figure 3.50 Top view of the grating waveguide shown in Figure 3.32

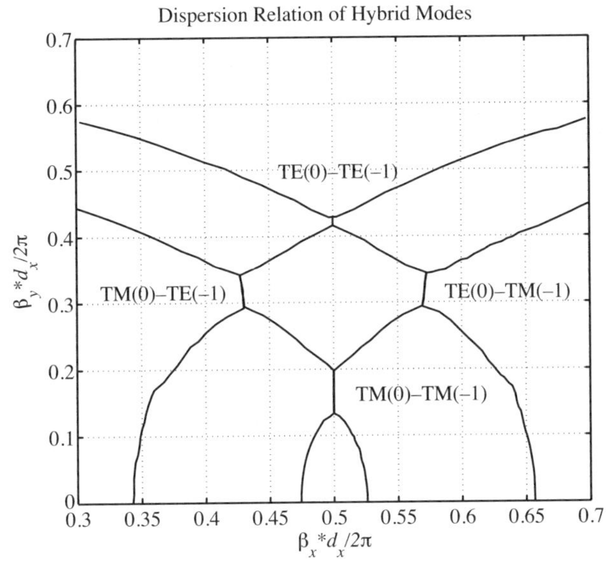

Figure 3.51 Phase relation of the guided modes in the grating waveguide: $d_x/\lambda = 0.4$

denoted TM(0)–TM(-1) corresponds to the coupling between the fundamental TM mode and its -1th-order space harmonic. Those two stop-bands are vertical with respect to $\beta_x d_x/2\pi$. Significantly, the stop-bands caused by the coupling between the TE- and TM-polarized modes are unusual, being slanted with respect to the horizontal axis. Within the slanted stop-band at the right-hand side, the TM-polarized -1th-order space harmonic reflects once the TE-polarized fundamental surface wave obliquely impinges on the grating. Contrarily, within the left-hand side slanted stop-band, the TE-polarized -1th-order space harmonic reflects when the TM-polarized fundamental surface wave is incident obliquely on the grating. Finally, although not shown, the propagation constant $k_x d_x/2\pi$ is a complex number in the vertical or slanted stop-bands.

References

Bertoni, H. L., Cheo, L.-h. S., and Tamir, T. (1989) Frequency-selective reflection and transmission by a periodic dielectric layer. *IEEE Transactions on Antennas and Propagation* **37**, 78–83.

Hall, R. C., Mittra, R., and Mitzner, K. M. (1988) Analysis of multilayered periodic structures using generalized scattering matrix theory. *IEEE Transactions on Antennas and Propagation* **36**, 511–517.

Hessel, A. and Oliner, A. A. (1965) A new theory of Wood's anomalies on optical gratings. *Applied Optics* **4**, 1275–1297.

Peng S.-T. 1989 Rigorous formulation of scattering and guidance by dielectric grating waveguides: general case of oblique incidence.*J. Opt. Soc. Am. A* **6**, 1869–1883.

Peng, S.-T., Tamir, T., and Bertoni, H. L. 1975 Theory of periodic dielectric waveguides. *IEEE Transactions on Microwave Theory and Techniques* **23**, 123–133.

Sharon, A., Rosenblatt, D., and Friesem, A. A. 1997 Resonant grating-waveguide structures for visible and near-infrared radiation.*Journal of the Optical Society of America A* **42**, 2985–2993.
Tamir, T. (ed.) (1979) *Integrated Optics: Topics in Applied Physics*, Springer, Berlin.

Further Readings

Born, M. and Wolf, E. (1964) *Principles of Optics*, Macmillan, New York, NY.
Elachi, C. (1989) Waves in active and passive periodic structures: a review. *Proceedings of the IEEE* **64**, 1666–1698.
Felsen, L. B. and Marcuvitz, N. (1973) *Radiation and Scattering of Waves*, Prentice-Hall, Englewood Cliffs, NJ.
Gaylord, T. K. and Moharam, M. G. (1985) Analysis and applications of optical diffraction by gratings. *Proceedings of the IEEE* **73**, 894–937.
Hunsperger, R. G. (1985) *Integrated Optics: Theory and Technology*, 2nd edn, Springer-Verlag.
Joannopoulous, J. D., Meade, R. D., and Winn, J. N. (1995) *Photonic Crystals: Modeling the Flow of Light*, Princeton University Press, Princeton, NJ.
Marcuse, D. (1974) *Theory of Dielectric Optical Waveguides*, Academic Press, New York, NY.

4

Two- and Three-Dimensional Periodic Structures

Wave propagating characteristics in periodic structures (gratings) have been intensively studied for several decades; most of the research results were focused on those of 1D periodic structures employed as diffraction gratings or grating waveguides. In the previous chapters the phase and dispersion relations of waves supported in a 1D periodic medium have been studied using the Fourier and modal transmission-line approaches, respectively. Moreover, the scattering of a plane wave by a 1D periodic structure (or grating) and the guiding characteristics of a dielectric waveguide incorporating 1D periodic structures have also been investigated. In this chapter we will extend the 1D periodic structure to two dimensions: one that is periodic in two directions and homogeneous in the third direction. Furthermore, the 3D periodic structure, which can be considered as a stack of 2D periodic layers, is also taken into account for evaluating the eigenwave and scattering characteristics. The rigorous mode-matching method will be employed for solving the electromagnetic boundary-value problem of the two types of periodic structures including the metallic and dielectric cylinders. Specifically, the Brillouin and phase diagrams will be exploited to understand the physical picture of the wave process involved in the structure under consideration.

4.1 Modal Transmission-Line Approach for a 2D Periodic Metallic Medium: In-Plane Propagation

A periodic structure consisting of metallic cylinders arranged in a 2D array is shown in Figure 4.1. Each rectangular cylinder is made from a PEC with widths w_x and w_y along the x- and y-directions, respectively, and is homogeneous along the z direction (we consider that the cylinder is infinite in extent along the z-axis). In addition, the

Periodic Structures: Mode-Matching Approach and Applications in Electromagnetic Engineering, First Edition. Ruey-Bing (Raybeam) Hwang.

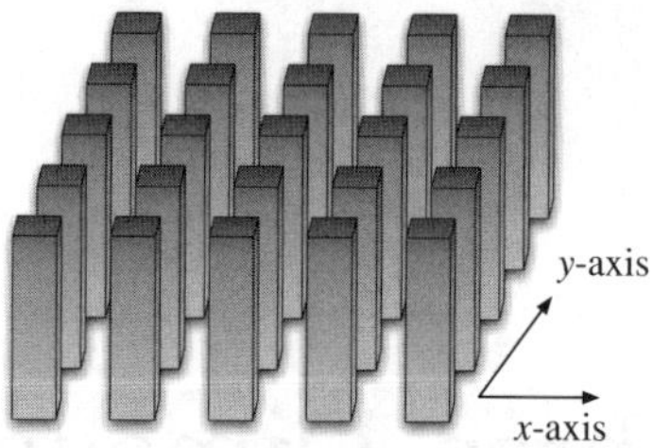

Figure 4.1 The 2D metallic periodic structure

periods along the x- and y-directions are d_x and d_y, respectively. The dielectric medium surrounding the metallic cylinders has the relative dielectric constant ε_S. In fact, the 2D periodic structure shown in Figure 4.1 can also be regarded as a stack of 1D periodic layers (Hwang 2004). For example, Figures 4.2 and 4.3 are top views of the 2D periodic structure. In Figure 4.2, each 1D metallic grating layer has a period along the x-direction, while the overall structure can be regarded as a stack of 1D gratings along the y-axis. Alternatively, in Figure 4.3 the 1D grating has its period along the y-axis. The cascade of the 1D gratings along the x-axis forms the same 2D periodic structure.

Moreover, since that the metallic cylinders array is assumed to be infinite in extent along the z-direction and there is no field variation along the z-axis (that is, $k_z = 0$), the electromagnetic fields in the structure can be individually treated as E_z and H_z modes without considering the coupling between them. The vanishment of the propagation constant along the z-axis, k_z, means that the incident wave is along the transverse plane, which is called the *in-plane* propagation.

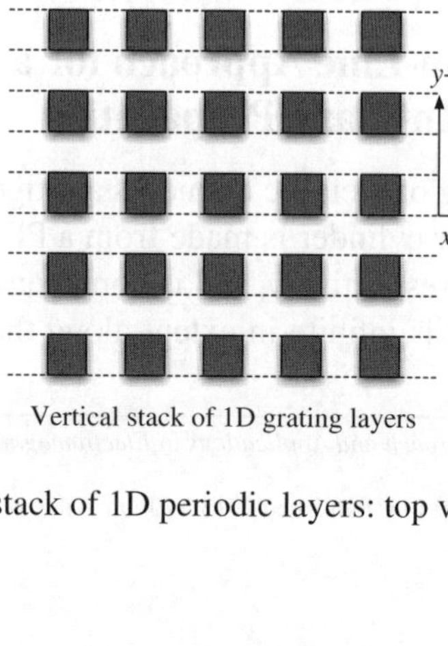

Figure 4.2 A vertical stack of 1D periodic layers: top view of a 2D periodic medium

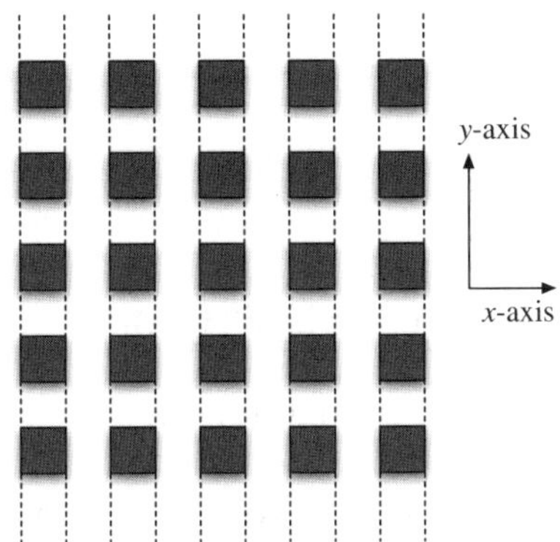

Horizontal stack of 1D grating layers

Figure 4.3 A horizontal stack of 1D periodic layers: top view of a 2D periodic medium

4.1.1 Generalized Scattering Matrix at the Interface between a 1D Periodic Metallic Medium and Uniform Medium

In Chapter 3, in the mathematical analysis of wave propagation in a 1D periodic medium composed of dielectric material we resorted to the method of Fourier series expansion (or plane-wave expansion). Namely, the 1D periodic dielectric function can be expressed in terms of the Fourier series expansion; the electric and magnetic fields in the periodic medium can also be expressed in terms of the superposition of the Floquet solution (plane-wave solution). Nevertheless, as far as the periodic medium consisting of PEC is concerned, the Fourier series expansion method fails because the relative dielectric constant of the PEC is an infinite number.

In this section we will introduce a rigorous approach based on the mode-matching technique for determining the eigenmode supported in a periodic structure consisting of a PEC. As mentioned previously, the 2D periodic medium can be regarded as a stack of 1D grating layers; see Figures 4.2 and 4.3. The input–output relation of the 1D metallic periodic layer (or grating) was treated in Chapter 3. Therefore, for a 2D periodic structure composed of a finite stack of 1D metallic gratings, the overall scattering characteristics can be easily obtained by cascading each of the input–output relations of the 1D metallic grating successively. This is called the building-block approach, commonly used in microwave engineering.

Regarding the source-free solution in an infinite 2D periodic medium, the eigenwave solutions can be determined by imposing the periodic boundary (Bloch–Floquet) condition on a period along the stacking direction (please refer to Figure 4.2), resulting in a generalized eigenvalue equation with the eigenvalue equal to $\exp(\pm jk_y d_y)$, where k_y and d_y are the propagation constant and period along the y-axis (we assume that the gratings are stacked up along the y-direction).

We begin with the input–output relation at the interface between a semi-infinite 1D periodic metallic medium and a uniform dielectric medium, which was derived in Section 3.4.1:

$$\underline{V}_{\rm p}(0) = \mathbf{R}\underline{V}_{\rm u}(0) \tag{4.1}$$

$$\underline{V}_{\rm u}(0) = \mathbf{R}^{H}\underline{V}_{\rm p}(0) \tag{4.2}$$

$$\underline{I}_{\rm p}(0) = \mathbf{R}\underline{I}_{\rm u}(0) \tag{4.3}$$

The subscripts "p" and "u" denote the voltage (current) wave in the parallel-plate region and the uniform medium, respectively. Parameter $\mathbf{R}$ is a matrix with its elements representing the coupling between the modes inside the parallel-plate region and uniform medium.

The voltage and current waves at the discontinuity are the transmission-line solutions, which can be expressed in terms of the superposition of forward- and backward-propagating waves $\underline{a}_{\rm p}$ and $\underline{b}_{\rm p}$ in the parallel-plate region and $\underline{a}_{\rm u}$ and $\underline{b}_{\rm u}$ in the uniform medium. Thus, Equations (4.1) and (4.3) can be rewritten as

$$\underline{b}_{\rm p}(0) + \underline{a}_{\rm p}(0) = \mathbf{R}[\underline{a}_{\rm u}(0) + \underline{b}_{\rm u}(0)] \tag{4.4}$$

$$\underline{a}_{\rm u}(0) + \underline{b}_{\rm u}(0) = \mathbf{R}^{H}[\underline{b}_{\rm p}(0) + \underline{a}_{\rm p}(0)] \tag{4.5}$$

$$\mathbf{Y}_{\rm p}[\underline{b}_{\rm p}(0) - \underline{a}_{\rm p}(0)] = \mathbf{RY}[\underline{a}_{\rm u}(0) - \underline{b}_{\rm u}(0)] \tag{4.6}$$

The diagonal matrices $\mathbf{Y}_{\rm u}$ and $\mathbf{Y}_{\rm p}$ are the admittance matrices in the uniform and parallel-plate regions, respectively. After processing Equations (4.4) and (4.6), we obtain the generalized scattering matrix (a scattering matrix including not only the propagating waves but also nonpropagating waves (Hall *et al.* 1988)) at the discontinuity $y = 0$ given below:

$$\begin{pmatrix} \underline{b}_{\rm u}(0) \\ \underline{b}_{\rm p}(0) \end{pmatrix} = \mathbf{S} \begin{pmatrix} \underline{a}_{\rm u}(0) \\ \underline{a}_{\rm p}(0) \end{pmatrix} \tag{4.7}$$

$$\mathbf{S} = \begin{pmatrix} -\mathbf{R} & \mathbf{I} \\ \mathbf{RY}_{\rm u} & \mathbf{Y}_{\rm p} \end{pmatrix}^{-1} \begin{pmatrix} \mathbf{R} & -\mathbf{I} \\ \mathbf{RY}_{\rm u} & \mathbf{Y}_{\rm p} \end{pmatrix} \tag{4.8}$$

Since the propagation constants of each space harmonic in the uniform and periodic mediums are known, the transitions of the waves in the two mediums only involve the propagation phase delay (propagation constant times the propagating distance). The generalized scattering matrix originally defined at $y = 0$ can be transformed into the new one containing two interfaces defined at $y = y_1$ ($y_1 = 0.5w_y$) and

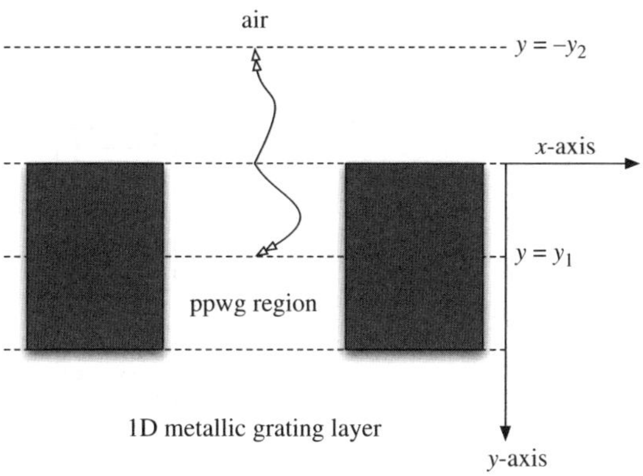

Figure 4.4 The new interfaces defined in a unit cell (1D grating layer) along the y-axis

$y = -y_2\,(y_2 = 0.5d_y - 0.5w_y)$, depicted in Figure 4.4. Therefore, the new generalized scattering matrix can be written as

$$\begin{pmatrix} \underline{b}_{\mathrm{u}}(-y_2) \\ \underline{b}_{\mathrm{p}}(y_1) \end{pmatrix} = \mathbf{S}^* \begin{pmatrix} \underline{a}_{\mathrm{u}}(-y_2) \\ \underline{a}_{\mathrm{p}}(y_1) \end{pmatrix} \tag{4.9}$$

$$\mathbf{S}^* = \mathbf{TST} \tag{4.10}$$

$$\mathbf{T} = \begin{pmatrix} \exp[-\mathrm{j}\mathbf{k}_{y,\mathrm{u}}(d_y - w_y)/2] & \mathbf{0} \\ \mathbf{0} & \exp[-\mathrm{j}\mathbf{k}_{y,\mathrm{p}}w_y/2] \end{pmatrix} \tag{4.11}$$

As shown in Figure 4.5, the two generalized scattering matrices are given in Equation (4.10). Consequently, the scattering matrix of the single grating layer demonstrated in this figure can be obtained by cascading the two scattering matrices:

$$\begin{pmatrix} \underline{b}_3 \\ \underline{b}_1 \end{pmatrix} = \begin{pmatrix} \mathbf{S}^*_{11}+\mathbf{S}^*_{12}\mathbf{T}\mathbf{S}^*_{22}\mathbf{S}^*_{21} & \mathbf{S}^*_{12}\mathbf{T}\mathbf{S}^*_{21} \\ \mathbf{S}^*_{12}\mathbf{T}\mathbf{S}^*_{21} & \mathbf{S}^*_{11}+\mathbf{S}^*_{12}\mathbf{T}\mathbf{S}^*_{22}\mathbf{S}^*_{21} \end{pmatrix} \begin{pmatrix} \underline{a}_3 \\ \underline{a}_1 \end{pmatrix} \tag{4.12}$$

$$\mathbf{T} = \left(\mathbf{I} - \mathbf{S}^*_{22}\mathbf{S}^*_{22}\right)^{-1} \tag{4.13}$$

where $\underline{a}_3$ and $\underline{b}_3$ are the forward- and backward-propagating vectors in the uniform medium at $y = 0.5(d_y + w_y)$ and $\underline{a}_1$ and $\underline{b}_1$ are the forward- and backward-propagating vectors in the uniform medium at $y = 0.5(w_y - d_y)$. In doing so, we could obtain the generalized scattering matrix of the single grating layer sandwiched by two uniform dielectric layers along the y-axis. With the generalized scattering matrix in Equation (4.12), we can calculate the input–output relation of a 2D periodic structure having a finite number of periods along the y-axis; however, it is infinite in extent along the x-axis, which is shown in Figure 4.6.

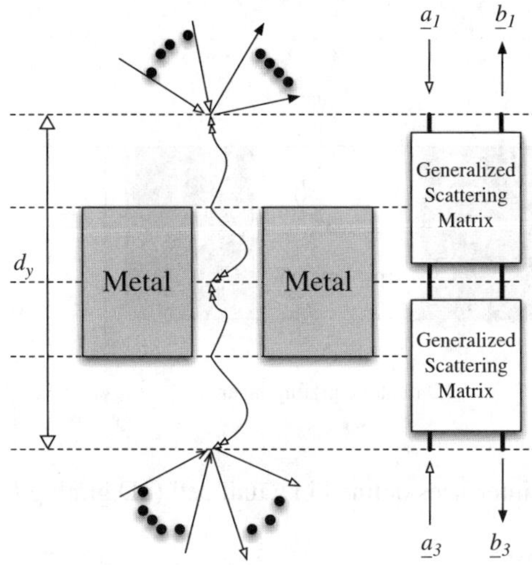

Figure 4.5 Generalized scattering matrix of the unit cell (1D grating layer)

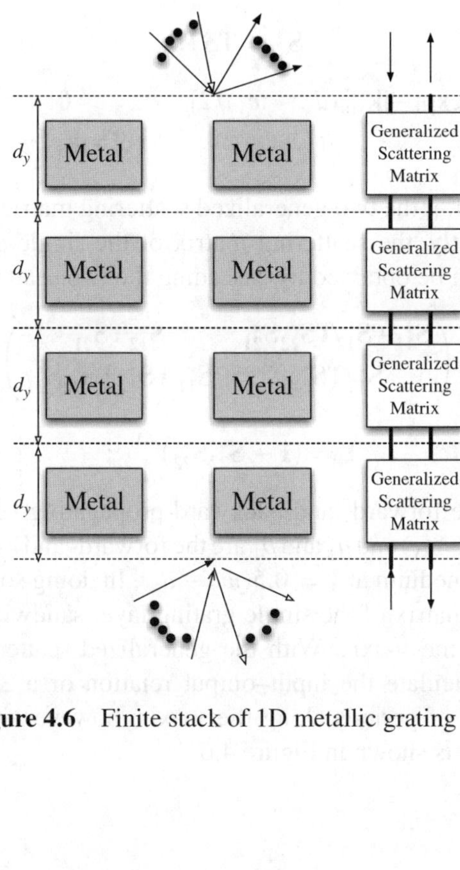

Figure 4.6 Finite stack of 1D metallic grating layers

Additionally, if the structure consists of an infinite number of periods along the y-axis, we can solve the source-free (eigen) solution supported in the 2D periodic medium. Those eigensolutions can be regarded as a basis for expanding the general field solution in the periodic medium when the excitation source is considered. In the following, we will exploit the generalized scattering matrix of the unit cell along the y-axis, which was derived previously, along with the Bloch–Floquet (periodic boundary) condition for building up a generalized eigenvalue equation for solving the propagation constant along the y-axis for given k_x and frequency.

4.1.2 Periodic Boundary Condition on the Unit Cell along the y-axis

Let us consider that the structure is infinite in extent along the y-direction, according to the Bloch–Floquet theorem, the wave traveling through a unit cell along the y-direction (see Figure 4.5) experiences a phase shift, denoted λ. Thus, the wave vectors propagating along the forward (backward) direction at the two interfaces separated by a distance d_y (period along the y-axis) must satisfy the following condition:

$$\underline{b}_3 = \lambda \underline{a}_1 \tag{4.14}$$

$$\underline{a}_3 = \lambda \underline{b}_1 \tag{4.15}$$

Substitution of Equations (4.14) and (4.15) into Equation (4.12) yields the following matrix equation:

$$\begin{pmatrix} \mathbf{S}_{21} & \mathbf{0} \\ \mathbf{S}_{11} & -\mathbf{I} \end{pmatrix} \begin{pmatrix} \underline{a}_1 \\ \underline{b}_1 \end{pmatrix} = \lambda \begin{pmatrix} \mathbf{I} & -\mathbf{S}_{22} \\ \mathbf{0} & -\mathbf{S}_{12} \end{pmatrix} \begin{pmatrix} \underline{a}_1 \\ \underline{b}_1 \end{pmatrix} \tag{4.16}$$

where matrices $\mathbf{S_{11}}$, $\mathbf{S_{12}}$, $\mathbf{S_{21}}$, and $\mathbf{S_{22}}$ are the sub-matrices of the generalized scattering matrix given in Equation (4.12). Equation (4.16) is a typical generalized eigenvalue equation, which can be solved by conventional numerical packages such as IMSL or MATLAB function eig(A,B). Here, the eigenvalue λ is related to the propagation constant (k_y) in the y-direction as well the period d_y, which can be written as

$$\lambda = \exp(\pm j k_y d_y) \tag{4.17}$$

Moreover, column vectors $\underline{a}_1$ and $\underline{a}_2$ are associated eigenvectors. With the eigenvector, we can determine the electric and magnetic fields inside the unit cell.

So far, we have formulated the eigenvalue equation for determining the source-free solutions in such a class of 2D metallic periodic medium. Because Equation (4.16) is a function of structure parameters k_x, k_y and k_o, we can determine the relationship among k_x, k_y and k_o by either of the two given parameters. For example, if the phase constant along the x-axis, k_x, is specified under a certain frequency of operation, we can obtain the value of k_y by solving the eigenvalue in Equation (4.16). The relationship between k_x and k_y is referred to as the phase relation, while relationship between k_o and k_y is termed the dispersion relation. In general, k_y is a complex number.

The real and imaginary parts represent the phase and attenuation constants of the wave propagating along the y-direction. The attenuation of the wave means that the wave decays exponentially along the propagation direction. The frequency band with nonvanishing attenuation constant is called the stop-band (or band-gap), which will become clear later on.

On the other hand, we can also calculate the value of k_x for a given phase constant along the y-axis, k_y; however, solving the propagation constant k_x requires resorting to resolve the nonlinear equation by searching the complex roots, which is more complicated than the previous case.

Through the rigorous analysis described previously, we can derive the phase relation as well as the dispersion relation of the eigenwave (source-free solution) supported in such a class of 2D metallic periodic mediums. Furthermore, the electromagnetic fields in the unit cell can be obtained accordingly from the associated eigenvector in Equation (4.16), which defines the amplitude of each space harmonic.

4.1.3 A Simple Graphical Method

Before embarking on the elaborate numerical calculation, I would first of all like to introduce a simple graphical method (Hwang 2000; Hwang and Peng 1999) for predicting the phase and dispersion relations of the eigenwave in a 2D metallic periodic medium. If we consider a 2D periodic medium with infinitesimal metallic cylinders (small perturbation problem), the field solution can be approximated by the plane-wave solution with periodic modulation (in fact, it is the Floquet solution). For the sake of simplicity, we assume that the magnetic field is along the z-axis (H_z mode); here, the propagation constant along the z-axis is supposed to be zero, $k_z = 0$. Therefore, the wave is propagating along the x–y plane (so-called *in-plane* propagation). In this situation, the TEM wave is the fundamental mode supported in this environment. Owing to the periodic variations along the x- and y-directions, the phase relation of each space harmonic of the TEM wave can be independently approximated by

$$\left(k_x + m\frac{2\pi}{d_x}\right)^2 + \left(k_y + n\frac{2\pi}{d_y}\right)^2 \approx k_o^2 \tag{4.18}$$

Parameters m and n individually are the indices of the space harmonic along the x- and y-axes, which are integers ranging from negative to positive infinity. Moreover, each set of (m, n) represents a circle with radius k_o centered at $k_x = -m2\pi/d_x$ and $k_y = -n2\pi/d_y$.

4.1.4 Phase Relation: The Relationship among k_x, k_y, and k_o

We first consider a 2D square lattice with equal period along the x- and y-axes, $d_x = d_y = 10$ mm. The dimension of the square metallic cylinder is 1 mm. The surrounding medium outside the metallic cylinders is air. We apply the formulation given in Section 4.1.2 to calculate the phase relation; that is, the relation among k_x, k_y, and k_o. In addition to the rigorous phase relation obtained by solving the generalized eigenvalue Equation (4.16), I also plot the unperturbed phase relation based on Equation (4.18) for ease of reference, since the dimension of the metallic cylinder is indeed much smaller than that of the period. Notably, the phase constants k_x and k_y were normalized to the period $k_x d_x/2\pi$ and $k_y d_x/2\pi$, so that each circle will center at the integer number and the radius becomes $k_o d_x/2\pi = d_x/\lambda$, and Equation (4.18) can be rewritten as

$$\left(\frac{k_x d_x}{2\pi} + m\right)^2 + \left(\frac{k_y d_x}{2\pi} + n\right)^2 \approx \left(\frac{k_o d_x}{2\pi}\right)^2 \tag{4.19}$$

We demonstrate the phase diagram at several frequencies for observing the in-plane wave-propagation characteristics. Figures 4.7, 4.8, 4.9, and 4.10 show the phase relation at 14 GHz, 15 GHz, 16 GHz, and 22 GHz, respectively. In the figures, each of the circles represents the space harmonic with its center located at $(-m, -n)$, where m and n are the indices of the space harmonic along the x- and y-axes, respectively. From these figures, we may observe that the unperturbed phase diagram does coincide with the exact one except for the region in the vicinity of the intersection points between

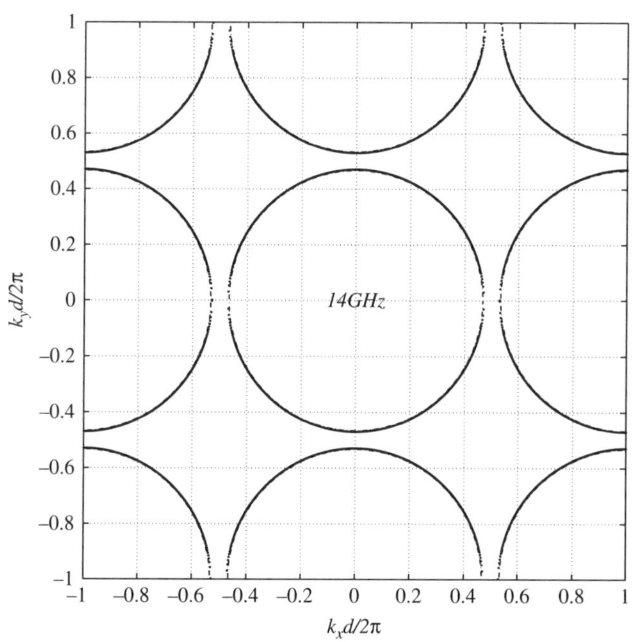

Figure 4.7 Phase relation at 14 GHz. The 2D square lattice has period 10 mm along the x and y-axes, and the dimension of the square metallic rod is 1 mm. The surrounding medium outside the metallic rods is air

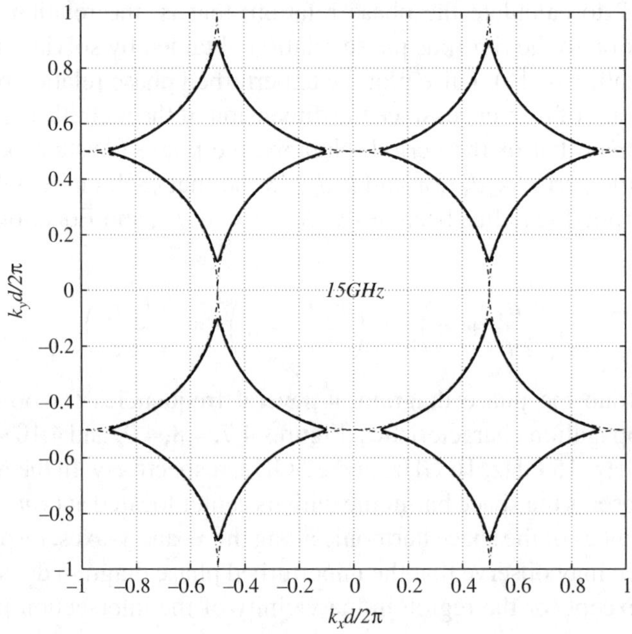

Figure 4.8 Phase relation at 15 GHz with the structure parameters and material properties indicated in Figure 4.7

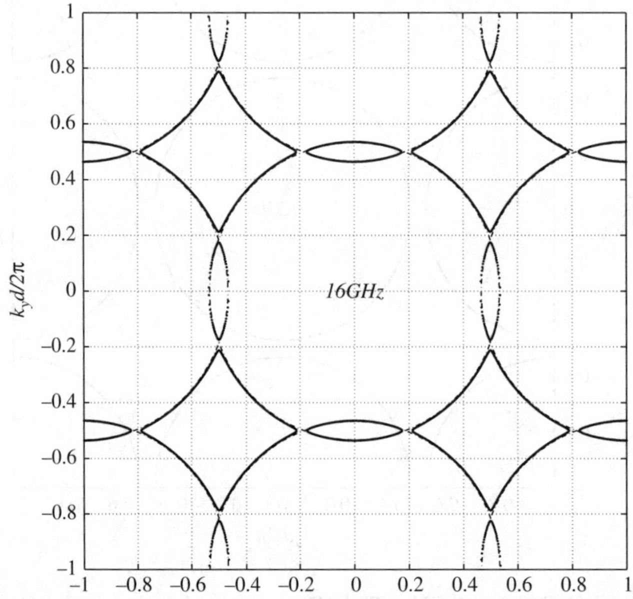

Figure 4.9 Phase relation at 16 GHz with the structure parameters and material properties indicated in Figure 4.7

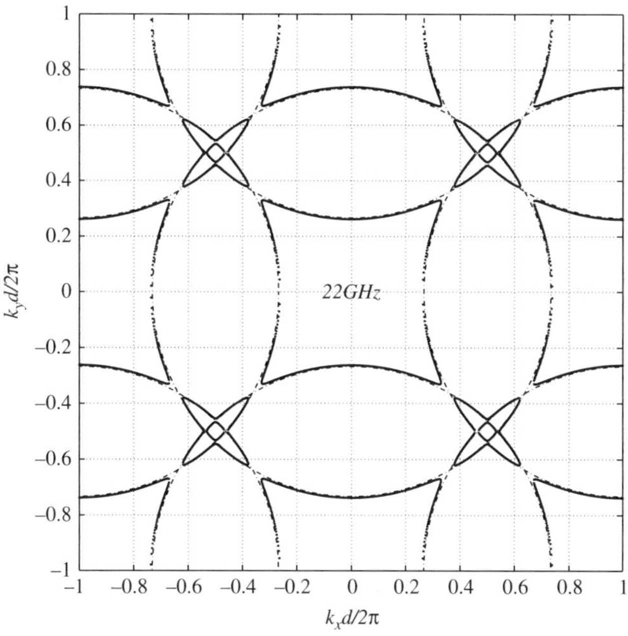

Figure 4.10 Phase relation at 22 GHz with the structure parameters and material properties indicated in Figure 4.7

the circles. Regarding the curves around the intersection points, owing to the phase matching condition, mode coupling between space harmonics (circles) is taking place and they definitely differ from the unperturbed ones. Thus, the unperturbed phase diagram can be taken as a reference for predicting the underlying physics of the wave process involved in a 2D periodic structure.

Looking at Figure 4.7, it is obvious that there are no intersection points between circles. In fact, we can easily figure out the condition of mode coupling between the space harmonics along the x- or y-axis. From the unperturbed diagram, we know that if the distance between the origins of the two adjacent circles along the x- or y-axis is less than or equal to the diameter of the circle (that is, $1 \leq 2d_x/\lambda$), then mode coupling occurs. Substituting the parameters into the above equation, the onset frequency of mode coupling is 15 GHz. At 15 GHz, the intersection point locating on the vertical or horizontal axis appears; see Figure 4.8.

In addition to the electromagnetic coupling along the vertical and horizontal directions, the coupling between the space harmonics along the diagonal direction, such as $(m = 0, n = 0)$ and $(m = -1, n = -1)$, can occur, as the distance between the origins of the above two circles is less than or equal to the diameter of the circle; that is, $\sqrt{2} \leq 2d_x/\lambda$. The onset frequency of the coupling along the diagonal space harmonics will be about 21.2132 GHz. In Figure 4.10, coupling taking place along the diagonal

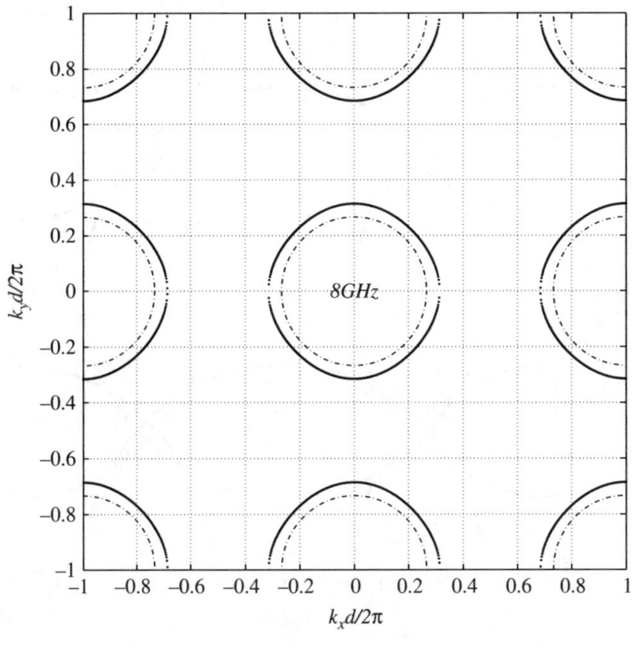

Figure 4.11 Phase relation at 8 GHz. The 2D square lattice has period 10 mm along the *x* and *y* axes, and the dimension of the square metallic rod is 5 mm. The surrounding medium outside the metallic rods is air. The dashed and solid lines represent the unperturbed and exact phase relations, respectively

direction is apparent. Significantly, the slanted stop-band, which is distributed along the diagonal direction rather than those along the vertical or horizontal directions, is found in this example.

In addition to the small perturbation case, we also study the case with large cylinder dimension but maintaining the lattice constant (period). In this example, the square cylinder has width 5 mm. Apparently, the exact phase diagram in Figures 4.11 and 4.12 has an effective refractive index greater than that of the surrounding medium (the radius of the unperturbed phase relation equals $\sqrt{\varepsilon_a}d_x/\lambda$). Additionally, the phase and group velocities are not identical, especially for the case shown in Figure 4.12. From the above two examples, we know that it would be viable to synthesize a material having an effective refractive index greater than the background medium, by immersing 2D metallic rods array in that medium. In addition, when the frequency increases from 10 to 12 GHz, the phase diagram has a dramatic change. The group velocity is along the direction normal to the contour of constant phase diagram. There are four stop-bands along the $\pm x$ and $\pm y$ directions, depicted in Figures 4.13. Along with the increase in frequency, the angular spectrum of the stop-band is enlarged, as was shown in Figure 4.14. As the frequency progressively increases to 16 GHz, only four small

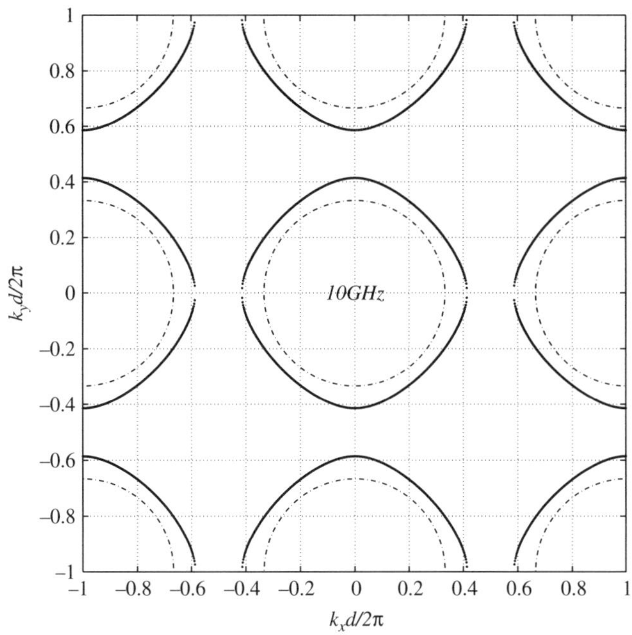

Figure 4.12 Phase relation at 10 GHz with the structure parameters and material properties indicated in Figure 4.11. The dashed and solid lines represent the unperturbed and exact phase relations, respectively

angular spectrums can allow the wave to pass through, which is shown in Figure 4.15. It is interesting to note that the stop-band region is again narrowing when the operating frequency is up to 17 GHz, shown in Figure 4.16.

4.1.5 Dispersion Relation: The Relationship between k_o and k_x (or k_y)

Returning to the previous example; however, this time we would like to investigate the relationship between k_o and k_y for a given k_x. As usual, we first plot the unperturbed dispersion diagram by considering the limiting case of infinitesimal metallic cylinders (or you may simply consider it as the plane-wave solution with periodic modulation). Figure 4.17 shows the dispersion diagram with the horizontal and vertical axes representing the normalized phase constant along the y-axis, $k_yd_x/2\pi$, and normalized frequency $k_od_x/2\pi(=d_x/\lambda)$, respectively. Notably, we have $d_x = d_y = d$ in the ensuing examples. Each of the curves can be easily plotted using Equation (4.19). Such a graph format is called a Brillouin diagram, commonly used in *solid-state physics*. In addition, the integers m and n respectively denote the index of the space harmonic along the x- and y-axes, which are attached to each curve for ease of reference. For $k_x = 0$, the dispersion curves with $m = 0$ are the fundamental space harmonics, which

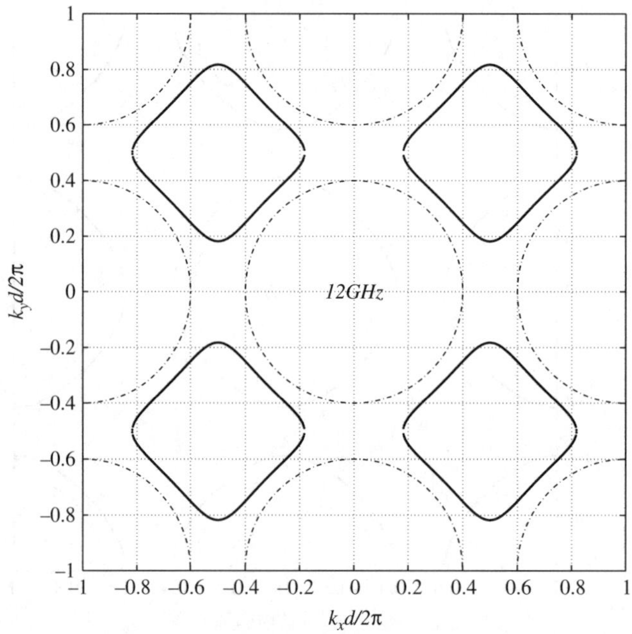

Figure 4.13 Phase relation at 12 GHz with the structure parameters and material properties indicated in Figure 4.11. The dashed and solid lines represent the unperturbed and exact phase relations, respectively

are straight lines. Contrarily, the curves with index $m \neq 0$ are hyperbolic. Additionally, the intersection points denoted in alphabetical order indicate that the electromagnetic coupling may occur therein.

Because $k_x = 0$, the curves $(m = +1, n = 0)$ and $(m = -1, n = 0)$ coincide with each other, as do those for $(m = +1, n = -1)$ and $(m = -1, n = -1)$; see Figure 4.17. Returning to the two straight lines, $(m = 0, n = 0)$ is the fundamental space harmonic, while $(m = 0, n = -1)$ is the first higher order one contributed by the periodicity along the y-axis. The two straight lines intersect at $(k_y d/2\pi = 0.5, k_o d/2\pi = 0.5)$. From the coupled-mode theory, at the intersection point, the two waves share the same phase constant but propagate in opposite directions ($k_y = \pi/d$ and $k_{y,-1} = -\pi/d$). Contraflow interaction (coupling) takes place, creating a stop-band around the intersection point. The phase-match condition can also be applied to explain the underlying physics at the other intersection points.

Figure 4.18 demonstrates the exact dispersion diagram; the vertical axis represents the normalized frequency (d_x/λ), while the horizontal axis is the normalized propagation constant along the y-axis $(k_y = \beta_y - j\alpha_y)$, including the phase (β) and attenuation (α) constants. Because of mirror reflection with respect to $k_y d/2\pi = 0.5$, we only plot half of the diagram. The negative sign of the normalized attenuation constant is plotted

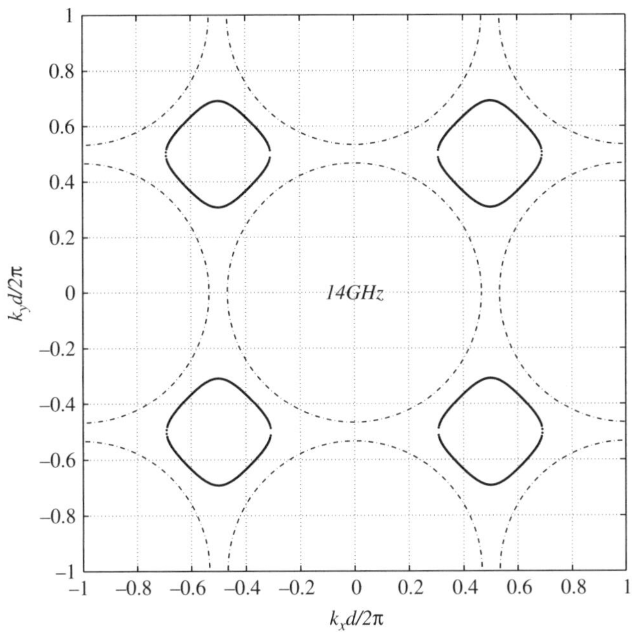

Figure 4.14 Phase relation at 14 GHz with the structure parameters and material properties indicated in Figure 4.11. The dashed and solid lines represent the unperturbed and exact phase relations, respectively

on the left-hand side of the real axis for ease of observation. As was indicated in the corresponding unperturbed dispersion diagram shown in Figure 4.17, the stop-bands due to the contraflow coupling of space harmonics are observed and indicated by the same letter. Moreover, the significant attenuation constant of the space harmonics ($m = \pm 1, n = 0$) and ($m = 0, n = +1$) represents the below-cutoff (nonpropagating) phenomenon.

In Figure 4.19, the normalized phase constant along the x-axis is changed to $k_x d_x/2\pi = 0.5$ for observing the variation on the dispersion relation. Notably, the dispersion curves are all hyperbolic when $k_x \neq 0$. As usual, the stop-bands caused by the contraflow interaction between space harmonics are present and indicated in alphabetic order. For identifying the space harmonics involved, the unperturbed dispersion diagram as well as the intersection points highlighted in the alphabetic order are depicted in Figure 4.20. Specifically, the slanted stop-bands (Hwang 2006) indicated by "B" and "E" are due to the coupling between the two space harmonics along the diagonal direction (please refer to the phase diagram in Figure 4.10).

The last example is the dispersion diagram plotted under the condition $k_x = \pi/d_x$, illustrated in Figure 4.21. Apparently, for such a large phase constant along the x-direction, the two space harmonics ($m = 0, n = 0$) and ($m = -1, n = 0$) are below

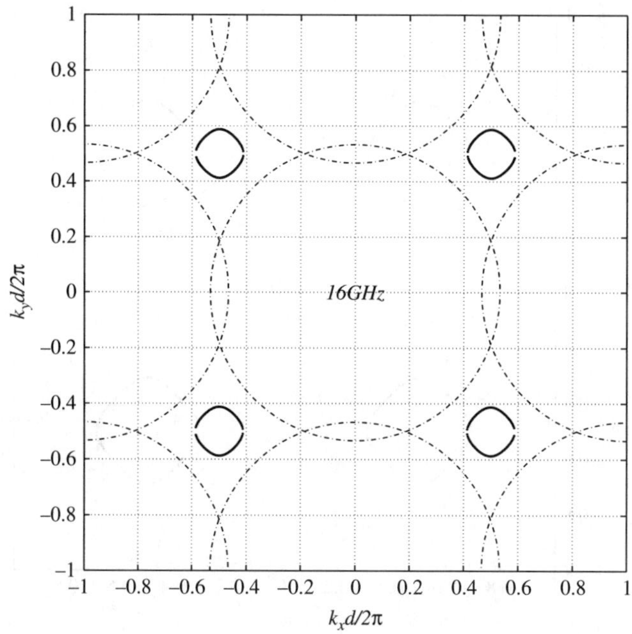

Figure 4.15 Phase relation at 16 GHz with the structure parameters and material properties indicated in Figure 4.11. The dashed and solid lines represent the unperturbed and exact phase relations, respectively

cutoff for a normalized frequency of less than about 0.5, which can be seen from their significant attenuation. Alternatively, it can also be observed from the corresponding unperturbed dispersion diagram shown in Figure 4.22.

4.1.6 Brillouin Zone and Band Structure

The phase diagram demonstrated previously can allow us to understand the wave propagation characteristics, which include the phase and group velocities and the forbidden angular spectrum, at a single frequency. However, the forbidden range in the spectrum cannot be understood from the single-frequency phase diagram. Intuitively, if we stack up the phase diagram with the vertical axis representing the frequency and the transverse plane denoted by k_x and k_y, we can plot the dispersion diagram in the format of a 3D graph. As long as the frequency increment is small enough, the 3D graph is "continuous" in some sense. We shall see some complex surfaces. Significantly, the discontinuity between two surfaces indicates the existence of complex dispersion roots and the wave cannot propagate therein. Such a spectral gap is termed the stop-band or band-gap.

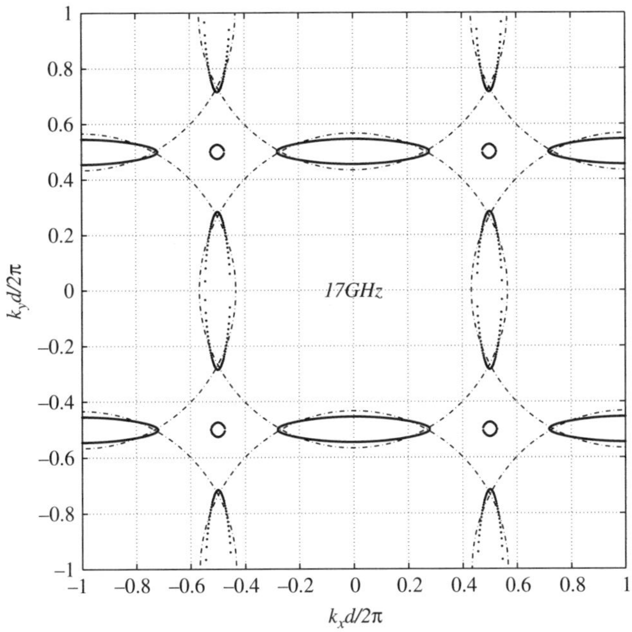

Figure 4.16 Phase relation at 17 GHz with the structure parameters and material properties indicated in Figure 4.11. The dashed and solid lines represent the unperturbed and exact phase relations, respectively

In Figure 4.23 we plot the 3D dispersion diagram using a number of 2D phase diagrams drawn at different frequencies. The structure under consideration is a 2D square lattice with the structure parameters illustrated in the figure caption. The ranges in k_x and k_y are both π/d, since the phase diagram has the basic unit cell pattern distributed in that region; the rest of the figure can be related to this fundamental one by translational symmetry. The 3D dispersion diagram is useful for observing the band-gap from every aspect; however, without 3D computer graphics technology it is hard to allocate the stop-band precisely. In the following, we will introduce a representation that is commonly used in solid-state physics for plotting the band-structure in a 2D diagram without invoking the complicated process of computer graphics.

Figure 4.24 is a Brillouin zone diagram of a square lattice ($d_x = d_y = d$) with reciprocal lattice vectors $\underline{x}_o 2\pi/d_x$ and $\underline{y}_o 2\pi/d_y$ attached. The area enclosed by the short dashed line is the first Brillouin zone, while that enclosed by the long dashed line is the second Brillouin zone. The first Brillouin zone of a square lattice exhibits a square pattern. The lattice center, face, and corner are denoted by Γ, X, and M, corresponding to ($k_x = 0, k_y = 0$), ($k_x = \pi/d, k_y = 0$), and ($k_x = \pi/d, k_y = \pi/d$), respectively. Notably, the irreducible Brillouin zone is the triangular wedge, which is one-eighth the area of the full Brillouin zone, surrounded by three points: Γ, X, and M. By rotational

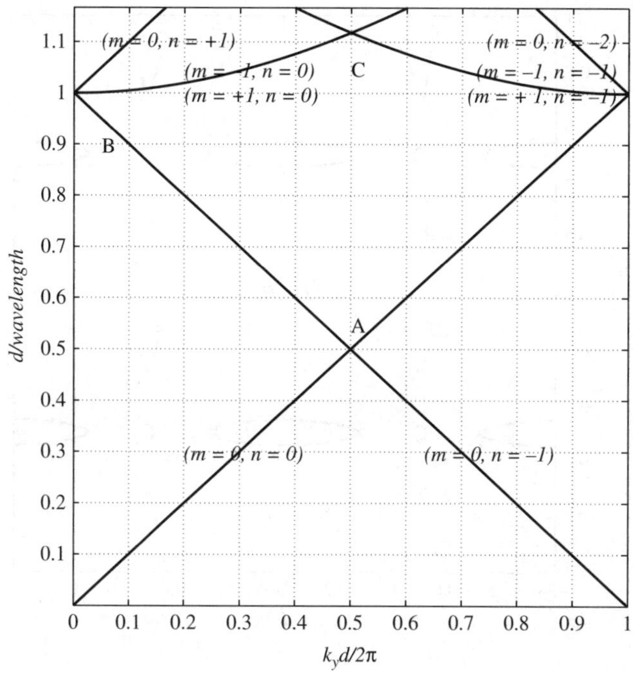

Figure 4.17 Unperturbed dispersion relation for $k_x = 0$. The structure parameters are shown in Figure 4.11: $d_x = d_y = d$

symmetry, the rest of the wedge in the Brillouin zone can be readily obtained. The band structure is plotted along the boundary of the triangular wedge starting from Γ through X and M and returning to Γ. Specifically, the three paths correspond to different incident conditions: (i) from Γ to X, the x-component of phase constant k_x is increasing from zero to π/d while the y-component of phase constant k_y is set to zero; (ii) from X to M, k_y is increasing from zero to π/d at a fixed x-component of phase constant $k_x = \pi/d$; (iii) from M to Γ, we keep $k_x = k_y$ and iterate it from π/d to zero along the diagonal segment.

Alternatively, if we trace the the boundary of the irreducible Brillouin zone (the triangular wedge), we can find the one-to-one correspondence between the boundary and the three planes enclosing the wedge indicated in Figure 4.25. As a consequence, if the 3D dispersion diagram has been obtained, the band structure plotted in the irreducible Brillouin zone can be easily determined by viewing the trajectories on the three cut planes. According to solid-state physics textbooks, the dispersion relation plotted on the boundary of an irreducible Brillouin zone reveals to us the band-gap behavior of the crystal; however, once the 3D dispersion diagram is determined, it can provide us not only the band-structure on the irreducible Brillouin zone boundary, but also allow us to know the wave propagation characteristics within the Brillouin zone.

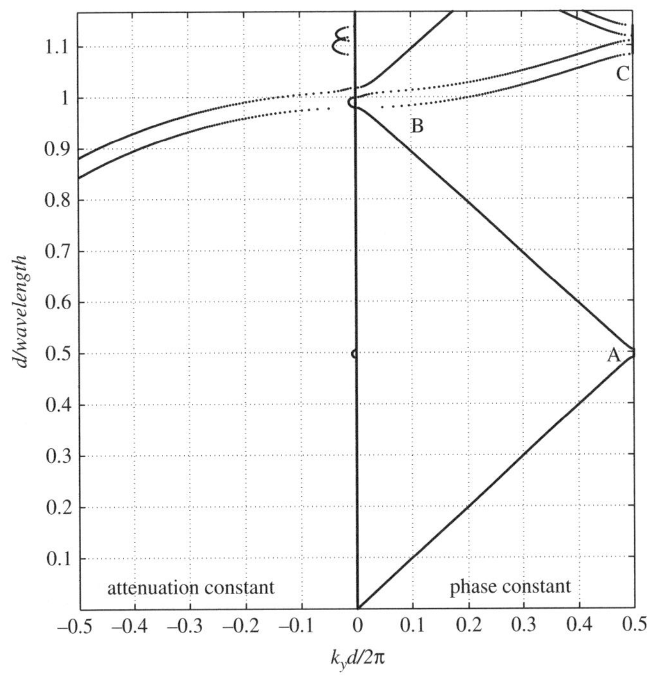

Figure 4.18 Exact dispersion diagram for $k_x = 0$. The structure parameters are shown in Figure 4.11: $d_x = d_y = d$

Note that the dispersion relation in Equation (4.16) is in the form of generalized eigenvalue equation, in which the frequency and k_x are given for solving the eigenvalue k_y. For a square lattice, the above-mentioned problem-solving scenario can deal with the band structure along the paths of Γ-to-X and X-to-M; however, the advantage of directly solving the eigenvalue is not able to be applied for determining the band structure along the path of M-to-Γ. As a consequence, searching the root (frequency) of the nonlinear equation, which is written below, for given k_x and k_y is inevitable:

$$\det\left[\begin{pmatrix} \mathbf{S}_{21} & \mathbf{0} \\ \mathbf{S}_{11} & -\mathbf{I} \end{pmatrix} - \exp(-\mathrm{j}k_y d_y)\begin{pmatrix} \mathbf{I} & -\mathbf{S}_{22} \\ \mathbf{0} & -\mathbf{S}_{12} \end{pmatrix}\right] = 0 \tag{4.20}$$

With the definition of irreducible Brillouin zone described previously, we again plot the dispersion relation in accordance with the new format. In the following examples, we present the band structure of the three examples with different dimensions of square column (1, 5, and 8 mm) while the lattice constant remains ($d = 10$ mm).

Let us start with the first example shown in Figure 4.26. From Γ to X, k_x is increasing from zero to π/d and k_y is set to be zero. Owing to the rotational symmetry of the square lattice, the result is the same as iterating k_y while k_x is zero. This is the reason

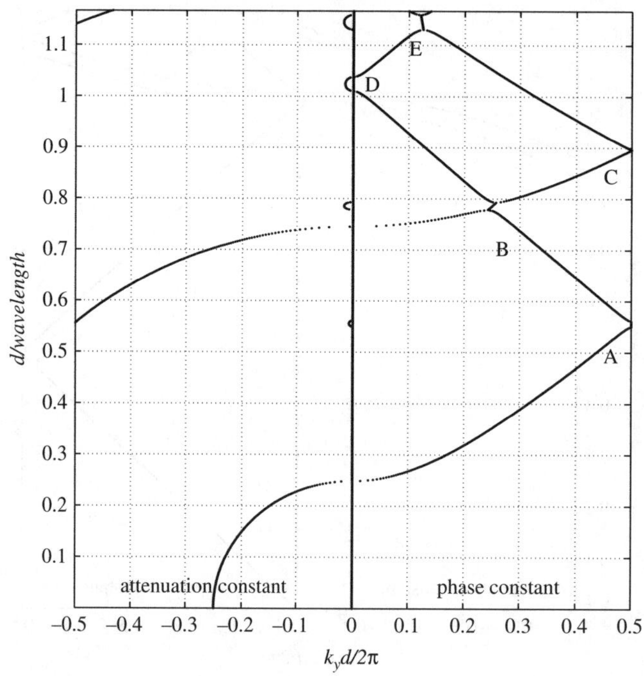

Figure 4.19 Exact dispersion diagram for $k_x = \pi/2d_x$. The structure parameters are shown in Figure 4.11: $d_x = d_y = d$

why this portion of band structure is the same as that shown in Figure 4.18. The band structure from X to M also corresponds to that depicted in Figure 4.21. Unfortunately, the band structure from M to Γ must invoke the root searching process, so that no reference data can be found for comparison.

In addition to the case of small perturbation depicted in Figure 4.26, the band structures of the periodic structure with 25% and 64% filling ratios (defined as the ratio of area between metallic cylinder and the unit cell) were calculated and are shown in Figures 4.27 and 4.28, respectively. It is worth emphasizing that each of the dispersion curves in the band-structure diagram can, in fact, be traced back to its corresponding space harmonic with indices (m, n), which can be figured out from the unperturbed dispersion curves. For the example with large filling ratio, the dispersion curves will differ from the unperturbed ones significantly; however, it can be deduced from the evolution of the dispersion curves by gradually increasing the filling ratio.

If there is a frequency band without any dispersion curves crossing over in an irreducible Brillouin zone, such a specific *spectral gap* is termed the *complete stop-band*. A single lattice pattern may not easily produce a complete stop-band; however, the composite periodic structure consisting of several types of lattice patterns, such as

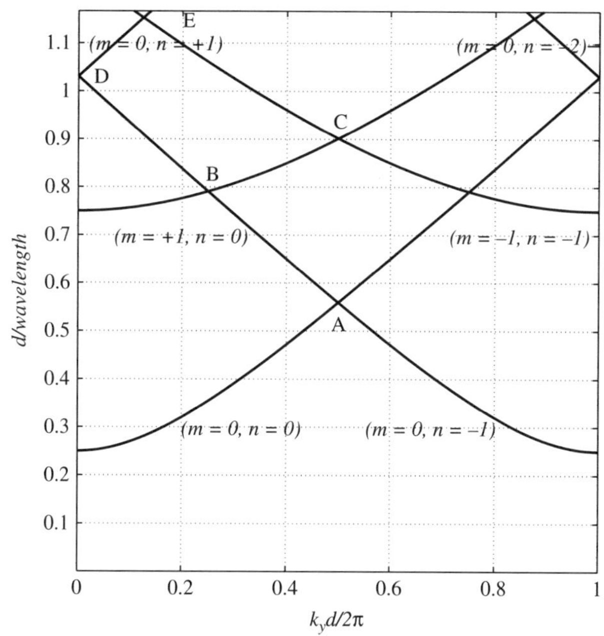

Figure 4.20 Unperturbed dispersion diagram for $k_x = \pi/2d_x$, which is taken as the reference of Figure 4.19: $d_x = d_y = d$

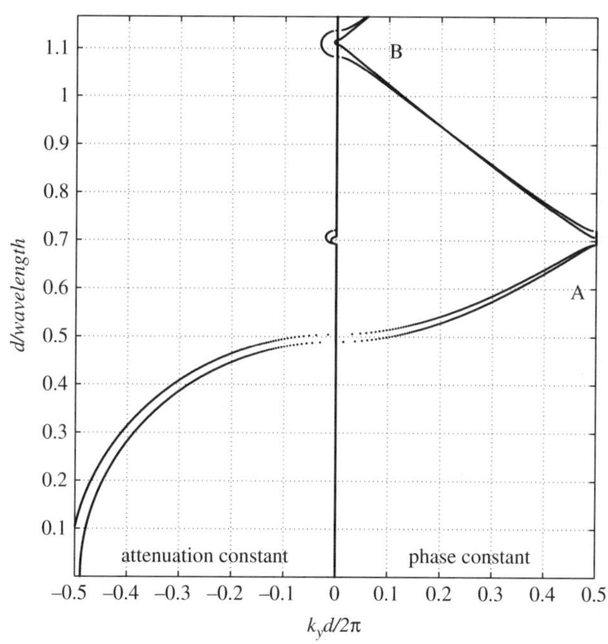

Figure 4.21 Exact dispersion diagram for $k_x = \pi/d_x$

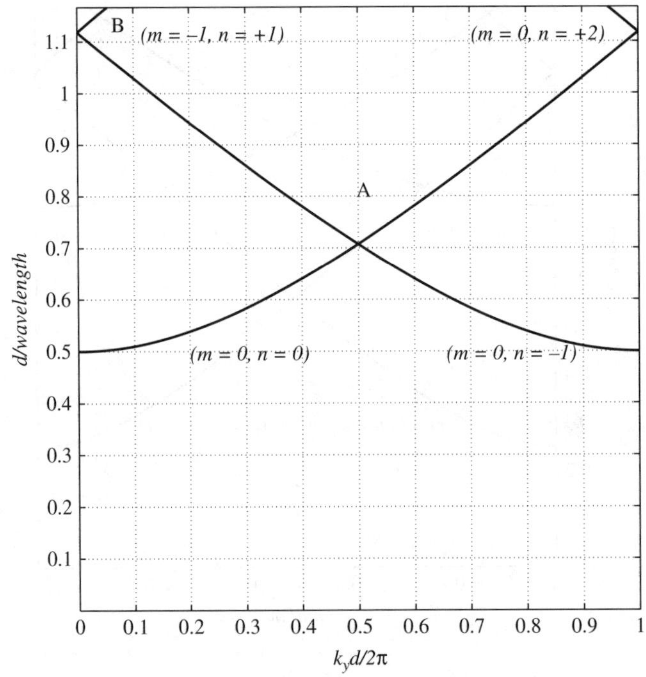

Figure 4.22 Unperturbed dispersion diagram for $k_x = \pi/d_x$, which corresponds to Figure 4.21

square and triangular lattices, can synthesize a complete stop-band by compensating several stop-bands (Hwang and Pu 2007).

4.2 Modal Transmission Line Approach for a 2D Periodic Dielectric Medium: In-Plane Propagation

In Section 4.1 we introduced a rigorous method for determining the eigenwave solution in an infinite 2D periodic metallic medium by stacking up an infinite numbers of 1D metallic gratings. By imposing a periodic boundary condition on the two interfaces of the unit cell along the stacking direction we can set up a generalized eigenvalue equation for solving the propagation constant and space harmonic amplitude of the mode supported in the unit cell. Notably, the eigenwave solutions in the 2D periodic metallic medium can be separately treated for TE and TM polarization since the propagation constant k_z along the length direction of the metal rod vanishes (the periodicity is along the x- and y-axes). Nevertheless, if $k_z \neq 0$, the superposition of the TE- and TM-polarized modes is needed to satisfy the electromagnetic boundary conditions at the discontinuity.

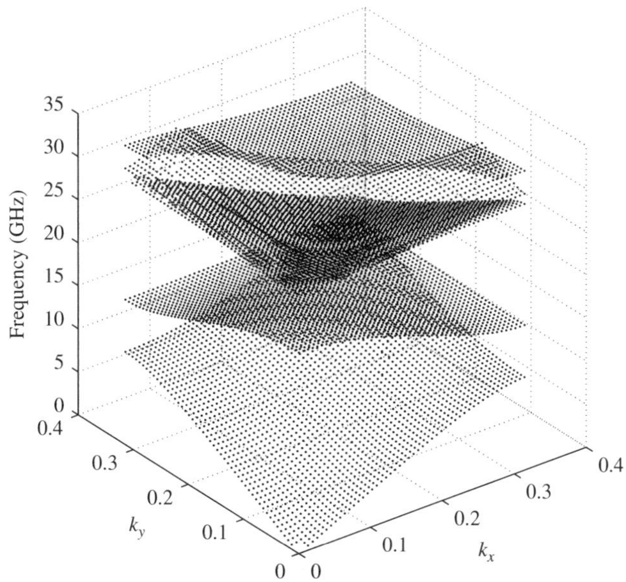

Figure 4.23 A 3D dispersion diagram plotted in the Brillouin zone with k_x and k_y in the range from zero to π/d. The period of the square lattice is 10 mm; the dimension of the square metallic column is 5 mm. Here, we have $k_z = 0$, so the wave is propagating strictly along the x–y plane (in-plane propagation)

In this section we will investigate the general field solutions in a 2D periodic dielectric medium consisting of 2D dielectric rods array (see Figure 4.29), particularly for the case with nonvanishing propagation constant along the length direction of the dielectric column. The scattering and guiding characteristics of a wave involved in a 2D dielectric periodic structure will be investigated. Significantly, the cross-coupling

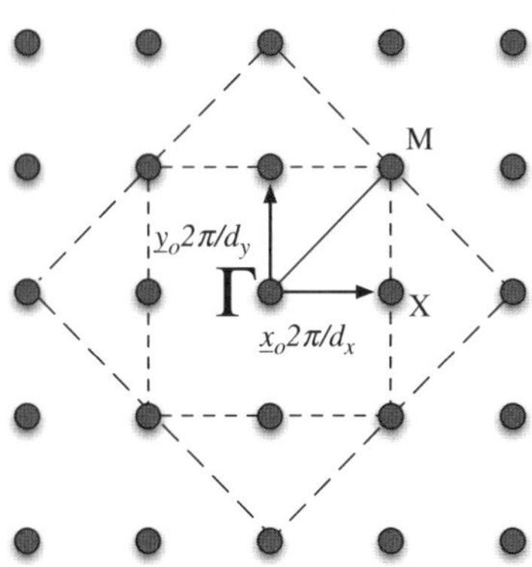

Figure 4.24 Brillouin zone of a square lattice

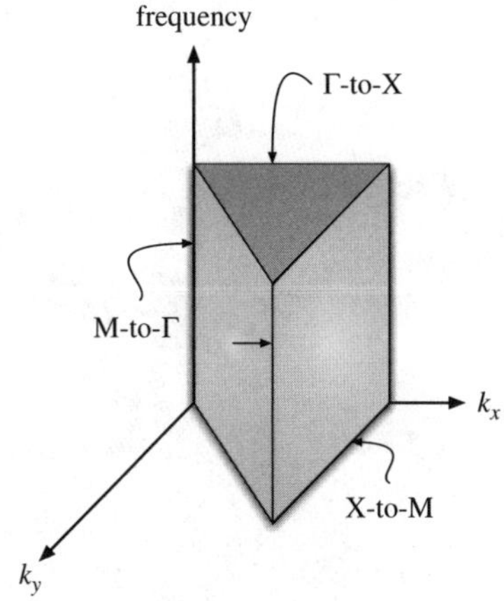

Figure 4.25 Irreducible Brillouin zone defined in a 3D dispersion diagram

between TE and TM modes (space harmonics of waveguide modes) will be studied in detail.

In Section 3.5, the hybrid-mode analysis of a 1D dielectric grating was rigorously formulated in the presence of the propagation constant k_y along the dielectric rod (the periodicity is along the x- and z-axes). Referring to Figure 4.29, each row in the

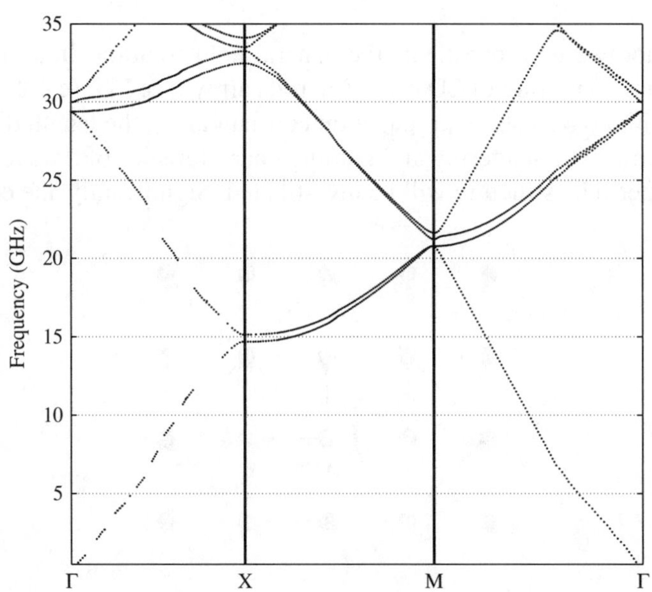

Figure 4.26 Band structure of the 2D square lattice. The lattice constant (period) and the dimension of the square metallic column are 10 mm and 1 mm, respectively

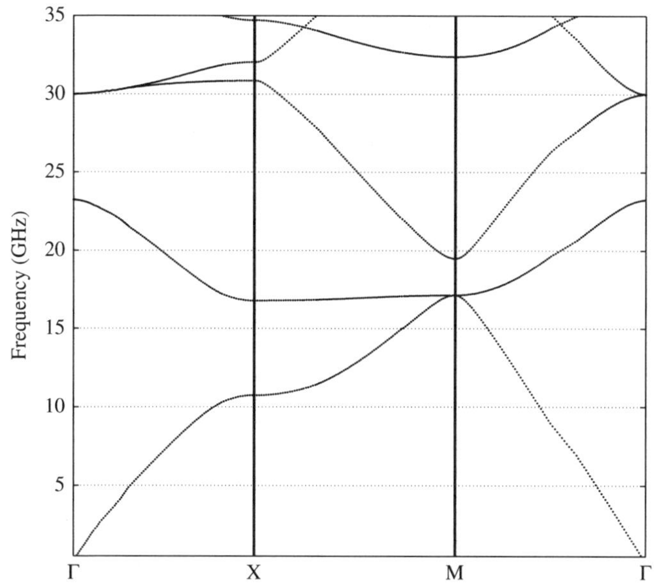

Figure 4.27 Band structure of the 2D square lattice. The lattice constant (period) and the dimension of the square metallic column are 10 mm and 5 mm, respectively

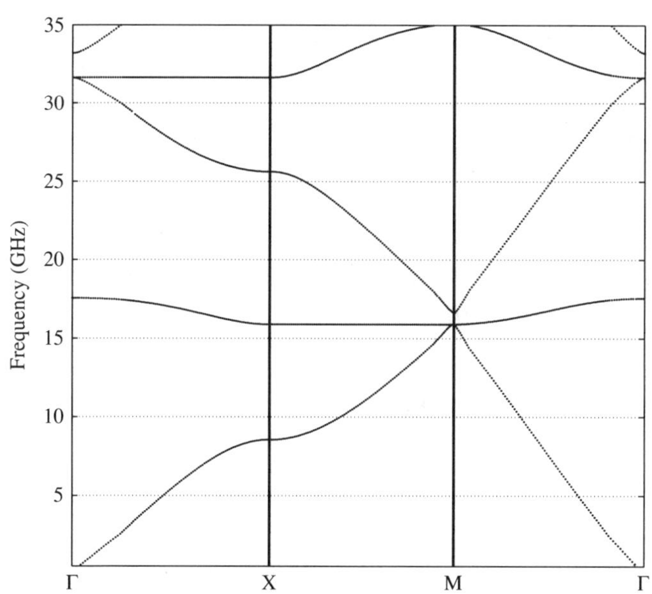

Figure 4.28 Band structure of the 2D square lattice. The lattice constant (period) and the dimension of the square metallic column are 10 mm and 8 mm, respectively

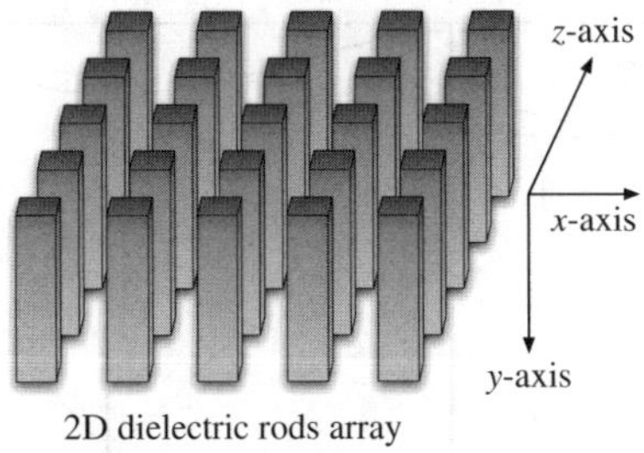

Figure 4.29 The 2D periodic dielectric medium composed of 2D dielectric rods array

figure is a 1D dielectric grating with period along the x-axis and infinite length along the y-axis. For ease of reference, we rewrite the equations concerning the tangential electric and magnetic fields in a 1D periodic dielectric medium:

$$\begin{bmatrix} -\underline{\mathcal{E}}_y(z) \\ \underline{\mathcal{E}}_x(z) \end{bmatrix} = \mathbb{Q}\left[\exp(-\mathrm{j}\mathbb{K}_z z)\underline{f} + \exp(+\mathrm{j}\mathbb{K}_z z)\underline{g}\right] \tag{4.21}$$

$$\begin{bmatrix} \underline{\mathcal{H}}_x(z) \\ \underline{\mathcal{H}}_y(z) \end{bmatrix} = \mathbb{P}\left[\exp(-\mathrm{j}\mathbb{K}_z z)\underline{f} - \exp(+\mathrm{j}\mathbb{K}_z z)\underline{g}\right] \tag{4.22}$$

Additionally, the tangential electric and magnetic fields in a uniform medium can be considered as the limiting case of a periodic medium with vanishing periodic variation. Consequently, their mathematical expression remains the same form as the above two equations by simply replacing $\mathbb{Q}$ with $\mathbf{A}$, and $\mathbb{P}$ with $\mathbf{AY}$, and $\mathbb{K}_z$ can be obtained from the phase relation of a plane wave in a uniform medium, where $\mathbf{A}$ and $\mathbf{Y}_u$ are defined and given in Section 3.6.

4.2.1 Input–Output Relation at the Interface: Generalized Scattering Matrix Representation

At the interface $z = 0$ between the 1D periodic medium and uniform dielectric medium depicted in Figure 4.30, the tangential electric and magnetic fields must be continuous across the interface. The equations are as follows:

$$\mathbb{Q}(\underline{f} + \underline{g}) = \mathbf{A}(\underline{a} + \underline{b}) \tag{4.23}$$

$$\mathbb{P}(\underline{f} - \underline{g}) = \mathbf{AY}_{\mathrm{u}}(\underline{a} - \underline{b}) \tag{4.24}$$

where $\underline{a}$ and $\underline{b}$ are respectively the forward- and backward-propagating wave vectors along the $\pm z$-axes in the uniform medium and $\underline{f}$ and $\underline{g}$ are those in the 1D periodic dielectric medium.

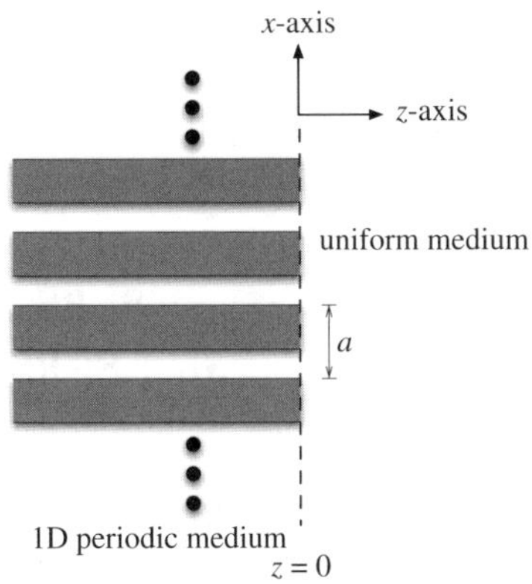

Figure 4.30 An interface between a semi-infinite 1D periodic dielectric medium (with period a along the x-axis) and a uniform medium

After rearranging the above two equations, we can obtain the following matrix equation:

$$\begin{pmatrix} \underline{a} \\ \underline{g} \end{pmatrix} = \mathbf{S}_{\mathrm{o}} \begin{pmatrix} \underline{b} \\ \underline{f} \end{pmatrix} \tag{4.25}$$

$$\mathbf{S}_{\mathrm{o}} = \begin{pmatrix} \mathbf{A} & -\mathbb{Q} \\ \mathbf{A}\mathbf{Y}_{\mathrm{u}} & \mathbb{P} \end{pmatrix}^{-1} \begin{pmatrix} -\mathbf{A} & \mathbb{Q} \\ \mathbf{A}\mathbf{Y}_{\mathrm{u}} & \mathbb{P} \end{pmatrix} \tag{4.26}$$

Through the presentation of the above matrix equation, the outgoing wave vector can be related to the incoming wave vector through the matrix $\mathbf{S}_{\mathrm{o}}$, which is usually termed the *scattering matrix* in microwave engineering. Moreover, since the modes considered here include not only the propagating modes but also the *nonpropagating* (evanescent) modes, the scattering matrix is usually termed the *generalized scattering matrix*. After determining the generalized scattering matrix at the interface, we may follow the mathematical procedures demonstrated in Section 4.1.1 for determining that of the unit cell along the z-axis. I just list the results and the detailed procedures are given as homework:

$$\begin{pmatrix} \mathbf{S}_{11} & \mathbf{S}_{12} \\ \mathbf{S}_{21} & \mathbf{S}_{22} \end{pmatrix} = \begin{pmatrix} \mathbf{T}_{\mathrm{u}} & \mathbf{0} \\ \mathbf{0} & \mathbf{T}_{\mathrm{g}} \end{pmatrix} \mathbf{S}_{\mathrm{o}} \begin{pmatrix} \mathbf{T}_{\mathrm{u}} & \mathbf{0} \\ \mathbf{0} & \mathbf{T}_{\mathrm{g}} \end{pmatrix} \tag{4.27}$$

$$\mathbf{T}_{\mathrm{u}} = \mathrm{diag}(\mathrm{e}^{-\mathrm{j}\mathbf{k}_z^{\mathrm{u}} \frac{t_{\mathrm{u}}}{2}}) \tag{4.28}$$

$$\mathbf{T}_{\mathrm{g}} = \mathrm{diag}(\mathrm{e}^{-\mathrm{j}\mathbf{k}_z^{\mathrm{g}} \frac{t_{\mathrm{g}}}{2}}) \tag{4.29}$$

where $\mathbf{T}_{\mathrm{u}}$ and $\mathbf{T}_{\mathrm{g}}$ are transition matrices of the modes in the uniform and periodic mediums, respectively; the two vectors, $\mathbf{k}_z^{\mathrm{u}}$ and $\mathbf{k}_z^{\mathrm{g}}$ contain the propagation constants of each of the space harmonics in a uniform medium and a 1D periodic dielectric medium, respectively.

The scattering matrix of the unit cell along the z-axis over a period $d_z = t_u + t_g$ is given as

$$\begin{pmatrix} \underline{b}_1 \\ \underline{b}_2 \end{pmatrix} = \begin{pmatrix} \mathbf{S}_{11} + \mathbf{S}_{12}\mathbf{R}\mathbf{S}_{22}\mathbf{S}_{21} & \mathbf{S}_{12}\mathbf{R}\mathbf{S}_{21} \\ \mathbf{S}_{12}\mathbf{R}\mathbf{S}_{21} & \mathbf{S}_{11} + \mathbf{S}_{12}\mathbf{R}\mathbf{S}_{22}\mathbf{S}_{21} \end{pmatrix} \begin{pmatrix} \underline{a}_1 \\ \underline{a}_2 \end{pmatrix} \tag{4.30}$$

where $\underline{a}_1$ and $\underline{b}_1$ are the incoming and outgoing wave vectors at the input interface of the unit cell and $\underline{a}_2$ and $\underline{b}_2$ are those at the output interface.

By invoking the *Bloch–Floquet* (periodic boundary) condition along the z-axis, the above scattering matrix over a unit cell can be rewritten as

$$\begin{pmatrix} \mathbf{S}_{21} & \mathbf{0} \\ -\mathbf{S}_{11} & \mathbf{1} \end{pmatrix} \begin{pmatrix} \underline{a}_1 \\ \underline{b}_1 \end{pmatrix} = \lambda \begin{pmatrix} \mathbf{1} & -\mathbf{S}_{22} \\ 0 & \mathbf{S}_{12} \end{pmatrix} \begin{pmatrix} \underline{a}_1 \\ \underline{b}_1 \end{pmatrix} \tag{4.31}$$

where $\lambda = \exp(\pm \mathrm{j} k_z d_z)$ is the eigenvalue of the generalized eigenvalue equation in Equation (4.31). The above equation can be solved by a traditionally used eigenvalue solver such as the MATLAB function of eig(A,B).

4.2.2 Brillouin Diagram and Phase Relation

In the first example we consider a 2D periodic medium consisting of a square dielectric rod array arranged in square lattice; the periods along the x and z-axes are both 1.0. The dimension of the square dielectric rod is 0.5, while it is of infinite extent along the length direction; that is, the y-axis. It is noted that all the dimensions are in arbitrary units. The relative dielectric constant of the dielectric rod and surrounding medium are $\varepsilon_s = 8.0$ and $\varepsilon_a = 1.0$, respectively.

Recalling the Brillouin diagrams of the 2D periodic metallic medium shown in Section 4.1.5, the stop-band can be clearly observed by its significant attenuation constant. Here, not only is the Brillouin diagram demonstrated, but also the scattering characteristics of the corresponding 2D dielectric periodic structure of finite thickness (can be simply considered as a finite stack of the 1D dielectric grating) will be shown to elucidate the nonpropagation of waves in the stop-band region. In the following numerical simulation, the structure contains a finite number of 1D dielectric grating layers. For the scattering analysis, the elevation angle θ and azimuth angle ϕ of the incident plane wave are considered simultaneously; therefore, k_y is nonvanished for the condition of $\phi \neq 0$; the hybrid-mode analysis has to be taken into account and the cross-polarization coupling takes place, which will become clear later on.

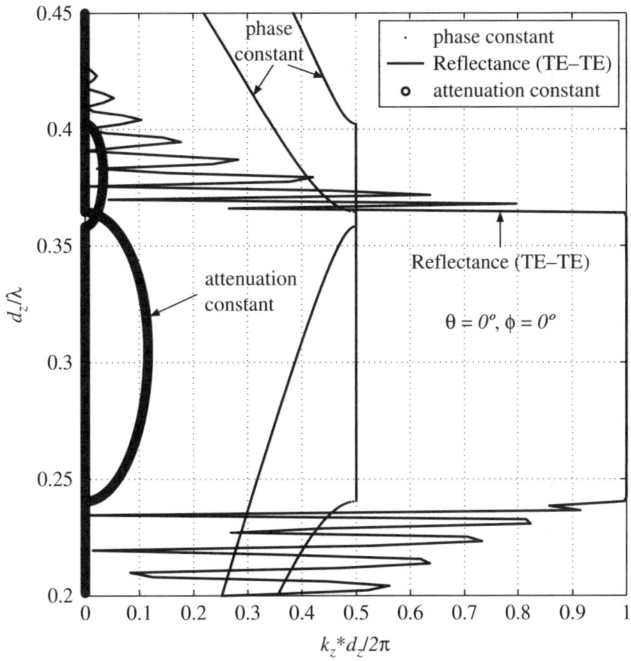

Figure 4.31 Brillouin diagram and the reflectance of zeroth-order space harmonic: TE-polarized incidence, $\theta = 0^\circ$ and $\phi = 0^\circ$

Figures 4.31 and 4.32 demonstrate the normalized phase constant along the z-axis $\beta_z d_z/2\pi$ against the normalized frequency d_z/λ in an infinite 2D periodic dielectric medium, together with the reflectance of a normal incident plane wave by the finite 2D periodic structure consisting of 10 1D periodic layers. The plane wave with TE polarization is normally impinging on the structure. The dotted line is the normalized phase constant; the line with circle symbols is the normalized attenuation constant; the solid line is the reflectance of TE-polarized space harmonic $n = 0$. It is apparent that the strong reflection region corresponds to the TE-polarized stop-band shown in Figure 4.31. Similarly, the TM-polarized stop-band can be identified by the strong reflection of the TM-polarized fundamental space harmonic for the TM-polarized incident plane wave, as shown in Figure 4.32. Notably, because the propagation constant along the y-axis vanishes ($k_y = k_o \sin\theta\cos\phi$), cross-polarization does not occur; the co-polarization reflection characteristic is shown in these two figures.

If the elevation and azimuth angles are both not zero, the propagation constants along the x- and y-axes are $k_x = k_o \sin\theta\cos\phi$ and $k_y = k_o \sin\theta\sin\phi$, respectively. From the mathematical formulation in this section, the cross-coupling between TE- and TM-polarized space harmonics certainly takes place. Figure 4.33 shows the dispersion

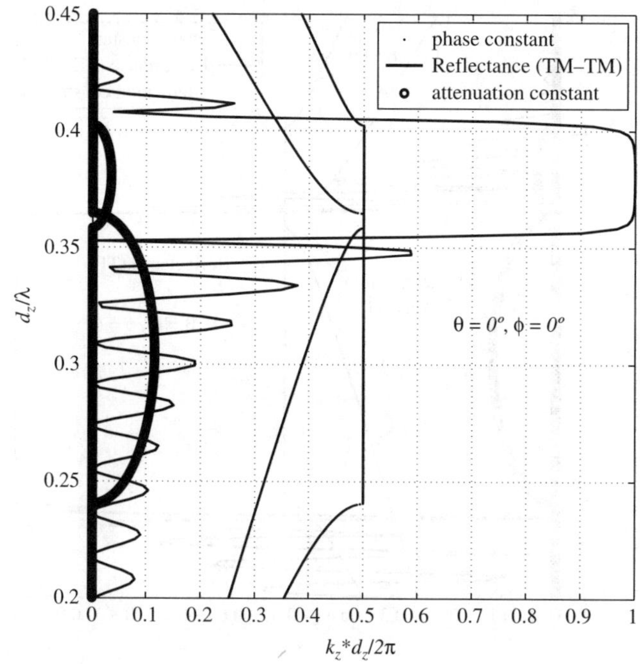

Figure 4.32 Brillouin diagram and the reflectance of zeroth-order space harmonic: TM-polarized incidence, $\theta = 0^\circ$ and $\phi = 0^\circ$

relation, which is shown in the format of a Brillouin diagram, and the reflectance of TE- and TM-fundamental space harmonic under the incidence of the TE plane wave as well. Obviously, the TE reflected wave exhibits a strong reflection in the first stop-band region, which is due to the contraflow coupling between TE space harmonics $(n = -1)$ and $(n = 0)$. Moreover, the contraflow coupling between the TM space harmonics $(n = -1)$ and $(n = 0)$ can cause the second vertical stop-band, enabling an obvious reflection for the TM wave incidence, which is shown in Figure 4.34. Significantly, the slanted stop-band in between the two vertical stop-bands also causes an extra reflection for both TE and TM wave incidence. Regarding the physics of such a slanted stop-band, we will explain this in the following example. It is noted that the index n of the space harmonic is due to the periodicity along the z-axis.

Figures 4.35 and 4.36 are the partial enlarged views of the dispersion relation shown in Figure 4.33, including the phase and attenuation constants, particularly in the slanted stop-band region. The slanted stop-band is due to the cross-coupling between TE$(n = 0)$ and TM$(n = -1)$, or TE$(n = -1)$ and TM$(n = 0)$. This is distinct from the slanted stop-band caused by the co-polarization coupling between the fundamental and higher order space harmonics along the x-axis that was shown in Figure 4.19.

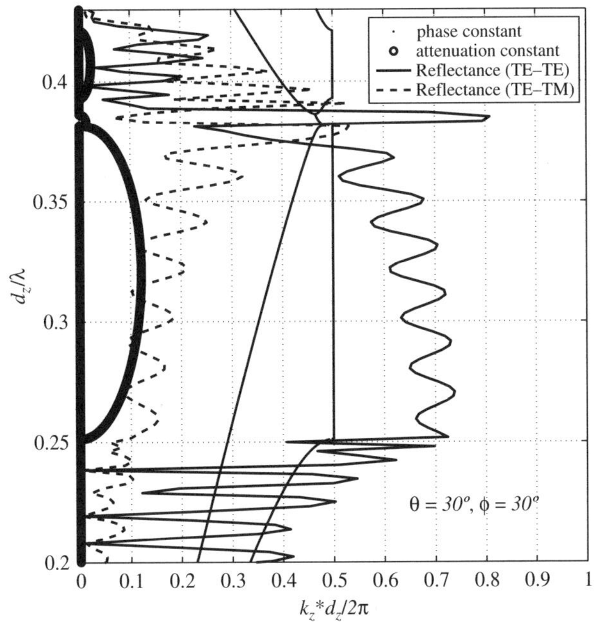

Figure 4.33 Brillouin diagram and the reflectance of zeroth-order space harmonic: TE incidence $\theta = 30^\circ$ and $\phi = 30^\circ$

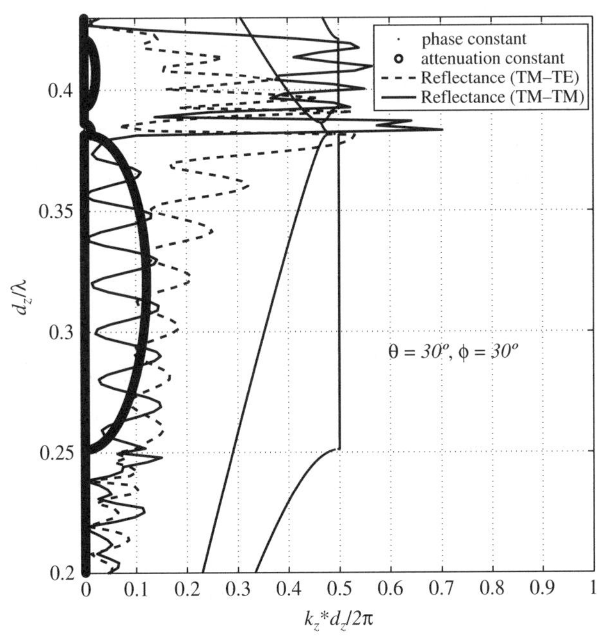

Figure 4.34 Brillouin diagram and the reflectance of zeroth-order space harmonic: TM incidence $\theta = 30^\circ$ and $\phi = 30^\circ$

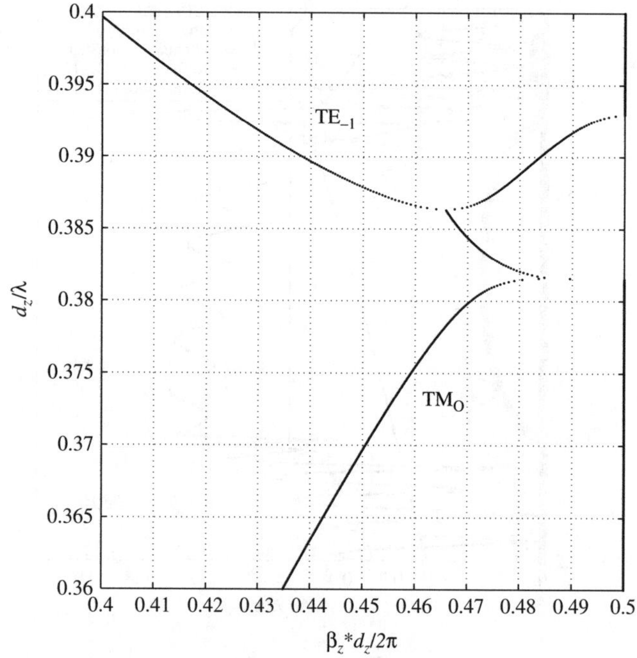

Figure 4.35 A partial enlarged view for the slanted stop-band shown in Figure 4.33: normalized phase constant against normalized frequency

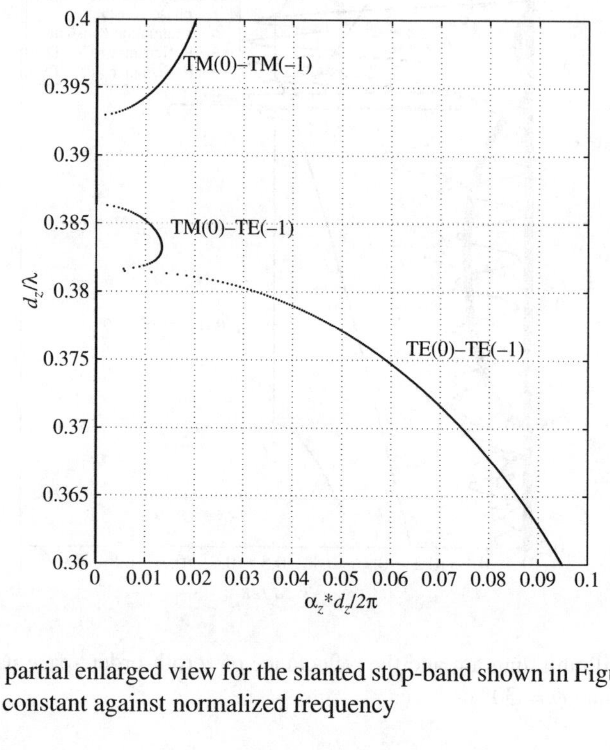

Figure 4.36 A partial enlarged view for the slanted stop-band shown in Figure 4.33: normalized attenuation constant against normalized frequency

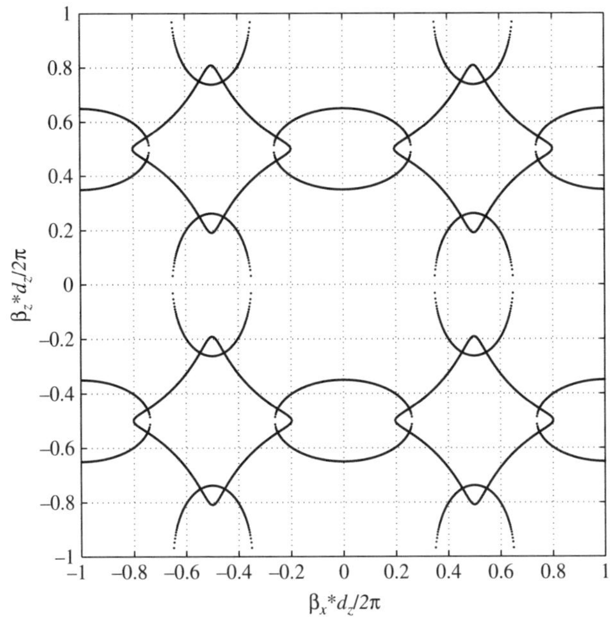

Figure 4.37 Phase diagram for the 2D periodic medium with the same structure parameters shown at the beginning of this section, $k_y = 0$ and $d_z/\lambda = 0.4$

Additionally, compared with the vertical stop-band, the slanted one has an asymmetric pattern on its attenuation constant; this is unlike the vertical stop-band, which has a symmetric pattern.

In addition to the dispersion relation, the phase relation was calculated; that is, the relationship between $k_x d_z/2\pi$ and $k_z d_z/2\pi$ for given normalized frequency d_z/λ and phase constant along the y-axis (k_y). Here, we change k_y for observing the variation on the phase diagram. Parameter k_y is assumed to be equal to $n\pi/10d_z$ and n is an integer. Figures 4.37–4.41 plot the phase diagrams for $n = 0$, 2, 5, 8 and 10, respectively. In Figure 4.37, the phase diagram in the shapes of diamonds and ellipses independently represent those of the two polarizations; therefore, no coupling occurred at the intersection point between them.

As k_y increases, coupling between the two polarization modes takes place; the slanted stop-band is present in the vicinity of the intersection point, which can be seen in Figures 4.38 and 4.39. For the examples with large k_y, such as those in Figures 4.40 and 4.41, the phase relation returns to a shape like concentric circles. It is worth pointing out that the effective refractive index n_{eff} of the two polarization modes for the former example are approximately 1.073 and 1.172 and are roughly 0.7235 and 0.8195 for the latter example. This means that we may synthesize an artificial medium with an effective refractive index smaller than unity.

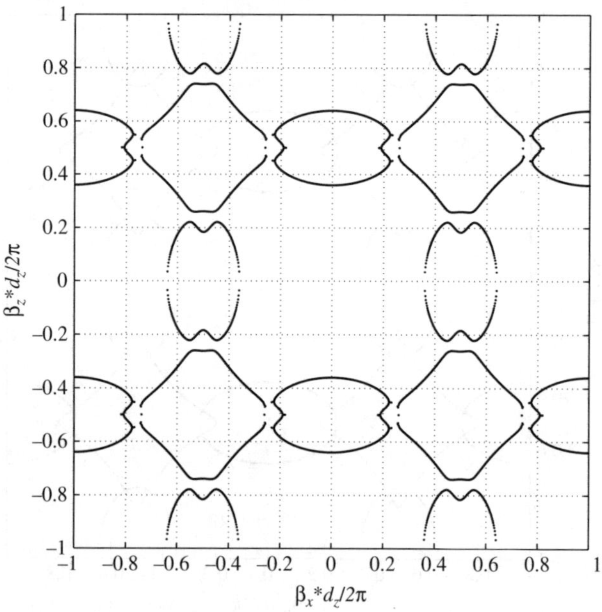

Figure 4.38 Phase diagram for the 2D periodic medium with the same structure parameters shown at the beginning of this section; $k_y = 2\pi/10d_z$ and $d_z/\lambda = 0.4$

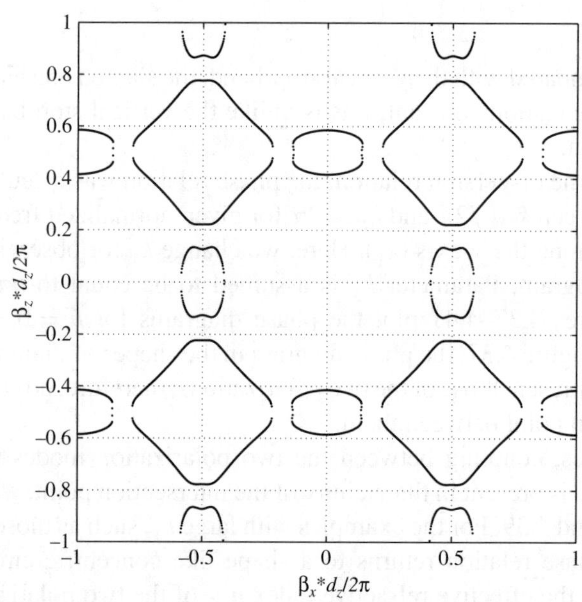

Figure 4.39 Phase diagram for the 2D periodic medium with the same structure parameters shown at the beginning of this section; $k_y = 5\pi/10d_z$ and $d_z/\lambda = 0.4$

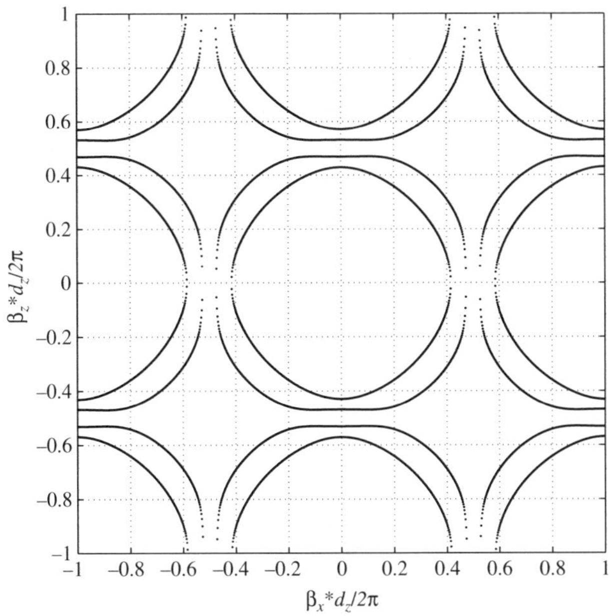

Figure 4.40 Phase diagram for the 2D periodic medium with the same structure parameters shown at the beginning of this section; $k_y = 8\pi/10d_z$ and $d_z/\lambda = 0.4$

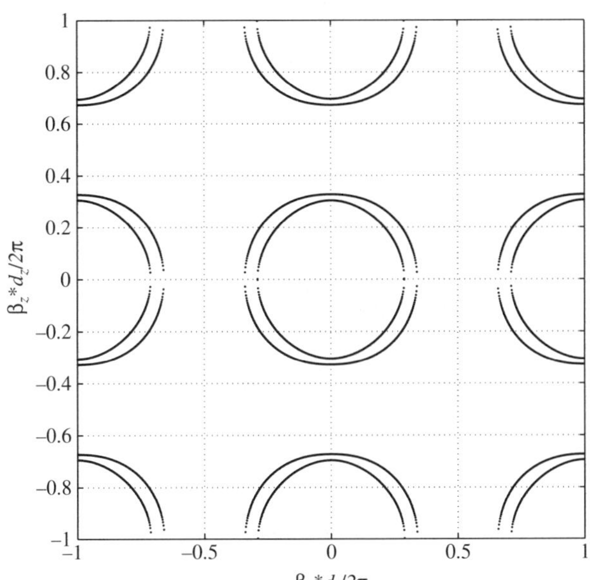

Figure 4.41 Phase diagram for the 2D periodic medium with the same structure parameters shown at the beginning of this section; $k_y = 10\pi/10d_z$ and $d_z/\lambda = 0.4$

4.3 Double Fourier-Modal Approach for a 2D Dielectric Periodic Structure: Out-of-Plane Propagation

In Section 4.2, the phase constant along the length direction of the dielectric column, the y-axis, is assumed to be known for solving the propagation constant along the periodicity directions, x- or z-axis. Therefore, the eigenwave is propagating on the x–z plane and is called the *in-plane* propagation. In this section we consider the medium has periodic variation on its permittivity and permeability along the transverse plane, x–y plane, as shown in Figure 4.42; the periods along the x- and y-axes are a and b, respectively, with the relative permittivity and permeability functions satisfying the equations given below:

$$\varepsilon(x + a, y + b) = \varepsilon(x, y) \tag{4.32}$$

$$\mu(x + a, y + b) = \mu(x, y) \tag{4.33}$$

The structure along the z-axis is uniform; however, the thickness is a finite number. Therefore, the propagation constant along the z-axis is not zero. For the plane-wave excitation problem, the phase constants k_x and k_y are given by the incident plane-wave, while k_z will be the unknown to be resolved. In this section we will carry out the mathematical procedures for formulating an eigenvalue equation with the eigenvalue k_z, for given k_x and k_y.

Let us begin with the following two Maxwell equations:

$$\nabla \times \underline{E} = -\mathrm{j}\omega\mu_0\mu(x, y)\underline{H} \tag{4.34}$$

$$\nabla \times \underline{H} = +\mathrm{j}\omega\varepsilon_0\varepsilon(x, y)\underline{E} \tag{4.35}$$

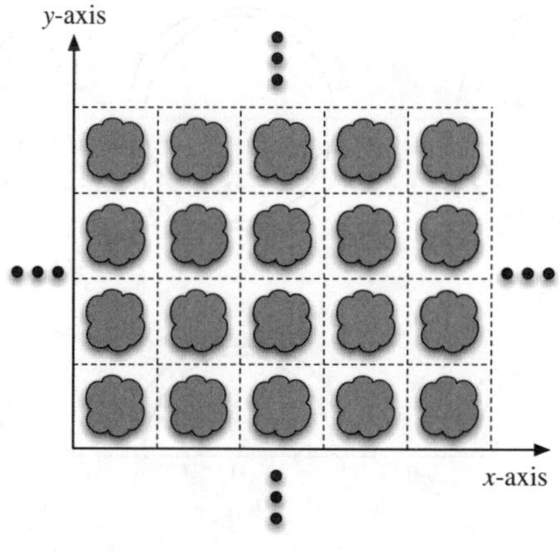

Figure 4.42 A 2D periodic medium with periodic variation on its permittivity and permeability along the x- and y-axes and homogeneous along the z-axis

Since the medium is nonhomogeneous along the transverse plane, a single polarization mode (E_z or H_z) cannot exist independently. The electric and magnetic fields must contain x, y, and z components simultaneously (Hwang and Peng 2000). By the Bloch theorem, the field component can be written as follows:

$$E_i(x, y, z) = \sum_m \sum_n E_{i,mn}(z) \exp(-\mathrm{j}\underline{k}_{\mathrm{t},mn} \cdot \underline{\rho}) \tag{4.36}$$

$$H_i(x, y, z) = \sum_m \sum_n H_{i,mn}(z) \exp(-\mathrm{j}\underline{k}_{\mathrm{t},mn} \cdot \underline{\rho}) \tag{4.37}$$

The function consisting of variables x and y shown in Equations (4.36) and (4.37) is the Floquet solution that satisfies the periodic boundary condition along the x and y-axes. However, the z-dependent function is the amplitude of the (m, n)th-order space harmonic to be resolved. The index i represents the x, y, or z component; vector $\underline{\rho} = x\underline{x}_{\mathrm{o}} + y\underline{y}_{\mathrm{o}}$ is the position of observation point; vector $\underline{k}_{\mathrm{t},mn} = k_{x,m}\underline{x}_{\mathrm{o}} + k_{y,n}\underline{y}_{\mathrm{o}}$, where $k_{x,m} = k_x + m2\pi/a$ and $k_{y,n} = k_y + n2\pi/b$.

Substituting Equations (4.36) and (4.37) into Equation (4.34), the three magnetic field components can be expressed in terms of the spatial derivative of the electric field components . The magnetic field component along the x-axis is written as

$$\frac{\partial E_z}{\partial y} - \frac{\partial E_y}{\partial z} = -\mathrm{j}\omega\mu_0\mu(x, y)H_x \tag{4.38}$$

Substituting E_z, E_y, and H_x in the form given in Equations (4.36) and (4.37), we obtain the following equation:

$$\begin{aligned} &\sum_m \sum_n \left[-\mathrm{j}k_{y,n}E_{z,mn}(z) - \frac{\mathrm{d}E_{y,mn}(z)}{\mathrm{d}z}\right] \exp(-\mathrm{j}\underline{k}_{\mathrm{t},mn} \cdot \underline{\rho}) \\ &\quad = \sum_m \sum_n -\mathrm{j}\omega\mu_0\mu(x, y)H_{x,mn}(z) \exp(-\mathrm{j}\underline{k}_{\mathrm{t},mn} \cdot \underline{\rho}) \end{aligned} \tag{4.39}$$

Since the corresponding Fourier amplitude of each harmonic (m, n) must be equal on both sides, we could multiply both sides by the complex conjugate of the Floquet solution with distinct space harmonic pair (p, q) and take the overlap integral over the unit cell, leading to

$$\begin{aligned} &\left\langle \exp(+\mathrm{j}\underline{k}_{\mathrm{t},pq} \cdot \underline{\rho}) \middle| \sum_m \sum_n \left[-\mathrm{j}k_{y,n}E_{z,mn}(z) - \frac{\mathrm{d}E_{y,mn}(z)}{\mathrm{d}z}\right] \exp(-\mathrm{j}\underline{k}_{\mathrm{t},mn} \cdot \underline{\rho}) \right\rangle \\ &= \left\langle \exp(+\mathrm{j}\underline{k}_{\mathrm{t},pq} \cdot \underline{\rho}) \middle| \sum_m \sum_n -\mathrm{j}\omega\mu_0\mu(x, y)H_{x,mn}(z) \exp(-\mathrm{j}\underline{k}_{\mathrm{t},mn} \cdot \underline{\rho}) \right\rangle \end{aligned} \tag{4.40}$$

In the above equation, the existence of the overlap integral allows us to interchange the operator of the double summation and overlap integral. The equation becomes

$$\begin{aligned}&\sum_m\sum_n\left[-\mathrm{j}k_{y,n}E_{z,mn}(z)-\frac{\mathrm{d}E_{y,mn}(z)}{\mathrm{d}z}\right]\left\langle\exp(+\mathrm{j}\underline{k}_{\mathrm{t},pq}\cdot\underline{\rho})\middle|\exp(-\mathrm{j}\underline{k}_{\mathrm{t},mn}\cdot\underline{\rho})\right\rangle\\&=\sum_m\sum_n-\mathrm{j}\omega\mu_0H_{x,mn}(z)\left\langle\exp(+\mathrm{j}\underline{k}_{\mathrm{t},pq}\cdot\underline{\rho})\middle|\mu(x,y)\exp(-\mathrm{j}\underline{k}_{t,mn}\cdot\underline{\rho})\right\rangle\end{aligned}\tag{4.41}$$

Owing to the orthogonality of the Floquet solution, the overlap integral on the left-hand side becomes

$$\iint \mathrm{e}^{+\mathrm{j}\underline{k}_{\mathrm{t},pq}\cdot\underline{\rho}}\,\mathrm{e}^{-\mathrm{j}\underline{k}_{\mathrm{t},mn}\cdot\underline{\rho}}\,\mathrm{d}x\,\mathrm{d}y=ab\delta_{mp}(x)\delta_{nq}(y)\tag{4.42}$$

The overlap integral on the right-hand side can be regarded the double Fourier series expansion of the periodic relative permeability function, given below:

$$\mu_{pq,mn}=\frac{1}{ab}\iint \mathrm{e}^{+\mathrm{j}\underline{k}_{\mathrm{t},pq}\cdot\underline{\rho}}\,\mathrm{e}^{-\mathrm{j}\underline{k}_{\mathrm{t},mn}\cdot\underline{\rho}}\mu(x,y)\,\mathrm{d}x\,\mathrm{d}y\tag{4.43}$$

In doing so, the (p, q)th-order space harmonic of the E_z and E_y components are related to every space harmonic of H_x, which is written as

$$\mathrm{j}k_{y,q}E_{z,pq}(z)+\frac{\mathrm{d}E_{y,pq}(z)}{\mathrm{d}z}=\sum_m\sum_n\mathrm{j}\omega\mu_0\mu_{pq,mn}H_{x,mn}(z)\tag{4.44}$$

Similarly, the magnetic field component along the y-axis can be written as

$$\frac{\partial E_x}{\partial z}-\frac{\partial E_z}{\partial x}=-\mathrm{j}\omega\mu_0\mu(x,y)H_y\tag{4.45}$$

By performing the same process of overlap integral mentioned previously, we can obtain

$$\frac{\mathrm{d}E_{x,pq}(z)}{\mathrm{d}z}+\mathrm{j}k_{x,p}E_{z,pq}(z)=\sum_m\sum_n-\mathrm{j}\omega\mu_0\mu_{pq,mn}H_{y,mn}(z)\tag{4.46}$$

Finally, the magnetic field component along the z-axis is expressed in terms of the derivative of E_y and Ex, given as

$$\frac{\partial E_y}{\partial x}-\frac{\partial E_x}{\partial y}=-\mathrm{j}\omega\mu_0\mu(x,y)H_z\tag{4.47}$$

Here, the detailed mathematical procedures are omitted (which is taken as homework for readers) and the relationship between the space harmonics of the electric- and magnetic-field components is written as

$$k_{x,p}E_{y,pq}(z)-k_{y,q}E_{x,pq}(z)=\sum_m\sum_n\omega\mu_0\mu_{pq,mn}H_{z,mn}(z)\tag{4.48}$$

For the shake of succinctness, we group the index pair (m, n) and (p, q) and denote them by new integers r and s, respectively; that is, $r = (p, q)$, and $s = (m, n)$. Equations(4.44), (4.46), and (4.48) can be rewritten in a vector-matrix form by collecting each of the space harmonic pairs and putting them into a vector or a matrix, which are written as

$$\mathrm{j}\mathbb{k}_y \mathbf{E}_z(z) + \frac{\mathrm{d}}{\mathrm{d}z}\mathbf{E}_y(z) = \mathrm{j}\omega\mu_0\overline{\overline{\mu}}\mathbf{H}_x(z) \tag{4.49}$$

$$\frac{\mathrm{d}\mathbf{E}_x(z)}{\mathrm{d}z} + \mathrm{j}\mathbb{k}_x \mathbf{E}_z(z) = -\mathrm{j}\omega\mu_0\overline{\overline{\mu}}\mathbf{H}_y(z) \tag{4.50}$$

$$\mathbb{k}_x \mathbf{E}_y(z) - \mathbb{k}_y \mathbf{E}_x(z) = \omega\mu_0\overline{\overline{\mu}}\mathbf{H}_z(z) \tag{4.51}$$

$$\mathbb{k}_x = \mathrm{diag}(k_{x,p}) \tag{4.52}$$

$$\mathbb{k}_y = \mathrm{diag}(k_{y,q}) \tag{4.53}$$

where “diag” means the diagonal matrix with each of the diagonal terms indicated within parentheses. Parameter $\overline{\overline{\mu}}$ is a full matrix with the (r, s)th entry given in Equation (4.43).

Similarly, substituting Equations (4.36) and (4.37) into Equation (4.35), the electric field components can be expressed in terms of the spatial derivative of magnetic fields. Following the same procedures described previously, we obtain the following three equations:

$$\mathrm{j}\mathbb{k}_y \mathbf{H}_z(z) + \frac{d}{dz}\mathbf{H}_y(z) = -\mathrm{j}\omega\varepsilon_0\overline{\overline{\varepsilon}}\mathbf{E}_x(z) \tag{4.54}$$

$$\frac{\mathrm{d}\mathbf{H}_x(z)}{\mathrm{d}z} + \mathrm{j}\mathbb{k}_x \mathbf{H}_z(z) = +\mathrm{j}\omega\varepsilon_0\overline{\overline{\varepsilon}}\mathbf{E}_y(z) \tag{4.55}$$

$$\mathbb{k}_x \mathbf{H}_y(z) - \mathbb{k}_y \mathbf{H}_x(z) = -\omega\varepsilon_0\overline{\overline{\varepsilon}}\mathbf{E}_z(z) \tag{4.56}$$

where $\overline{\overline{\varepsilon}}$ is a full matrix with its (r, s)th element given by

$$\varepsilon_{pq,mn} = \frac{1}{ab}\iint \mathrm{e}^{+\mathrm{j}\underline{k}_{t,pq}\cdot\underline{\rho}}\,\mathrm{e}^{-\mathrm{j}\underline{k}_{t,mn}\cdot\underline{\rho}}\varepsilon(x, y)\,\mathrm{d}x\,\mathrm{d}y \tag{4.57}$$

For ease of reference, we rearrange the six matrix equations obtained previously and list them as follows:

$$\mathrm{j}\mathbb{k}_y \mathbf{E}_z(z) + \frac{\mathrm{d}}{\mathrm{d}z}\mathbf{E}_y(z) = \mathrm{j}\omega\mu_0\overline{\overline{\mu}}\mathbf{H}_x(z) \tag{4.58}$$

$$\frac{\mathrm{d}\mathbf{E}_x(z)}{\mathrm{d}z} + \mathrm{j}\mathbb{k}_x \mathbf{E}_z(z) = -\mathrm{j}\omega\mu_0\overline{\overline{\mu}}\mathbf{H}_y(z) \tag{4.59}$$

$$\mathbb{k}_x \mathbf{E}_y(z) - \mathbb{k}_y \mathbf{E}_x(z) = -\omega\mu_0\overline{\overline{\mu}}\mathbf{H}_z(z) \tag{4.60}$$

$$\mathrm{j}\mathbb{k}_y \mathbf{H}_z(z) + \frac{\mathrm{d}}{\mathrm{d}z}\mathbf{H}_y(z) = -\mathrm{j}\omega\varepsilon_0\overline{\overline{\varepsilon}}\mathbf{E}_x(z) \tag{4.61}$$

$$\frac{\mathrm{d}\mathbf{H}_x(z)}{\mathrm{d}z}+\mathrm{j}\mathbb{k}_x\mathbf{H}_z(z)=+\mathrm{j}\omega\varepsilon_0\overline{\overline{\varepsilon}}\mathbf{E}_y(z) \tag{4.62}$$

$$\mathbb{k}_x\mathbf{H}_y(z)-\mathbb{k}_y\mathbf{H}_x(z)=-\omega\varepsilon_0\overline{\overline{\varepsilon}}\mathbf{E}_z(z) \tag{4.63}$$

Substituting Equation (4.63) into Equations (4.58) and (4.59) to eliminate the z-component electric field, we obtain the following two equations:

$$-\frac{\mathrm{d}\mathbf{E}_y(z)}{\mathrm{d}z}=-\mathrm{j}\omega\mu_0\left[\left(\overline{\overline{\mu}}-\frac{1}{k_0^2}\mathbb{k}_y\overline{\overline{\varepsilon}}^{-1}\mathbb{k}_y\right)\mathbf{H}_x(z)+\frac{1}{k_0^2}\mathbb{k}_y\overline{\overline{\varepsilon}}^{-1}\mathbb{k}_x\mathbf{H}_y(z)\right] \tag{4.64}$$

$$\frac{\mathrm{d}\mathbf{E}_x(z)}{\mathrm{d}z}=-\mathrm{j}\omega\mu_0\left[\left(\overline{\overline{\mu}}-\frac{1}{k_0^2}\mathbb{k}_x\overline{\overline{\varepsilon}}^{-1}\mathbb{k}_x\right)\mathbf{H}_y(z)+\frac{1}{k_0^2}\mathbb{k}_x\overline{\overline{\varepsilon}}^{-1}\mathbb{k}_y\mathbf{H}_x(z)\right] \tag{4.65}$$

The above two matrix equations can be integrated into a super-matrix equation, with the tangential electrical field components put into super-vector $\mathbf{E}_\mathrm{t}(z)=[-\mathbf{E}_y(z)\ \mathbf{E}_x(z)]^\mathrm{t}$ and the magnetic field components into $\mathbf{H}_\mathrm{t}(z)=[\mathbf{H}_x(z)\ \mathbf{H}_y(z)]^\mathrm{T}$, where the symbol "T" stands for the transpose of the vector. Consequently, the derivative of the tangential electric-field vector with respect to z is related to the tangential magnetic field vector by a super-matrix $\mathbb{Z}$, which is given as follows:

$$\frac{\mathrm{d}}{\mathrm{d}z}\begin{pmatrix}-\mathbf{E}_y(z)\\ \mathbf{E}_x(z)\end{pmatrix}=-\mathrm{j}\omega\mu_0\mathbb{Z}\begin{pmatrix}\mathbf{H}_x(z)\\ \mathbf{H}_y(z)\end{pmatrix} \tag{4.66}$$

$$\mathbb{Z}=\begin{bmatrix}\overline{\overline{\mu}}-\frac{1}{k_0^2}\mathbb{k}_y\overline{\overline{\varepsilon}}^{-1}\mathbb{k}_y & \frac{1}{k_0^2}\mathbb{k}_y\overline{\overline{\varepsilon}}^{-1}\mathbb{k}_x\\ \frac{1}{k_0^2}\mathbb{k}_x\overline{\overline{\varepsilon}}^{-1}\mathbb{k}_y & \overline{\overline{\mu}}-\frac{1}{k_0^2}\mathbb{k}_x\overline{\overline{\varepsilon}}^{-1}\mathbb{k}_x\end{bmatrix} \tag{4.67}$$

Similarly, by substituting Equation (4.60) into Equations (4.61) and (4.62) to eliminate $\mathbf{H}_z(z)$, the derivative of the tangential magnetic-field vectors with respect to z can be expressed in terms of the superposition of tangential electric field vectors as

$$\frac{\mathrm{d}\mathbf{H}_x(z)}{\mathrm{d}z}=\mathrm{j}\omega\varepsilon_0\left[\left(\overline{\overline{\varepsilon}}-\frac{1}{k_0^2}\mathbb{k}_x\overline{\overline{\mu}}^{-1}\mathbb{k}_x\right)\mathbf{E}_y(z)+\frac{1}{k_0^2}\mathbb{k}_x\overline{\overline{\mu}}^{-1}\mathbb{k}_y\mathbf{E}_x(z)\right] \tag{4.68}$$

$$\frac{\mathrm{d}\mathbf{H}_y(z)}{\mathrm{d}z}=-\mathrm{j}\omega\varepsilon_0\left[\left(\overline{\overline{\varepsilon}}-\frac{1}{k_0^2}\mathbb{k}_y\overline{\overline{\mu}}^{-1}\mathbb{k}_y\right)\mathbf{E}_x(z)+\frac{1}{k_0^2}\mathbb{k}_y\overline{\overline{\mu}}^{-1}\mathbb{k}_x\mathbf{E}_y(z)\right] \tag{4.69}$$

The derivative of the tangential magnetic-field super-vector is related to the tangential electric-field super-vector by super-matrix $\mathbb{Y}$ as follows:

$$\frac{\mathrm{d}}{\mathrm{d}z}\begin{pmatrix}\mathbf{H}_x(z)\\ \mathbf{H}_y(z)\end{pmatrix}=-\mathrm{j}\omega\varepsilon_0\mathbb{Y}\begin{pmatrix}-\mathbf{E}_y(z)\\ \mathbf{E}_x(z)\end{pmatrix} \tag{4.70}$$

$$\mathbb{Y}=\begin{bmatrix}\overline{\overline{\varepsilon}}-\frac{1}{k_0^2}\mathbb{k}_x\overline{\overline{\mu}}^{-1}\mathbb{k}_x & -\frac{1}{k_0^2}\mathbb{k}_x\overline{\overline{\mu}}^{-1}\mathbb{k}_y\\ -\frac{1}{k_0^2}\mathbb{k}_y\overline{\overline{\mu}}^{-1}\mathbb{k}_x & \overline{\overline{\varepsilon}}-\frac{1}{k_0^2}\mathbb{k}_y\overline{\overline{\mu}}^{-1}\mathbb{k}_y\end{bmatrix} \tag{4.71}$$

By differentiating Equation (4.66) with respect to z on both sides and substituting Equation (4.70) into it, we obtain the second-order matrix differential equation

$$\frac{\mathrm{d}^2}{\mathrm{d}z^2}\mathbf{E}_{\mathrm{t}}(z) = -\mathbb{A}\mathbf{E}_{\mathrm{t}}(z) \tag{4.72}$$

where $\mathbb{A} = k_{\mathrm{o}}^2\mathbb{Z}\mathbb{Y}$.

Matrix $\mathbb{A}$ can be factorized as $\mathbb{Q}\mathbb{D}\mathbb{Q}^{-1}$, where $\mathbb{Q}$ is the square matrix whose ith column is the eigenvector $\mathbf{q}_i$ of matrix $\mathbb{A}$ and where $\mathbb{D}$ is the diagonal matrix whose diagonal elements are the corresponding eigenvalues; that is, $D_{ii} = \lambda_{ii}$. The above equation can be rewritten as

$$\frac{\mathrm{d}^2}{\mathrm{d}z^2}\mathbf{E}_{\mathrm{t}}(z) = -\mathbb{Q}\mathbb{D}\mathbb{Q}^{-1}\mathbf{E}_{\mathrm{t}}(z) \tag{4.73}$$

Therefore, the general solution of the tangential electric-field vector $\mathbf{E}_{\mathrm{t}}(z)$ can be written as

$$\mathbf{E}_{\mathrm{t}}(z) = \mathbb{Q}[\exp(-\mathrm{j}\mathbf{K}_z z)\underline{v}_+ + \exp(+\mathrm{j}\mathbf{K}_z z)\underline{v}_-] \tag{4.74}$$

Vectors $\underline{v}_+$ and $\underline{v}_-$ are the forward- and backward-propagating vectors to be determined by the electromagnetic boundary condition at the interface of the periodic medium. Each element in vector $\underline{v}$ stands for the voltage amplitude of the corresponding space harmonic. Parameter $\mathbf{K}_z$ is a diagonal matrix with the element representing $\sqrt{\lambda_{ii}}$. Notably, the matrix $\exp(\pm\mathrm{j}\mathbf{K}_z z)$ is a diagonal matrix with its diagonal element representing $\exp(\pm\mathrm{j}K_{z,ii}z)$, where ii is the index of the (i, i)th entry.

By differentiating the tangential electric-field in Equation (4.74) with respect to z and substituting it into Equation (4.66) we obtain the tangential magnetic-field vector given below:

$$\mathbf{H}_{\mathrm{t}}(z) = \mathbb{P}[\exp(-\mathrm{j}\mathbf{K}_z z)\underline{v}_+ - \exp(+\mathrm{j}\mathbf{K}_z z)\underline{v}_-] \tag{4.75}$$

where matrix $\mathbb{P}$ can be determined by the following equation:

$$\mathbb{P} = \frac{\mathbb{Z}^{-1}\mathbb{Q}\mathbf{K}_z}{\omega\mu_{\mathrm{o}}} \tag{4.76}$$

4.3.1 Scattering Analysis of a 2D Grating: Out-of-Plane Propagation

With the tangential electric and magnetic fields given in Equations (4.74) and (4.75), we can apply the mathematical procedures described in Section 3.7 for formulating the reflectance and transmittance of a structure composed of multiple 2D periodic layers (or gratings). The numerical results reported by Peng and Morris (1996) have been taken as an example for verifying my computer code. A 2D dielectric grating consisting of square dielectric columns immersed in a dielectric medium is shown in

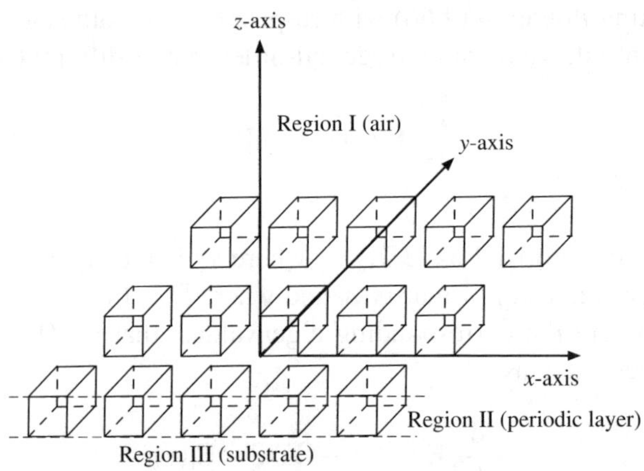

Figure 4.43 A 2D dielectric grating: regions I, II, and III are respectively air, grating, and substrate

Figure 4.43. A plane wave with electric field along the y-axis is normally incident on the structure. The detailed structure parameters and relative dielectric constants of the media employed are in the caption to Figure 4.44. Obviously, the behavior of frequency-selective reflection can be observed in this figure. Although not shown here, our numerical results exhibit a perfect agreement with those shown in Peng and Morris (1996), which can validate the correctness of my code.

4.4 Three-Dimensional Periodic Structures

In addition to the scattering analysis for a single 2D grating, I will demonstrate the numerical example concerning a 3D periodic structure shown in Figure 4.45, which can be regarded as a finite stack of 2D periodic layers. Furthermore, the eigenwave solutions in a 3D periodic medium, which is periodic along x-, y-, and z-axes and is infinite in extent along the three axes, will be determined by solving the generalized eigenvalue equation. Fortunately, the methodology for building up the generalized eigenvalue equation has been introduced in detail previously. For the sake of succinctness, we are going to review the mathematical procedures instead of repeating the mathematical derivation.

First of all, we consider an interface between the 2D semi-infinite periodic medium with periods along the x- and y-axes. Using the tangential electric and magnetic field solutions in Equations (4.74) and (4.75) and taking into account the electromagnetic boundary conditions (continuous tangential components of electric and magnetic fields across the interface), the input–output relation expressed in terms of the generalized scattering matrix can be obtained with the same form shown in Equation (4.27). Moreover, by invoking the periodic boundary condition, the generalized eigenvalue equation given in Equation (4.31) can be obtained. Notably, all the mathematical formulations in Section 4.2 can be directly applied here by simply replacing the $\mathbb{Q}$

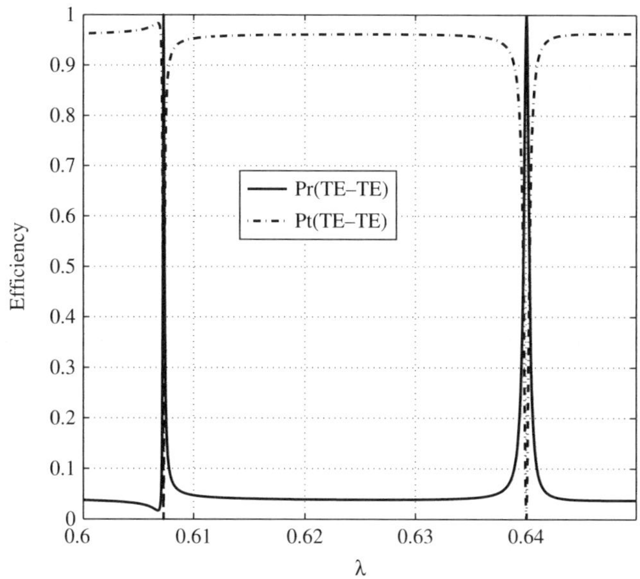

Figure 4.44 Reflectance and transmittance of a plane wave normally incident on the structure shown in Figure 4.43. The 2D grating has the same period (0.4) along the x- and y-axes; the dielectric blocks ($\varepsilon_r = 4.0$) arranged in a square lattice are immersed in the dielectric medium with $\varepsilon_r = 3.42$. The duty cycle of the dielectric block is $f_x = f_y = 0.5$ along the x- and y-axes. The refractive index in the substrate region is $n = 1.46$. The first nine space-harmonic pairs (m, n) are employed for the numerical simulation

and $\mathbb{P}$ given in Equations (4.74) and (4.75). Regarding the coordinate transfer matrix $\mathbf{A}$, the diagonal elements of the sub-matrices $\mathbf{A}_x$ and $\mathbf{A}_y$ are replaced by $k_{x,m}/k_{t,mn}$ and $k_{y,n}/k_{t,mn}$. The eigenvalue λ is equal to $\exp(\pm jk_z z)$; thus, for given structure parameters, wavelength, and phase constants along the x- and y-axes, k_x and k_y, the

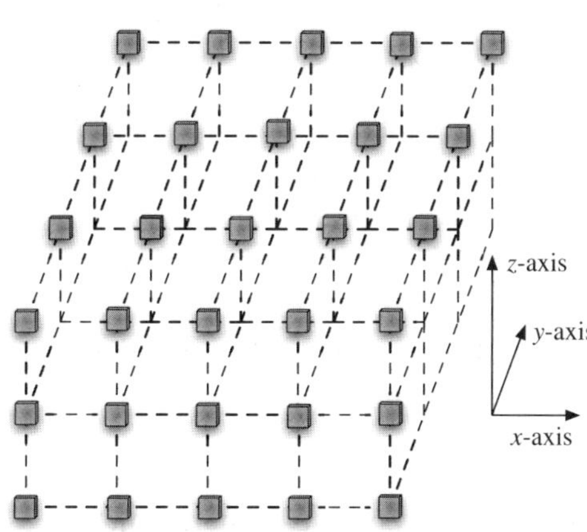

Figure 4.45 A 3D periodic structure consisting of three periods, along the x-, y-, and z-axes

propagation constant along the z-axis, k_z, can be readily determined by the MATLAB function eig(A,B). Generally, k_z is a complex number with its real and imaginary parts representing the phase and attenuation constants, respectively.

4.4.1 Scattering Analysis of a 3D Periodic Structure

We consider an infinite 3D periodic medium with the cubic lattice shown in Figure 4.45. The dimension of the dielectric cube is 0.5 with the relative dielectric constant 4.0; the 3D array is immersed in air. The periods along the x-, y-, and z-axes are all 1.0. Notably, all the dimension are in arbitrary units. For the scattering analysis, the 3D periodic structure consists of a finite stack of 2D periodic layers, each of which is infinite in extent along the x- and y-axes, but with finite thickness along the z-axis.

Notably, in the presence of a 2D periodic nature along the x- and y-axes, the tangential components of the electric and magnetic fields of the (m, n)th space harmonic in free space can be written as

$$\begin{pmatrix} -E_{y,mn}(x, y, z) \\ E_{x,mn}(x, y, z) \end{pmatrix} = \begin{pmatrix} \frac{k_{x,m}}{k_{t,mn}} & -\frac{k_{y,n}}{k_{t,mn}} \\ \frac{k_{y,n}}{k_{t,mn}} & \frac{k_{x,m}}{k_{t,mn}} \end{pmatrix} \begin{pmatrix} V'_{mn}(z) \\ V''_{mn}(z) \end{pmatrix} \mathrm{e}^{-\mathrm{j}\underline{k}_{t,mn}\cdot\underline{\rho}} \tag{4.77}$$

The above equation means that a coordinate transformation matrix exists between the spatial- and eigen-coordinate systems of the transverse electromagnetic-field vector for each space harmonic; m and n are the space harmonic indices along the x- and y-axes, respectively.

Consider a plane wave having incident angles θ and ϕ; such a uniform plane wave can be viewed as the fundamental space harmonic ($m = 0, n = 0$). From Equation (4.77), the uniform plane wave is able to be expressed in terms of the superposition of two orthogonal linearly polarized waves (TE^z and TM^z); the electric- and magnetic-field components over the transverse (x–y) plane can be written as

$$E_x = (\sin\phi V' + \cos\phi V'')\exp(-\mathrm{j}\underline{k}_t \cdot \underline{\rho}) \tag{4.78}$$
$$-E_y = (\cos\phi V' - \sin\phi V'')\exp(-\mathrm{j}\underline{k}_t \cdot \underline{\rho}) \tag{4.79}$$

where V' and V'' are the voltage amplitudes of the TE and TM waves. The propagation vector along the transverse plane $\underline{k}_t$ and the position vector $\underline{\rho}$ are equal to $k_x\underline{x}_o + k_y\underline{y}_o$ and $x\underline{x}_o + y\underline{y}_o$. In the following numerical calculation, we assume that the incident plane wave is a TE- or TM-polarized wave with the voltage amplitudes respectively being ($V' = 1$, $V'' = 0$) and ($V' = 0$, $V'' = 1$).

Contrarily, for the condition of principal-plane incidence, $\phi = 0$, the electric-field component is in the x-axis for a TM-polarized wave and in the y-axis for a TE-polarized wave.

Figure 4.46 shows the transmittance and reflectance (normalized to the incident power) of the TE-polarized fundamental space harmonic ($m = 0, n = 0$) against $1/\lambda$. The 3D periodic structure under consideration is shown in Figure 4.45. The dielectric

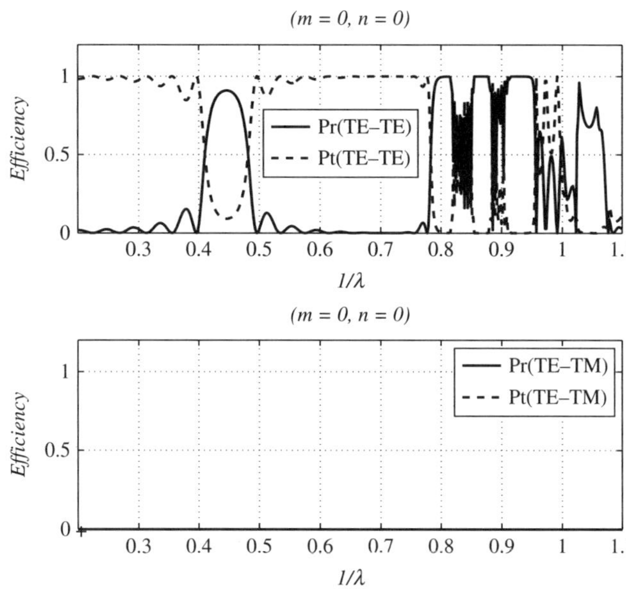

Figure 4.46 Scattering characteristics of the space harmonic ($m = 0, n = 0$) by the 3D periodic structure consisting of 10 2D periodic layers, each of which consists of a 2D square lattice with lattice constant 1.0 along the x- and y-axes. The dielectric cubes with dimension 0.5 and relative dielectric constant 4.0 are immersed in air. The separation distance between two 2D periodic layer is 0.5. The incident angles of the TE-polarized plane wave are $\theta = 0^\circ$ and $\phi = 0^\circ$. The upper and lower figures are the co- and cross-polarization scattering efficiencies including reflectance and transmittance, respectively. The legend TE–TM represents the reflected or transmitted TM wave under TE-polarized plane-wave incidence

cubes with dimension $w = 0.5$ and relative dielectric constant $\varepsilon_s = 4.0$ are immersed in air. Here, the 3D periodic structure can be regarded as a stack of 10 2D periodic layers with layer thickness 0.5. The separation distance between two successive layers also is 0.5. The lattice constants (or period) along the three axes are 1.0. The TE-polarized plane wave is normally incident on the 3D periodic structure. Notably, the dimensions of the structure are in the arbitrary units. As shown in this figure, the reflection is significant in the vicinity of $1/\lambda = 0.45$. This stop-band, which is commonly seen in 1D and 2D periodic structures, as discussed in the previous chapters, is attributed to the periodicity along the z-axis. Furthermore, such a stop-band is similar to that in a 1D periodic structure by considering the 2D periodic layer as an effective medium. Different from those depicted in Figure 2.13, the higher order stop-bands squeeze together. The physical consequence will become clear later on.

The lower graph in Figure 4.46 is the cross-polarization response; that is, the scattering space harmonic with electric field along the x-axis while the incident one is in

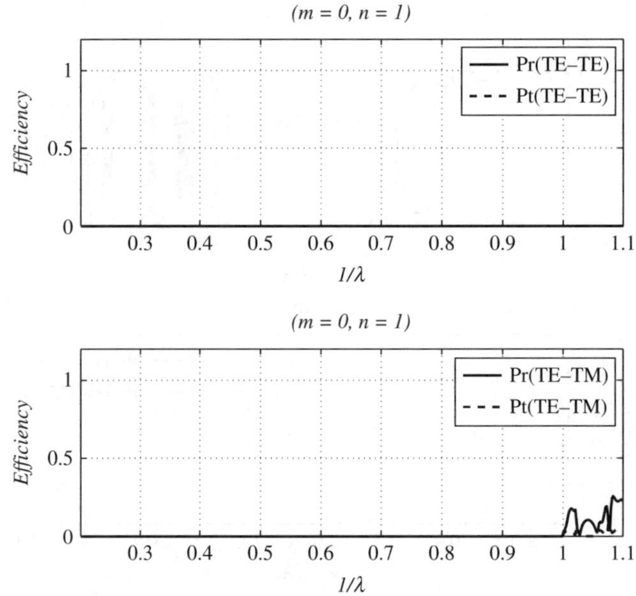

Figure 4.47 Transmittance and reflectance of the space harmonic ($m = 0, n = 1$) by the same 3D periodic structure indicated in Figure 4.46

the y-axis. It is apparent that cross-polarization scattering does not occur, since the y-direction momentum of the incident photon cannot induce the x-direction component for the square cube under the normally incident condition.

Figure 4.47 shows the scattering characteristics of the space harmonic ($m = 0, n = 1$). At first glance, it is strange that cross-polarization takes place and co-polarization disappears. However, it is noted that each space harmonic has its own eigen-coordinate system that is not coincident with the spatial coordinate system. For the space harmonic ($m = 0, n \neq 0$), we have $k_{x,m}/k_{t,mn} = 0$ and $k_{y,n}/k_{t,mn} = 1$; the x-axis electric field is contributed by the TE-polarized voltage wave while the y-axis component is contributed by the TM-polarized voltage wave; namely, $E_{y,0n} = V''_{0n}$ and $E_{x,0n} = -V'_{0n}$. Consequently, the TE space harmonic ($m = 0, n = 1$) has the electric field along the x-axis, which is perpendicular to the direction of the incident electric field. The top graph in Figure 4.47 is, in fact, the cross-polarization; however, the lower one is co-polarization scattering space harmonic with the same direction of incident electric field. Although not shown here, owing to the normal incidence, the space harmonic ($m = 0, n = -1$) shares the same response with Figure 4.47. The onset wavelength $\lambda = 1.0$ corresponds to the cutoff wavelength of the space harmonic ($m = 0, n = 1$).

In addition, the scattering properties of the space harmonic ($m = 1, n = 0$) including the reflectance and transmittance are shown in Figure 4.48. For normal incidence,

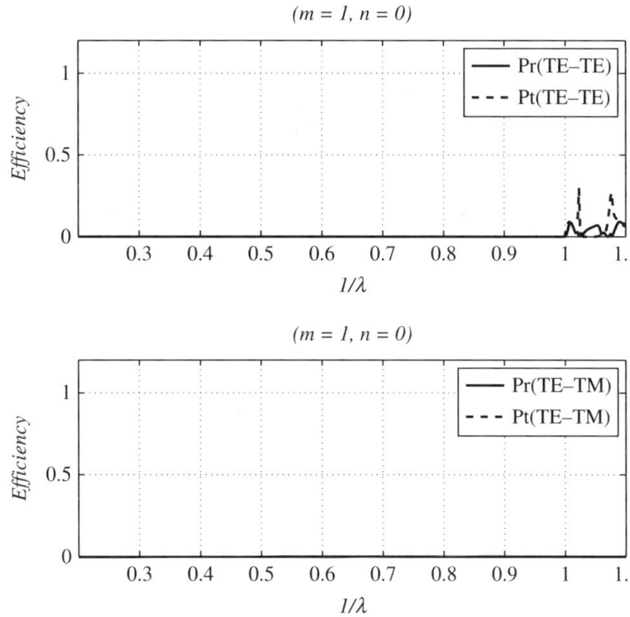

Figure 4.48 Transmittance and reflectance of space harmonic ($m = 1, n = 0$) by the same 3D periodic structure indicated in Figure 4.46

from Equation (4.77) we have $E_{y,10}(z) = -V'_{1,0}(z)$ and $E_{x,10}(z) = V''_{1,0}(z)$. Thus, the lower figure reveals that the cross-polarization of the space harmonic ($m = 1, n = 0$) does not occur. The onset wavelength of the reflection and transmission for that space harmonic is at its cutoff wavelength. Additionally, the scattering characteristics of the space harmonic ($m = -1, n = 0$) remain the same as shown in this figure.

In addition to the normal incidence case, in the following numerical experiment I will demonstrate the scattering characteristics of the same 3D periodic structure illuminated by a obliquely incident plane wave with nonzero elevation and azimuth angles. The occurrence of the cross-polarization space harmonics will be addressed.

Here, I assume that the incident angles of the TM-polarized plane wave are $\theta = 10^\circ$ and $\phi = 60^\circ$; the structure parameters remain the same as depicted in the last example. Notably, for the oblique incidence the propagation constants along the x- and y-axes are present simultaneously. It may be conjectured that, in general, each space harmonic has both the x- and y-components of the vector electric field.

We demonstrate the reflectance and transmittance of the space harmonics containing the fundamental ($m = 0, n = 0$) and the four higher order ones, ($m = 1, n = 0$), ($m = -1, n = 0$), ($m = 0, n = 1$), and ($m = 0, n = -1$), from Figures 4.49–4.53, respectively.

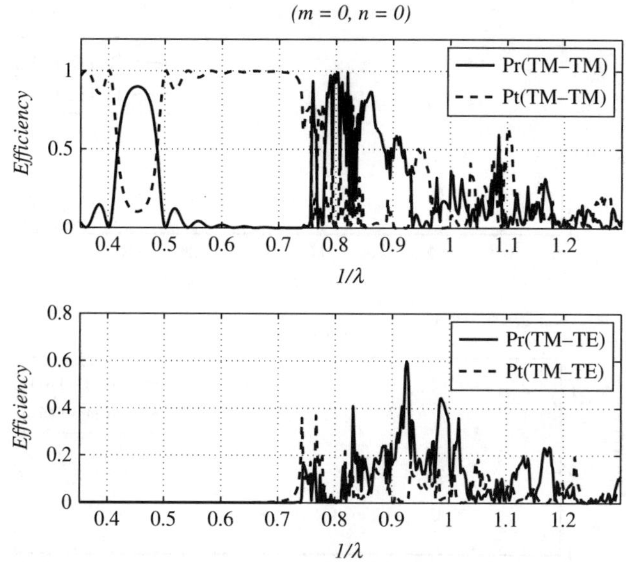

Figure 4.49 Transmittance and reflectance of space harmonic ($m = 0, n = 0$) by the same 3D periodic structure indicated in Figure 4.46; the TM-polarized plane wave incidence with $\theta = 10^\circ$ and $\phi = 60^\circ$

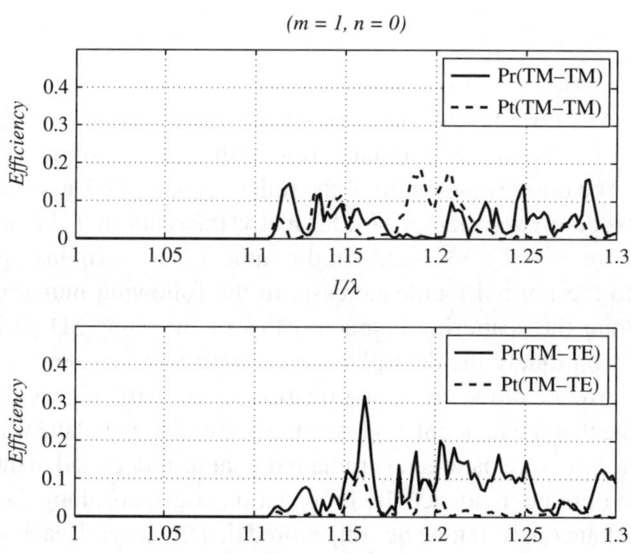

Figure 4.50 Transmittance and reflectance of space harmonic ($m = 1, n = 0$) by the same 3D periodic structure indicated in Figure 4.46; the TM-polarized plane wave incidence with $\theta = 10^\circ$ and $\phi = 60^\circ$

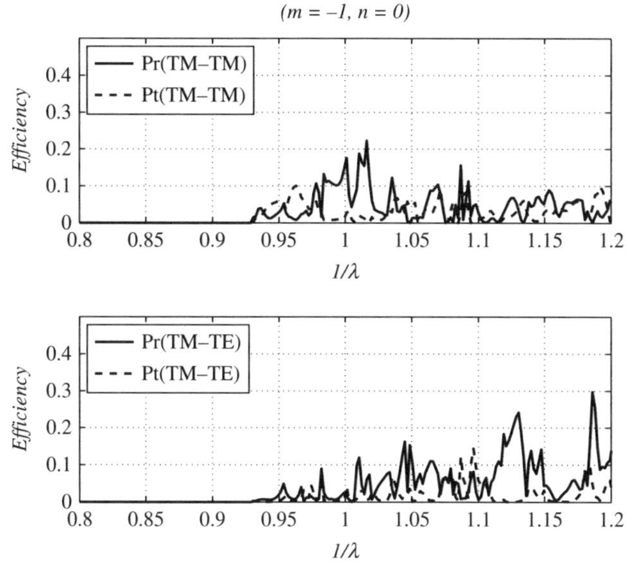

Figure 4.51 Transmittance and reflectance of space harmonic ($m = -1, n = 0$) by the same 3D periodic structure indicated in Figure 4.46; the TM-polarized plane wave incidence with $\theta = 10^\circ$ and $\phi = 60^\circ$

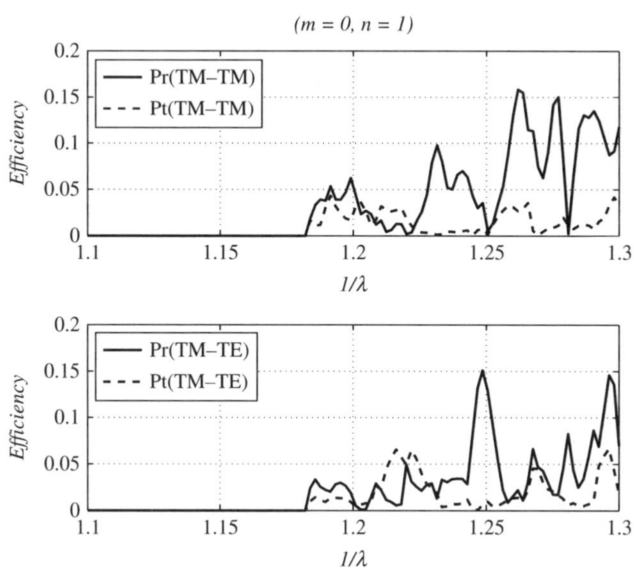

Figure 4.52 Transmittance and reflectance of space harmonic ($m = 0, n = 1$) by the same 3D periodic structure indicated in Figure 4.46; the TM-polarized plane wave incidence with $\theta = 10^\circ$ and $\phi = 60^\circ$

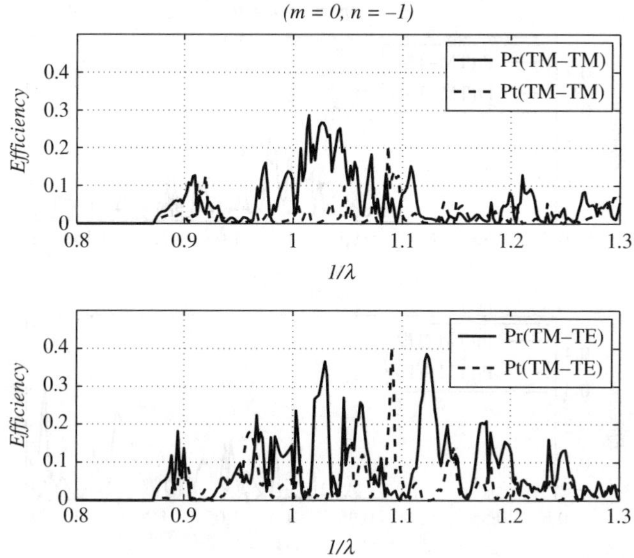

Figure 4.53 Transmittance and reflectance of space harmonic ($m = 0, n = -1$) by the same 3D periodic structure indicated in Figure 4.46; the TM-polarized plane wave incidence with $\theta = 10^\circ$ and $\phi = 60^\circ$

Equation (4.80) is the formula for calculating the cutoff wavelength of the space harmonic (m, n). The periods along the x- and y-axes are denoted a and b; the elevation and azimuth angles of the incident plane wave are θ and ϕ.

$$\frac{1}{\lambda_{c,mn}} = \frac{1}{\cos^2\theta}\left\{\left(\frac{m}{a}f + \frac{n}{b}g\right) + \sqrt{\left(\frac{m}{a}f + \frac{n}{b}g\right)^2 + \cos^2\theta\left[\left(\frac{m}{a}\right)^2 + \left(\frac{n}{b}\right)^2\right]}\right\}$$
$$f = \sin\theta\cos\varphi$$
$$g = \sin\theta\sin\varphi \tag{4.80}$$

From the above equation, the cutoff wavelengths of the four higher order space harmonics are respectively $\lambda_{1,0} = 1/1.1089$, $\lambda_{-1,0} = 1/0.9298$, $\lambda_{0,1} = 1/1.1823$, and $\lambda_{0,-1} = 1/0.8721$, which can be seen Figures 4.50–4.53.

4.4.2 *Eigenwave Analysis of a 3D Periodic Medium*

The scattering characteristics of a 3D periodic structure consisting of finite periods along the energy propagation direction, the z-axis, have been calculated and demonstrated in Section 4.4.1. As far as a periodic structure is concerned, the significant reflection in the stop-band region shall be the research highlight. In the last examples, extra stop-bands additional to those contributed by the periodicity along the z-axis were found that are not present in the 1D periodic structure. In fact, in Section 4.2.2 regarding the propagation characteristics of eigenwaves in a 2D periodic medium we demonstrated such a class of stop-band, called a slanted stop-band, contributed by the contraflow coupling between the fundamental space harmonic along the longitudinal direction and the higher order space harmonic along the other periodicity in the transverse plane. For the 3D periodic structure of interest in this subsection, owing to the two periodicities on the transverse plane and one in the longitudinal direction, the space harmonics along the three periods must be taken into account and the coupling mechanism among the space harmonics becomes much more complex. In the following we will first introduce the phase relation of a 3D periodic medium for realizing the interaction between space harmonics. Furthermore, the dispersion relation in the format of a Brillouin diagram will be presented to illustrate the stop-bands, including the vertical one commonly seen in the 1D periodic structure and the slanted ones caused by the periodicity along the transverse plane.

4.4.2.1 Phase Relation

As usual, before attacking the complex problem of solving the phase relation of the eigenwave in a 3D periodic medium, we begin by investigating the unperturbed phase relation of a plane wave solution in a uniform medium by considering the 3D periodic nature. If the periods along the x-, y-, and z-axes are denoted a, b, and c, respectively, then the relationship among k_x, k_y, and k_z can be simply written as

$$(k_x + m2\pi/a)^2 + (k_y + n2\pi/b)^2 + (k_z + p2\pi/c)^2 \approx k_o^2\varepsilon_{\text{eff}} \tag{4.81}$$

where m, n, and p are the indices of the space harmonic along the x-, y-, and z-axes respectively. Parameter ε_{eff} is the effective dielectric constant of the medium. Specifically, the above equation forms a sphere in Cartesian coordinate system with radius $k_o\sqrt{\epsilon_{eff}}$. If we consider a cubic lattice with period a, the above equation can be rewritten as

$$(k_xa/2\pi + m)^2 + (k_ya/2\pi + n)^2 + (k_za/2\pi + p)^2 \approx (a/\lambda)^2\varepsilon_{\text{eff}} \tag{4.82}$$

Therefore, the origin of each sphere locates at (m, n, p) with radius $\sqrt{\epsilon_{eff}}a/\lambda$ in the new rectangular coordinate system with three axes, $k_xa/2\pi$, $k_ya/2\pi$, and $k_za/2\pi$; that is, they locate at the integer grid of the new rectangular coordinate system. With the spheres in mind, this will facilitate us figuring out the electromagnetic coupling among the space harmonics that may occur in the vicinity of the intersections between spheres.

We make use of the same 3D cubic lattice with the structure parameters given in Figure 4.46 as an example to calculate the *exact* phase relation by solving the generalized eigenvalue problem. Since we know that the radius of each sphere is proportional to a/λ (a is the period), we may progressively alter the operation wavelength for observing the change in the 3D phase relation.

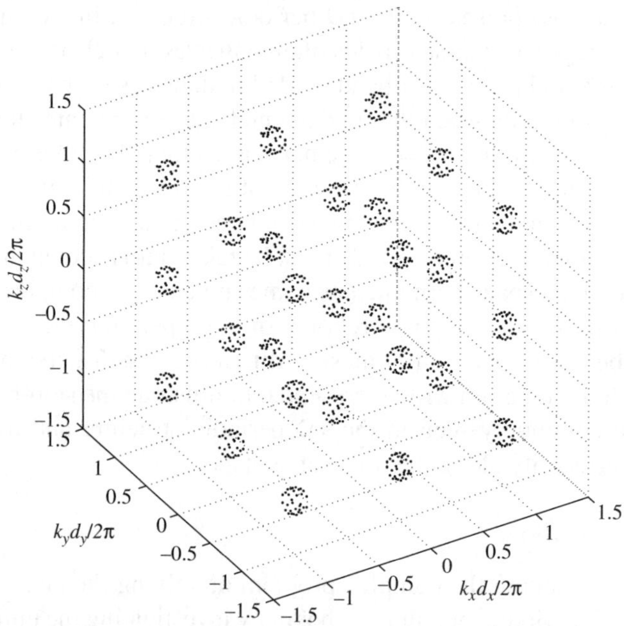

Figure 4.54 Phase relation of eigenwave in the same 3D periodic structure indicated in Figure 4.46: wavelength $\lambda = 10$; $d_x = d_y = d_z$

In Figure 4.54, the operational wavelength is 10 (in arbitrary units); the radius of each sphere is small enough that no interaction occurs between two adjacent two spheres. When the operational wavelength is decreased to 3.0, as shown in Figure 4.55, the radius of each sphere increases, but also without interaction. At wavelength $\lambda = 2.5$, the radius of each sphere is around 0.5, which is the onset wavelength that the interaction starts to happen, as shown in Figure 4.56. We can imagine that coupling will take place as the wavelength is smaller than 2.5. Figure 4.57 shows the phase relation for $\lambda = 2.0$. Obviously, the interaction between spheres occurs along the x-, y-, and z-axes; the overlap between two adjacent spheres enables the occurrence of the deformation shown in this figure. Moreover, because the radius of each sphere is greater than 0.5 and smaller than $\sqrt{3}/2$, interaction between the spheres along the nearest diagonal direction (inside the *irreducible Brillouin zone*) does not take place. I have also carried out the calculation for the case of $\lambda \ll 0.2$; however, owing to the complex interaction between spheres, it is hard to visualize clearly and the phase relation is not demonstrated here.

4.4.2.2 Dispersion Relation

The dispersion relation drawn in the format of a Brillouin diagram is carried out for given $k_x = k_o \sin\theta\cos\phi$ and $k_y = k_o \sin\theta\sin\phi$; here, we assume that the

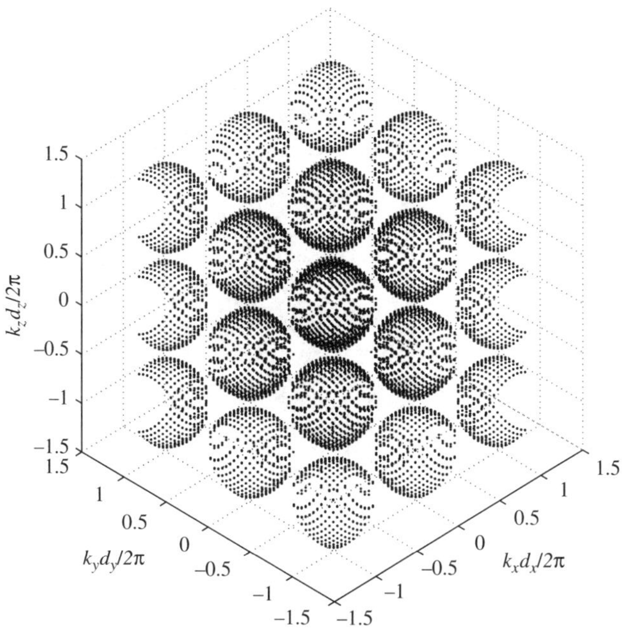

Figure 4.55 Phase relation of eigenwave in the same 3D periodic structure indicated in Figure 4.46: wavelength $\lambda = 3$; $d_x = d_y = d_z$

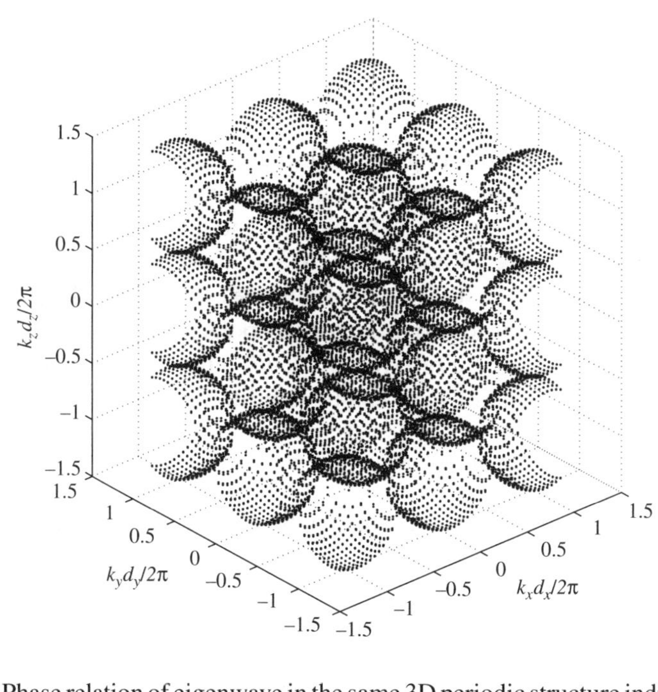

Figure 4.56 Phase relation of eigenwave in the same 3D periodic structure indicated in Figure 4.46: wavelength $\lambda = 2.5$; $d_x = d_y = d_z$

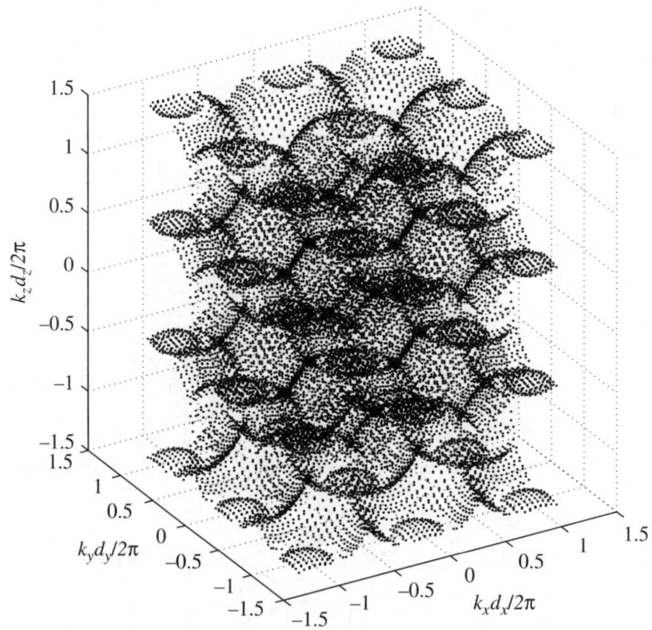

Figure 4.57 Phase relation of eigenwave in the same 3D periodic structure indicated in Figure 4.46: wavelength $\lambda = 2.0$; $d_x = d_y = d_z$

prescribed phase constants are assigned by the plane-wave excitation with incident angles θ and ϕ.

Figure 4.58 is the dispersion relation with the horizontal and vertical axes representing the normalized phase constant along the z-axis and the normalized frequency $\beta_z d_z/2\pi$ and d_z/λ, respectively. Since the electric and magnetic fields in a 3D periodic medium generally contain x-, y-, and z-components, these cannot be separated into two independent groups of TE- or TM- polarized modes. The general field solution consists of a superposition of these two orthogonally polarized modes. Consequently, the two sets of dispersion curves are referred as *TE-like* and *TM-like* modes, respectively.

For given k_x and k_y, Equation 4.82 can be rewritten as

$$(a/\lambda)^2\varepsilon_{\text{eff}} - (k_z a/2\pi + p)^2 \approx (k_x a/2\pi + m)^2 + (k_y a/2\pi + n)^2 \tag{4.83}$$

Equation (4.83), in general, exhibits a hyperbolic curve for each space harmonic (m, n, p), except for the case $k_x = k_y = 0$ and $m = n = 0$, representing a straight line.

The dispersion diagram shown in Figure 4.58 contains two types of stop-bands; namely, the vertical and slanted ones mentioned previously. Recalling the unperturbed dispersion diagram shown in Section 4.1.4, we know that the vertical stop-band is due

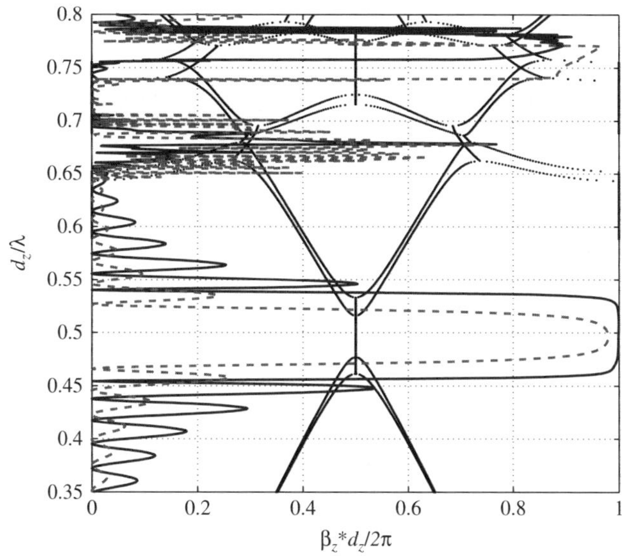

Figure 4.58 Dispersion relation shown in the format of a Brillouin diagram; this is the same 3D periodic structure profile (but with 20 2D periodic layers) indicated in Figure 4.46 with the plane wave having incident angles $\theta = 30^\circ$ and $\phi = 60^\circ$; the phase constant is drawn as a dotted line; the solid gray line is for the reflectance of the TE-polarized fundamental space harmonic, ($m = 0, n = 0$), with the TE-polarized plane wave incidence, while the dashed line is for the reflectance of the TM-polarized space harmonic with the TM-polarized plane wave incidence

to the coupling between the space harmonics along the $k_z d_z/2\pi$-axis (in the 3D phase diagram), while the slanted stop-band is contributed by the space harmonics along the diagonal direction. In the vicinity of $d_z/\lambda = 0.5$, as shown in Figure 4.58, the two stop-bands (which are due to *TE-like* and *TM-like* modes, respectively) overlap each other, each of which is caused by the contraflow coupling between the fundamental and -1th-order space harmonic along the z-axis. In addition, the reflectance of the fundamental space harmonic ($m = 0, n = 0$) is plotted together with the Brillouin diagram to facilitate the understanding of the band-structure properties. In this figure, the dotted line represents the dispersion relation, the solid line stands for the reflectance of TE-polarized fundamental space harmonic under the oblique incidence of the TE-polarized plane wave, and the dashed line is the TM-polarized space harmonic under the illumination of the TM-polarized plane wave.

Because the eigenvalue approach can directly solve the complex propagation constant, the attenuation constant in addition to the phase constant can be determined simultaneously. Figure 4.59 demonstrates the normalized attenuation constant against the normalized frequency. As is well known, the attenuation constant distribution

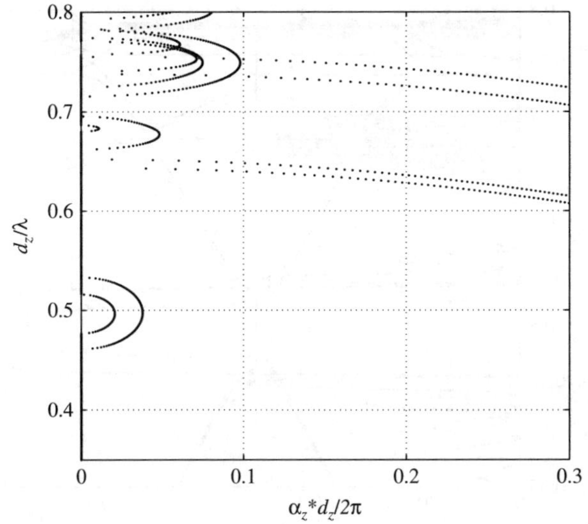

Figure 4.59 Dispersion relation shown in the format of a Brillouin diagram; this is the same 3D periodic structure indicated in Figure 4.46 with the plane wave having incident angles $\theta = 30°$ and $\phi = 60°$; the distribution of the normalized attenuation constant (horizontal axis) against the normalized frequency (vertical axis)

exhibits a parabolic shape in the stop-band region; the vertical stop-bands have a symmetric pattern while the slanted ones have an asymmetric pattern, which are shown in this figure. Moreover, the significant attenuation constant distribution of the four curves represents the below-cutoff of the higher order space harmonics.

In order to clearly observe the band structure in Figure 4.58, we partially enlarge the second and third stop-bands and redraw them in Figures 4.60 and 4.61, respectively. In the first slanted stop-band region shown in Figure 4.60, the reflectance of TE- (solid line) and TM-polarized waves (dashed line) are significant compared with those in the pass-band region. The case of TM-polarized has a wider stop-band than that of the TE-polarized one shown in this figure, so that the wide slanted stop-band may be designated as a *TM-like* stop-band while the narrow one is characterized as a *TE-like* stop-band.

In Figure 4.61, in comparison with those in the last figure, the two slanted stop-bands are able to be observed clearly. Since 20 2D periodic layers are employed in the scattering analysis, the band edge can be identified clearly. The wide stop-band is caused by the *TM-like* mode, while the narrow one is by the *TE-like* mode, which can be conjectured from the reflection property.

From the attenuation constant distribution shown in Figure 4.59, we know that the first slanted stop-band has a smaller attenuation constant than the others. As a

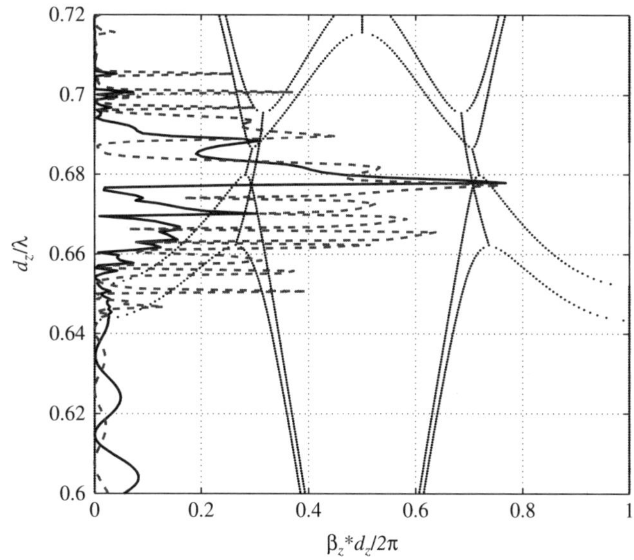

Figure 4.60 Partial enlargement of the dispersion diagram shown in Figure 4.58; the solid and dashed lines are for the TE–TE and TM–TM reflectances, respectively

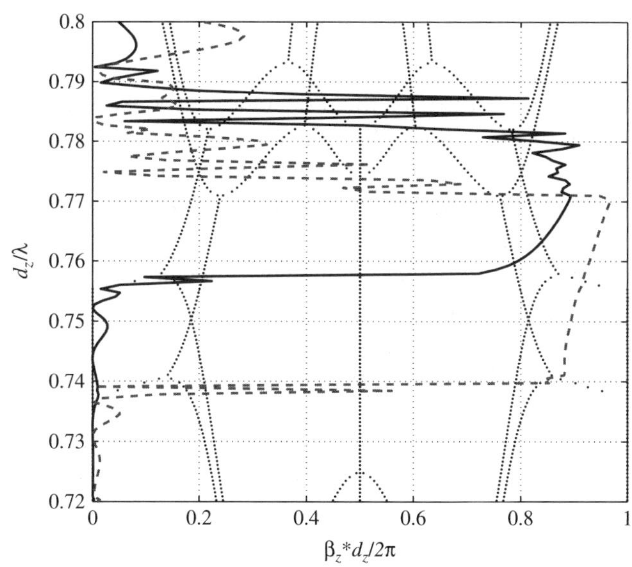

Figure 4.61 Partial enlargement of the dispersion diagram shown in Figure 4.58; the solid and dashed lines are for the TE–TE and TM–TM reflectances, respectively

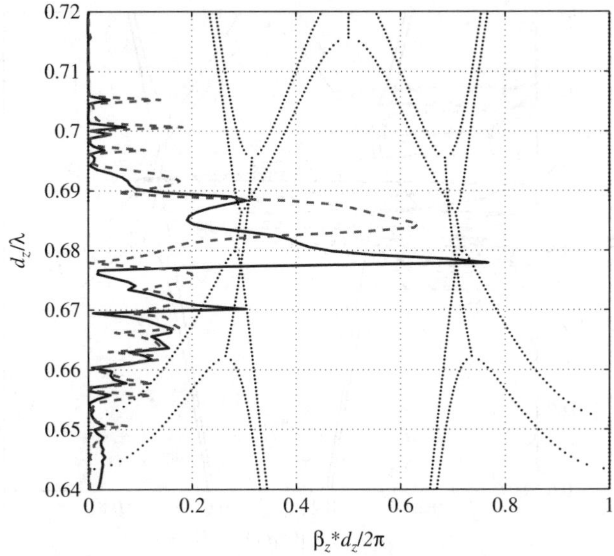

Figure 4.62 Partial enlargement of the dispersion diagram shown in Figure 4.58; the reflectance response of TE–TM is drawn as a solid line while the TE–TE is plotted as a dashed line

consequence, we may conjecture that the reflectance response in that stop-band is less significant. However, this seems not to be the case, as shown in Figure 4.60, since the narrower slanted stop-band has a stronger reflectance than that of the wider stop-band. Actually, in the previous figures, only co-polarization reflectance of the fundamental space harmonic is taken into account. For the oblique incidence condition presented in this example, cross-polarization does occur. The reflectance of the cross-polarization terms – that is, both the TE–TM (reflectance of TM-polarized space harmonic for the TE-polarized plane-wave incidence) and TM–TE (reflectance of TE-polarized space harmonic for the TM-polarized plane-wave incidence) – have to be taken into account.

In Figure 4.62, the co-polarization (denoted TE–TE) reflectance of the fundamental space harmonic is drawn as a solid line while the cross-polarization (TE–TM) term is plotted as a dashed line. Since the reflectance must take into account the contribution from both the co- and cross-polarizations, the overall reflectance is the sum of these two terms. The superposition of the reflectance responses allows us to have clear band edges for such a slanted stop-band. A similar explanation can be applied to Figure 4.63 for filling the "missing piece" in the wide slanted stop-band contributed by the *TM-like* mode.

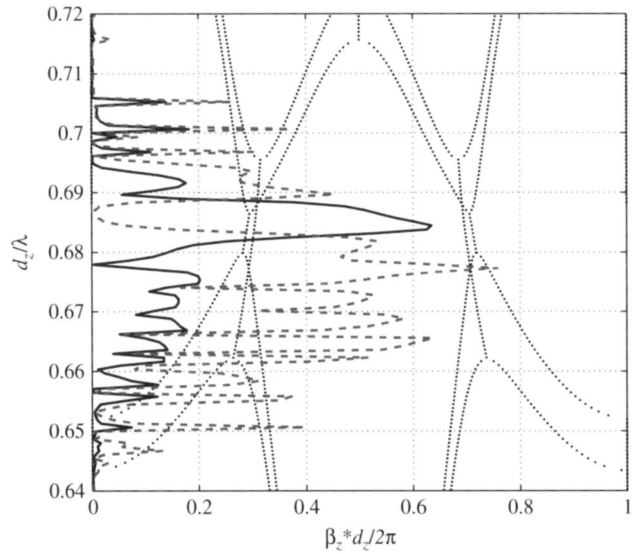

Figure 4.63 Partial enlargement of the dispersion diagram shown in Figure 4.58; the reflectance response of TM–TE is drawn as a solid line while the TM–TM is plotted as a dashed line

Appendix: Closed-Form Solution of $\varepsilon_{pq,mn}$ and $\mu_{pq,mn}$

In this appendix, the closed-form solution of the $\varepsilon_{pq,mn}$ and $\mu_{pq,mn}$ will be formulated. Here, we assume that the unit cell is in the rectangular coordinate system. The host medium of the unit cell is characterized by $(\varepsilon_{\mathrm{h}}, \mu_{\mathrm{h}})$. There are N filling materials doped into the host medium. Each filling material has its own relative permittivity and permeability, ε_j and μ_j, distributed in the region, $x \in [x_1^{(j)}, x_2^{(j)}]$ and $y \in [y_1^{(j)}, y_2^{(j)}]$, where $j = 1, 2, 3, \ldots, N$. The relative permittivity and permeability functions can be written as follows:

$$\varepsilon(x, y) = \varepsilon_{\mathrm{h}} + \sum_{j=1}^{N} (\varepsilon_j - \varepsilon_h)\Omega_j(x)\Omega_j(y) \tag{4A.1}$$

$$\mu(x, y) = \mu_{\mathrm{h}} + \sum_{j=1}^{N} (\mu_j - \mu_h)\Omega_j(x)\Omega_j(y) \tag{4A.2}$$

where $\Omega_j(x) = 1$ for $x \in [x_j^{(1)}, x_j^{(2)}]$ and is zero otherwise.

$$\varepsilon_{pq,mn} = \frac{1}{ab}\iint \mathrm{e}^{+\mathrm{j}\underline{k}_{\mathrm{t},pq}\cdot\underline{\rho}}\,\mathrm{e}^{-\mathrm{j}\underline{k}_{\mathrm{t},mn}\cdot\underline{\rho}}\varepsilon(x, y)\,\mathrm{d}x\,\mathrm{d}y \tag{4A.3}$$

$$\mu_{pq,mn} = \frac{1}{ab}\iint \mathrm{e}^{+\mathrm{j}\underline{k}_{\mathrm{t},pq}\cdot\underline{\rho}}\,\mathrm{e}^{-\mathrm{j}\underline{k}_{\mathrm{t},mn}\cdot\underline{\rho}}\mu(x, y)\,\mathrm{d}x\,\mathrm{d}y \tag{4A.4}$$

where $e^{+j\underline{k}_{t,pq}\cdot\underline{\rho}} \cdot e^{-j\underline{k}_{t,mn}\cdot\underline{\rho}}$ is equal to $e^{j(p-m)2\pi/a} \cdot e^{j(q-n)2\pi/b}$. For the sake of succinctness, we define $\alpha = p - m$ and $\beta = q - n$. After performing some mathematical procedures, the closed-form solution of the integration in Equations (4A.3) and (4A.4) is written as

$$\varepsilon_{pq,mn} = \varepsilon_h \delta_\alpha \delta_\beta + \sum_{j=1}^{N} (\varepsilon_j - \varepsilon_h) \frac{s_j}{s} e^{j2\pi\alpha x_c^{(j)}/a} e^{j2\pi\beta y_c^{(j)}/b} sinc \frac{\alpha\pi a^{(j)}}{a} \cdot sinc \frac{\beta\pi b^{(j)}}{b} \tag{4A.5}$$

where $x_c^{(j)}$ and $y_c^{(j)}$ are the center positions of the jth dielectric block along the x- and y-axes with $x_c^{(j)} = (x_1^{(j)} + x_2^{(j)}/2)$ and $y_c^{(j)} = (y_1^{(j)} + y_2^{(j)})/2$, respectively. Parameter s_j is the area of the jth filling material, which is given as $s_j = a_j b_j$; s is the area of the unit cell, which is given as $s = ab$. Moreover, δ is the Kronecker delta function.

Similarly, we have the expansion coefficient of relative permeability given as

$$\mu_{pq,mn} = \mu_h \delta_\alpha \delta_\beta + \sum_{j=1}^{N} (\mu_j - \mu_h) \frac{s_j}{s} e^{j2\pi\alpha x_c^{(j)}/a} e^{j2\pi\beta y_c^{(j)}/b} \,\mathrm{sinc}\, \frac{\alpha\pi a^{(j)}}{a} \cdot \mathrm{sinc}\, \frac{\beta\pi b^{(j)}}{b} \tag{4A.6}$$

References

Hall, R. C., Mittra, R., and Mitzner, K. M. (1988) Analysis of multilayered periodic structures using generalized scattering matrix theory. *IEEE Transactions on Antennas and Propagation* **36**, 511–517.

Hwang, R.-B. (2000) Scattering characteristics of two-dimensionally periodic impedance surface. *IEEE Transactions on Antennas and Propagation* **48**, 1521–1527.

Hwang R.-B. (2004) Relations between the reflectance and band structure of 2-D metallodielectric electromagnetic crystals. *IEEE Transactions on Antennas and Propagation* **52**, 1454–1464.

Hwang, R.-B. (2006) Correlation between a negative group velocity and a slanted stop band in two-dimensionally periodic structures. *Radio Science* **41**, RS1004. DOI: 10.1029/2005RS003262.

Hwang, R.-B. and Peng, S.-T. (1999) Guidance characteristics of two-dimensionally periodic impedance surface. *IEEE Transactions on Microwave Theory and Techniques* **47**, 2503–2511.

Hwang, R.-B. and Peng, S.-T. (2000) Guided waves on 2D periodic structures and their relation to planar photonic band gap structures. *IEICE Trans. on Electronics* **E83**, 705–712.

Hwang, R.-B. and Pu, T.-C. (2007) Omnidirectional stop band by using composite two-dimensionally artificial crystal. *Microwave and Optical Technology Letters* **49**, 1914–1917.

Peng, S. and Morris, G. M. (1996) Resonant scattering from two-dimensional gratings. *Journal of the Optical Society of America A* **47**, 993–1005.

5

Introducing Defects into Periodic Structures

During the early 1990s, the optical physics community *rediscovered* the idea that a wave can be guided in a uniform medium between a pair of periodic structures, in effect forming a waveguide with side walls composed of periodic structures, if the periodic structures are operated in their stop-bands (called photonic bandgaps by them). That community was unaware that the *identical concept* was published many years ago in the microwave community (Larsen and Oliner 1967). They designed a specific structure, and rigorously calculated the dispersion characteristics of waveguide modes, and also confirmed the guided-wave characteristics through the measurement. In this chapter, we will start from revisiting such a simple but important structure. In addition to the *ppwg*-like mode, the *surface state* supported by the asymmetry semi-infinite 1D periodic stratified medium will also be discussed. Secondly, the dispersion relation of waveguide, particularly on the leaky-wave phenomenon, in the channel sandwiched by a pair of 2D dielectric or metallic periodic structures will be discussed. Finally, the correlation between the scattering and guiding characteristics of the above-mentioned structure will be elucidated for interpreting the resonant transmission.

5.1 A Parallel-Plane Waveguide having a Pair of 1D Semi-Infinite Periodic Structures as its Side Walls

As shown in Figure 5.1, a parallel-plane waveguide consists of side walls made up of semi-infinite 1D periodic layers. The two dielectric mediums with relative dielectric constant and thickness, (ε_1,d_1) and (ε_2,d_2), are included in a unit cell. The structure along the y- and z-axes are assumed to be uniform and infinite; the fields have no variation along the y-direction. The relative dielectric constant and the thickness of the

Periodic Structures: Mode-Matching Approach and Applications in Electromagnetic Engineering, First Edition. Ruey-Bing (Raybeam) Hwang.

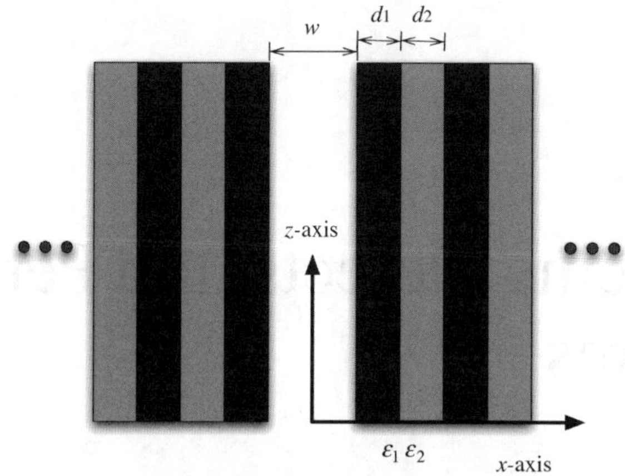

Figure 5.1 A parallel-plane waveguide consists of side walls made up of semi-infinite 1D periodic layers

uniform medium between the two semi-infinite 1D periodic structures is characterized by (ε_a, w).

5.1.1 Bloch Impedance

Regarding the two-tone 1D periodic medium, the input–output relation of the unit cell was rigorously developed in Section 2.3.3. In the presence of periodic boundary condition over the unit cell along the x-axis, the voltage and current at the input terminal satisfies the relation written below (which is the dispersion relation of Bloch wave in the 1D periodic medium).

$$\mathbb{T}\begin{bmatrix} V(0) \\ I(0) \end{bmatrix} = \lambda \begin{bmatrix} V(0) \\ I(0) \end{bmatrix} \tag{5.1}$$

where $\mathbb{T}$ is the 2×2 transfer matrix and $\lambda = \exp(\pm \mathrm{j}\kappa d)$. In addition to the above-mentioned eigenvalue equation, the eigenvalue can also be resolved by

$$\cos \kappa d = \cos k_{x,1} d_1 \cos k_{x,2} d_2 - Q \sin k_{x,1} d_1 \sin k_{x,2} d_2 \tag{5.2}$$

$$Q = \frac{1}{2}(Z_1/Z_2 + Z_2/Z_1) \tag{5.3}$$

$$k_{x,i} = k_o \sqrt{\varepsilon_i - \varepsilon_{\text{eff}}} \tag{5.4}$$

where the index $i = 1, 2$ and paramater ε_{eff} is the effective dielectric constant of the guided-wave propagating along the z-axis, which is defined as $\varepsilon_{\text{eff}} = (k_z/k_o)^2$, where k_z remains the same in each layer. Parameters Z_1 and Z_2 are the characteristic impedances in medium 1 and medium 2, respectively.

Once the eigenvalue is determined, the eiegnvectors can be readily obtained from Equation (5.1). Moreover, the ratio of $V(0)$ to $I(0)$ is defined as the *Bloch impedance*, which can be regarded as the characteristic impedance of a Bloch wave propagating in the 1D periodic medium. The Bloch impedance (Z_B) can be expressed explicitly as follows:

$$Z_\mathrm{B} = \frac{V(0)}{I(0)} = \frac{t_{1,2}}{\lambda - t_{1,1}} \tag{5.5}$$

$$t_{1,1} = \cos k_{x,1}d_1 \cos k_{x,2}d_2 - \frac{Z_1}{Z_2} \sin k_{x,1}d_1 \sin k_{x,2}d_2 \tag{5.6}$$

$$t_{1,2} = -\mathrm{j}Z_2 \sin k_{x,2}d_2 \cos k_{x,1}d_1 - \mathrm{j}Z_1 \sin k_{x,1}d_1 \cos k_{x,2}d_2 \tag{5.7}$$

Problem 5.1

Prove Equations (5.6) and (5.7).

5.1.2 Surface States Supported at the Interface of a Semi-Infinite 1D Periodic Structure

So far, we have determined the Bloch impedance of a 1D two-tone periodic medium and expressed it in a closed-form solution. Before exploring the guiding characteristics of the waveguide with periodic side walls, we first investigate a waveguide sidewall made up of a semi-infinite 1D periodic structure. According to the paper by Yeh *et al.* (1977), a semi-infinite periodic stratified medium consisting of alternating layers of different indices of refraction can support surface waves (or a so-called surface state) that decay in both directions (evanescent wave in the uniform medium and decaying Bloch wave in the stratified medium).

Considering the semi-infinite 1D periodic structure with relative dielectric constant distribution shown in Figure 5.2, the mode dispersion relation is written as

$$Z_\mathrm{B} + Z_\mathrm{a} = 0 \tag{5.8}$$

where Z_B is defined in Equation (5.5),and Z_a is the characteristic impedance in the uniform medium, such as the air. The above equation is also termed the transverse resonance equation, elucidated previously. The TE- and TM-mode characteristic impedance can be written as

$$Z_\mathrm{a} = \begin{cases} \frac{\omega\mu_\mathrm{o}}{k_\mathrm{o}\sqrt{\varepsilon_\mathrm{a}-\varepsilon_\mathrm{eff}}}; & \text{TE} \\ \frac{k_\mathrm{o}\sqrt{\varepsilon_\mathrm{a}-\varepsilon_\mathrm{eff}}}{\omega\varepsilon_\mathrm{o}\varepsilon_\mathrm{a}}; & \text{TM} \end{cases} \tag{5.9}$$

The mode dispersion relation in Equation (5.8) is a function of the structure parameters k_o and the unknown ε_eff to be determined. Since the Bloch impedance is an implicit function of the unknown ε_eff, there is no closed-form solution to

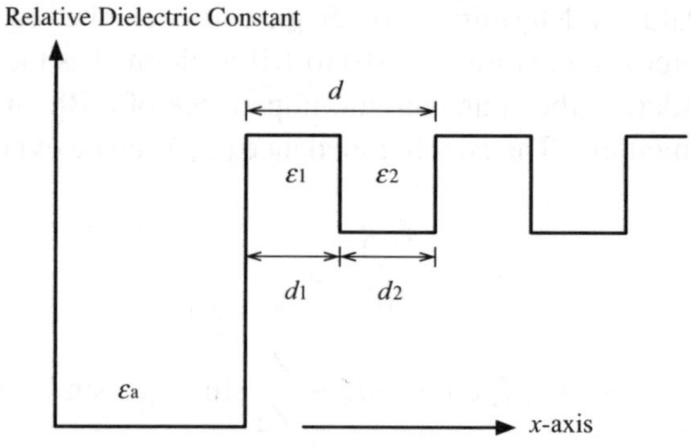

Figure 5.2 Relative dielectric constant distribution of a semi-infinite periodic stratified medium

Equation (5.9). The numerical method has to be exploited to search for the dispersion roots. However, before the brute-force approach, we may systematically allocate the regions that the roots ε_{eff} may exist along the real axis. First of all, we consider a general case that the relative dielectric constants are different for the three mediums and satisfy the condition $\varepsilon_2 > \varepsilon_1 > \varepsilon_{\text{a}}$.

For a lossless structure, the surface state must have an effective dielectric constant ε_{eff} greater than ε_{a}. Therefore, the possible dispersion root ranges from ε_{a} to infinity. We first check the region: $\varepsilon_{\text{eff}} \in [\varepsilon_2, \infty]$. Within this region, the propagation constants along the x-axis in medium 1 ($k_{x,1}$), medium 2 ($k_{x,2}$) and air ($k_{x,\text{a}}$) are all pure imaginary numbers, becoming evanescent (nonpropagating) waves, which are $-\text{j}\alpha_{x,1}$, $-\text{j}\alpha_{x,2}$, and $-\text{j}\alpha_{x,\text{a}}$, respectively. Variables $\alpha_{x,1}$, $\alpha_{x,2}$, and $\alpha_{x,\text{a}}$ are all positive real numbers. Under this situation, the dispersion relation in Equation (5.2) yields

$$\cos\kappa d = \cosh\alpha_{x,1}d_1 \cosh\alpha_{x,2}d_2 + Q\sinh\alpha_{x,1}d_1 \sinh\alpha_{x,2}d_2 \tag{5.10}$$

where

$$Q = \frac{1}{2}(\alpha_{x,2}/\alpha_{x,1} + \alpha_{x,1}/\alpha_{x,2}) \geq \sqrt{(\alpha_{x,2}/\alpha_{x,1})(\alpha_{x,1}/\alpha_{x,2})} = 1$$

for TE polarization and

$$Q = \frac{1}{2}(\varepsilon_2\alpha_{x,1}/\alpha_{x,2}\varepsilon_1 + \varepsilon_1\alpha_{x,2}/\alpha_{x,1}\varepsilon_2) \geq 1$$

for TM polarization.

Since $\cosh\alpha_{x,1}d_1$ and $\cosh\alpha_{x,2}d_2$ are both greater than unity, together with the positive real number of the rest of the terms on the right-hand side of Equation (5.10), we have $\cos\kappa d > 1.0$. This means that the Bloch wave decays along the x-axis in the periodic stratified medium; the wave is in the stop-band condition. The propagation constant κ is a pure imaginary number, $\kappa = -\text{j}\alpha$ ($\alpha > 0$).

Returning to the Bloch impedance in Equation (5.5), considering the attenuation waves in medium 1 and medium 2, the parameters $t_{1,1}$ and $t_{1,2}$ become

$$t_{1,1} = \cosh\alpha_{x,1}d_1 \cosh\alpha_{x,2}d_2 + \frac{Z_1}{Z_2}\sinh\alpha_{x,1}d_1 \sinh\alpha_{x,2}d_2 \tag{5.11}$$

where $Z_1/Z_2 = \alpha_2/\alpha_1$ for TE polarization and $Z_1/Z_2 = \alpha_1\varepsilon_2/\alpha_2\varepsilon_1$ for TM polarization. Parameter $t_{1,1}$ is a real number and is greater than unity.

$$t_{1,2}^{\mathrm{TE}} = -\mathrm{j}\frac{k_{\mathrm{o}}}{Z_{\mathrm{o}}}(\sinh\alpha_{x,2}d_2 \cosh\alpha_{x,1}d_1/\alpha_{x,2} + \sinh\alpha_{x,1}d_1 \cosh\alpha_{x,2}d_2/\alpha_{x,1}) \tag{5.12}$$

$$t_{1,2}^{\mathrm{TM}} = \frac{\mathrm{j}}{k_{\mathrm{o}}Z_{\mathrm{o}}}\left(\frac{\alpha_{x,2}}{\varepsilon_2}\sinh\alpha_{x,2}d_2 \cosh\alpha_{x,1}d_1 + \frac{\alpha_{x,1}}{\varepsilon_1}\sinh\alpha_{x,1}d_1 \cosh\alpha_{x,2}d_2\right) \tag{5.13}$$

Substitution of $t_{1,1}$ and $t_{1,2}$ together with the eigenvalue λ into Equation (5.5). In the stop-band, the wave is exponentially decaying in the periodic medium; the eigenvalue $\lambda = \exp(-\alpha d)$ is smaller than unity. Therefore, Z_{B} is a purely *inductive* reactance for TE polarization and a purely *capacitive* reactance for TM polarization.

Additionally, the input impedance looking into the uniform medium is equal to the characteristic impedance of the uniform medium, which is $\mathrm{j}\omega\mu_{\mathrm{o}}/\alpha_{\mathrm{a}}$ for TE polarization and $-\mathrm{j}\alpha_{\mathrm{a}}/\omega\varepsilon_{\mathrm{o}}\varepsilon_{\mathrm{a}}$ for TM polarization. Therefore, it is a purely *inductive* reactance for TE polarization and a purely *capacitive* reactance for TM polarization.

From the concept of transverse resonance, the resonance can occur only when the inductive and capacitive reactances cancel each other out (Figure 5.3). However, the Bloch impedance and the input impedance looking into the uniform medium share the same polarity of reactance; the transverse resonance condition does not take place. We conclude that although the semi-infinite periodic medium is operated in the stop-band when $\varepsilon_{\mathrm{eff}} > \max[\varepsilon_1, \varepsilon_2]$, the surface state is not present. Therefore, we have to search the dispersion roots in the region $\varepsilon_{\mathrm{eff}} \in (\varepsilon_{\mathrm{a}}, \varepsilon_2]$ subject to the necessary and sufficient condition that the periodic medium is operated in the stop-band condition.

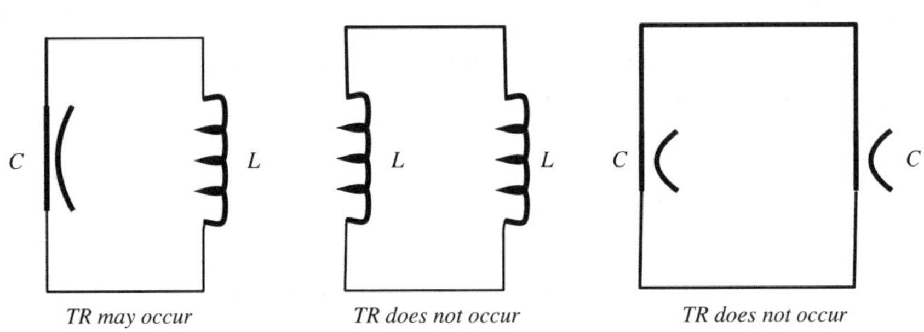

Figure 5.3 Condition of the existence of transverse resonance (TR)

Problem 5.2

Prove that the resistive part of the Bloch impedance exists when the Bloch wave in the periodic medium is operated in the pass-band condition (with the effective dielectric constant greater than ε_a).

Example 5.1

Consider the semi-infinite 1D periodic stratified medium with relative dielectric constant distribution sketched in Figure 5.2. The relative dielectric constants of the two mediums are respectively $\varepsilon_1 = 6.0$ and $\varepsilon_2 = 3.0$, and the uniform medium is assumed to be air. The period along the x-axis is 1.0 (arbitrary units) and the duty cycle is 50%. Draw the dispersion diagram of the waveguide together with the band structure of the 1D periodic medium; the TM mode is considered in this example.

Solution

As described previously, the surface state occurs in the region with the relative dielectric constant ε_{eff}, located between ε_a and ε_1. By solving the transverse resonance equation given in Equation (5.8), we may determine the mode dispersion relation. Additionally, the band edge of the 1D periodic medium can be obtained by solving the equation

$$\cos k_{x,1}d_1 \cos k_{x,2}d_2 - Q \sin k_{x,1}d_1 \sin k_{x,2}d_2 = \pm 1 \tag{5.14}$$

The region with the value on the left-hand side greater than one is denoted as the forbidden (stop-band) region.

Figure 5.4 shows the dispersion diagram of the guided waves together with the band structure of the corresponding 1D periodic medium of infinite extent along the $\pm x$-axis. The heavy dotted line is the mode dispersion relation, while the shaded area is the forbidden region with the Bloch wave decaying in the periodic medium. The dispersion curves with $\varepsilon_{\text{eff}} > 1.0$ belong to the *surface state*, which is an evanescent wave in a uniform medium and is a decaying Bloch wave in a semi-infinite periodic medium. Contrarily, a wave with $\varepsilon_{\text{eff}} < 1.0$ is a leaky wave. Notably, if the dispersion root is inside the stop-band, the leaky wave radiates its power into the uniform medium only and decays in the semi-infinite periodic medium; otherwise, it leaks its power into both the uniform and semi-infinite periodic mediums if the leaky dispersion root is in the pass-band region. Although not shown here, the leaky wave, in general, has a complex propagation constant with nonvanished attenuation constant α_z.

In addition to the dispersion diagram, we also plot the tangential electric field component along the x-axis for the surface state. The three dispersion roots on the different branches are chosen for plotting their field distribution. In Figure 5.5, the field pattern with ε_{eff} greater than but around unity is chosen. Since the dispersion root is near the band edge (inside the stop-band), the attenuation in the periodic medium is insignificant. The attenuation rate of the Bloch wave in the semi-infinite 1D periodic medium is not fast enough, so that it can survive for several periods, as shown in this figure.

In the second case, we chose the dispersion root with the effective dielectric constant near 5.0. Because this dispersion root is away from the band edge, it has a considerable attenuation constant in the 1D periodic medium along the $+x$-axis, and thus the tangential field component is decaying significantly therein. As indicated in Figure 5.6, the field strength almost concentrates in the first period and decays rapidly in the periodic medium.

Finally, the case with $\varepsilon_{\text{eff}} = 5.757$ at $k_o = 12.0$ is shown in Figure 5.7. In light of the large effective dielectric constant, the wave decays in both the $\pm x$-directions. Consequently, the electric field strength concentrates in the vicinity of the interface between the semi-infinite 1D periodic medium and air. Away from the interface, the wave decays rapidly to zero in both mediums, which is extremely similar to the TM-polarization surface plasmon supported at the interface between air and plasma with relative dielectric constant smaller than -1.0. Distinct with the TM-polarized surface plasmon, the same structure shown in Figure 5.2 with the structure parameters given in the last example can also support the surface state (or wave) for TE polarization, which is depicted in Figure 5.8.

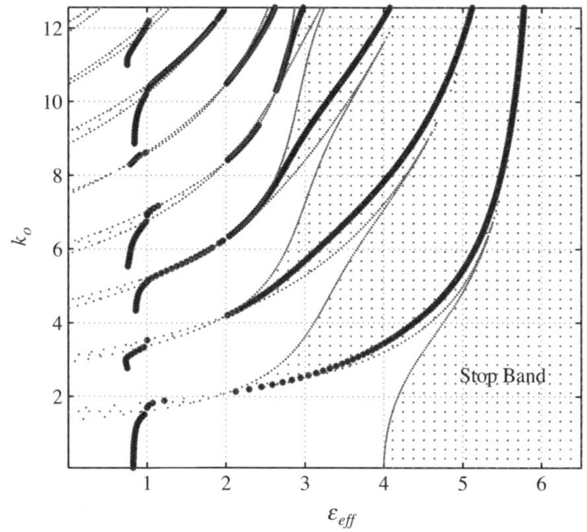

Figure 5.4 Dispersion diagram of the TM-polarized guided modes in a semi-infinite periodic stratified medium; the shaded region is the stop-band with the band edge attached; the heavy dotted lines are the mode dispersion curves

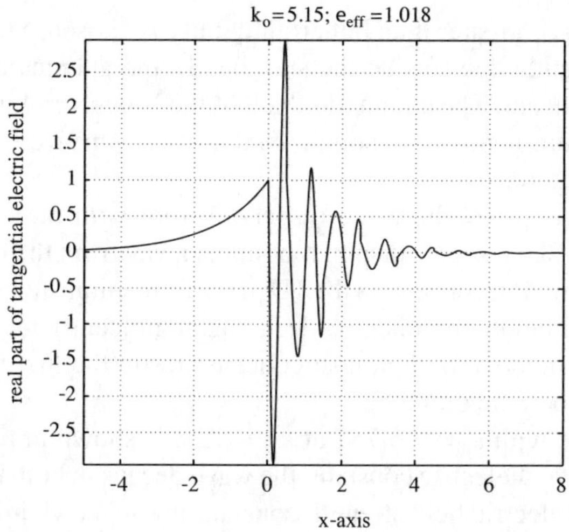

Figure 5.5 Variation of the real part of the tangential electric field component against the position along the x-axis; the interface is at $x = 0$ and $x \geq 0$ denotes the semi-infinite 1D periodic medium; this plot is drawn for the condition $k_o = 5.25$ and $\varepsilon_{\text{eff}} = 1.018$

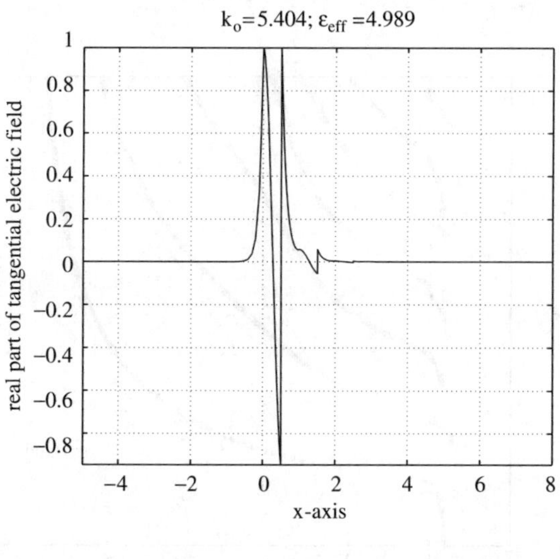

Figure 5.6 Variation of the real part of the tangential electric field component against the position along the x-axis; the interface is at $x = 0$ and $x \geq 0$ denotes the semi-infinite 1D periodic medium; this plot is drawn for the condition $k_o = 5.404$ and $\varepsilon_{\text{eff}} = 4.989$

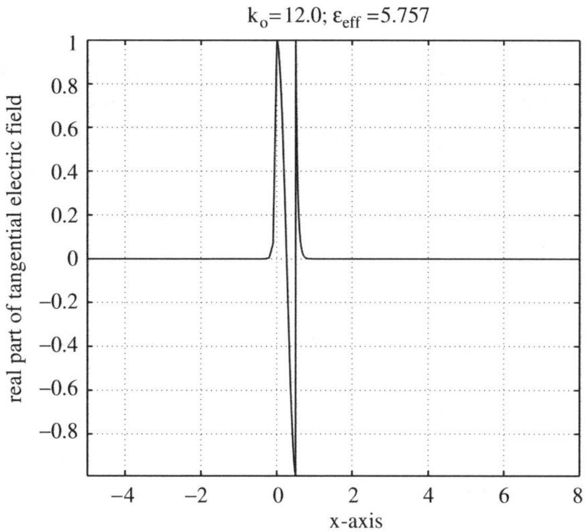

Figure 5.7 Variation of the real part of the tangential electric field component against the position along the x-axis; the interface is at $x = 0$ and $x \geq 0$ denotes the semi-infinite 1D periodic medium; this plot is drawn for the condition $k_o = 12.0$ and $\varepsilon_{eff} = 5.757$

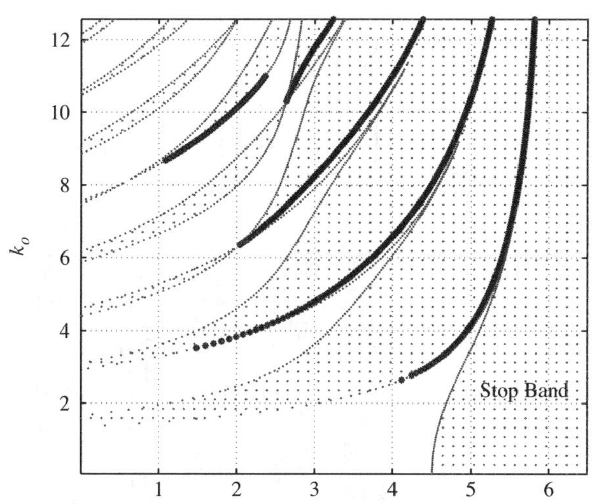

Figure 5.8 Dispersion diagram of the TE-polarized guided modes in a semi-infinite periodic stratified medium; the shaded region is the stop-band with the band edge attached; the heavy dotted lines are the surface-mode dispersion curves

Problem 5.3

If the semi-infinite 1D periodic stratified medium is replaced by finite-thickness 1D periodic layers, in addition to the surface waves, the *volume waves* will also be supported in an asymmetric structure. Write a computer program to calculate the dispersion diagram and observe the types of modes supported in the structure.

5.1.3 A Semi-Infinite 1D Periodic Structure Consisting of Symmetric Dielectric Waveguides

We consider a special case that the uniform medium shares the same dielectric constant with one of the dielectric slabs in the unit cell; thus, the overall structure can be regarded as a coupled channel waveguide immersed in a uniform medium such as air; the distribution of the relative dielectric constant of the coupled waveguide system is shown in Figure 5.9. It is interesting to note that, different from the similar structure having a different dielectric material in the uniform medium presented previously, such

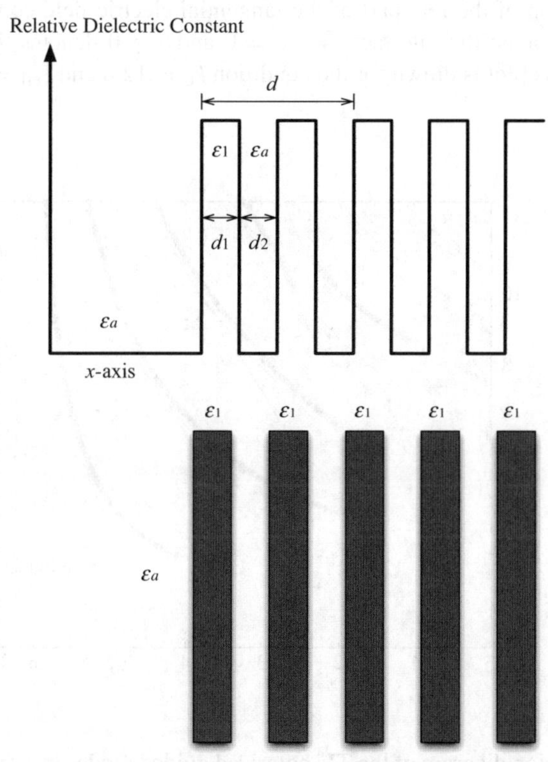

Figure 5.9 Distribution of the relative dielectric constant of a coupled dielectric waveguide system immersed in a uniform medium, such as air, which can also be regarded as a semi-infinite periodic stratified medium

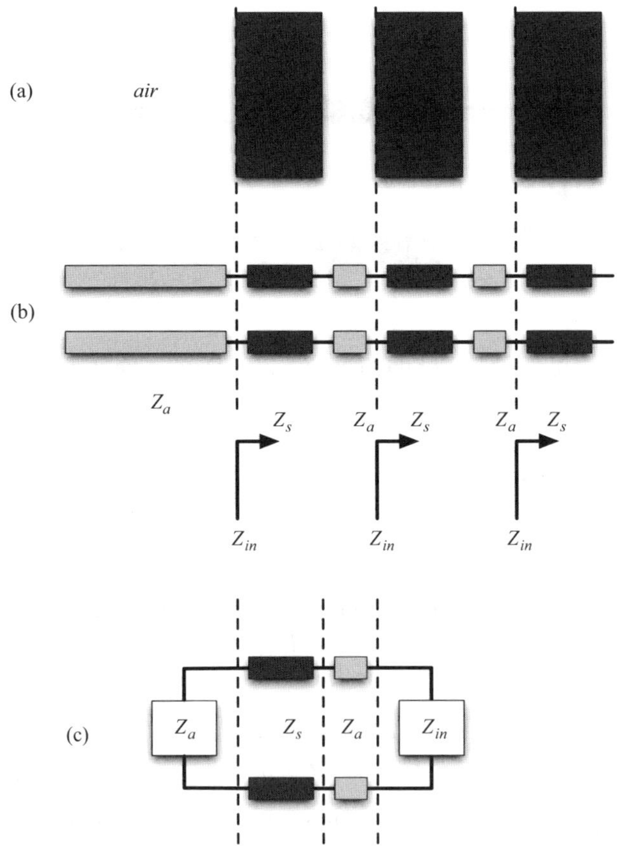

Figure 5.10 Schematic diagram of a semi-infinite 1D periodic structure: (a) dielectric slabs immersed in the air; (b) transmission-line network representation; (c) circuit diagram for evaluating the transverse resonance

a structure *cannot* support the surface state. The detailed mathematical interpretation follows.

Figure 5.10 shows a schematic diagram of a semi-infinite 1D periodic structure; (a) is the dielectric slabs immersed in the air and (b) is the transmission-line network representation of the electromagnetic fields in the multilayered structure. Here, we assume that the thicknesses of the dielectric slab and air slab are respectively t_s and t_a. The characteristic impedances in air and the dielectric substrate are individually denoted as Z_a and Z_s. In the presence of the periodic boundary condition, the input impedance looking into each unit cell (period) shares the same one; that is, Z_{in}.

If the guided-wave is supported in the semi-infinite 1D periodic structure, the transverse resonance condition at the interface yields

$$Z_a + Z_{in} = 0 \tag{5.15}$$

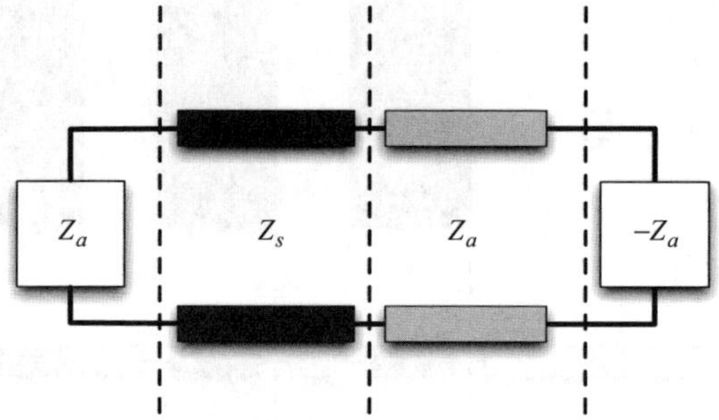

Figure 5.11 Schematic diagram of the semi-infinite 1D periodic structure for evaluating the transverse resonance

The transmission-line network in (b) of Figure 5.10 can be converted into (c); that is, the unit cell is terminated by Z_a and Z_{in} at both its ends. The following step is to evaluate the transverse resonance condition of the circuit depicted in (c). From the transverse resonance condition given in Equation (5.15), we have $Z_{in} = -Z_a$. Therefore, the circuit diagram in (c) can be transformed into Figure 5.11.

Referring to the circuit diagram shown in Figure 5.11, the input impedance looking into the input terminal of the air line is equal to

$$Z_{in} = Z_a \frac{-Z_a + jZ_a \tan k_{x,a} t_a}{Z_a + j(-Z_a) \tan k_{x,a} t_a} = -Z_a \tag{5.16}$$

where $k_{x,a}$ is the propagation constant along the x-axis in the air slab. Such an input impedance is the load impedance of the dielectric layer.

With the load impedance of the dielectric layer $-Z_a$, the input impedance looking into this dielectric layer is

$$Z_{in} = Z_s \frac{-Z_a + jZ_s \tan k_{x,s} t_s}{Z_s + j(-Z_a) \tan k_{x,s} t_s} \tag{5.17}$$

The transverse resonance condition at the input terminal of the dielectric layer is

$$Z_a + Z_s \frac{-Z_a + jZ_s \tan k_{x,s} t_s}{Z_s + j(-Z_a) \tan k_{x,s} t_s} = 0 \tag{5.18}$$

The above equation can be rearranged to:

$$\frac{j(Z_s^2 - Z_a^2) \tan k_{x,s} t_s}{Z_s + j(-Z_a) \tan k_{x,s} t_s} = 0 \tag{5.19}$$

The existence of the transverse resonance condition is $Z_s = \pm Z_a$, leading to a trivial solution; that is, the dielectric layer shares the same relative dielectric constant with that of air, or the dielectric layer is a plasma medium. Through the mathematical proof, we prove that such a structure depicted in Figure 5.9 cannot support any waveguide modes as well as the surface state.

Problem 5.4

Prove that for a TE wave the condition causing $Z_s = \pm Z_a$ is $\varepsilon_a = \varepsilon_s$, while it is $\varepsilon_a + \varepsilon_s < 0$ for a TM wave.

5.2 Dispersion Relation of a Parallel-Plane Waveguide with Semi-Infinite 1D Periodic Structures as Waveguide Side Walls

Let us now return to the structure shown in Figure 5.1; because of the symmetry of the wave-guiding structure along the x-axis, the electric field distribution along the x-axis contains both symmetry and anti-symmetry distribution with respect to $x = 0$. Thus, the anti-symmetry one can be simplified as the SCB which places a PEC at $x = 0$, while the symmetry one can be modeled by the OCB with a PMC placed at $x = 0$. With the OCB and SCB shown in Figure 5.12, the computation domain can be further reduced to half of the original one.

In the presence of the termination condition at $x = 0$, the transverse resonance condition in Equation (5.8) is modified as

$$Z_B + Z_{in} = 0 \tag{5.20}$$

where $Z_{in} = jZ_a \tan k_{x,a}w/2$ for SCB and $Z_{in} = -jZ_a \cot k_{x,a}w/2$ for OCB, and Z_B is the Bloch impedance of the 1D periodic medium.

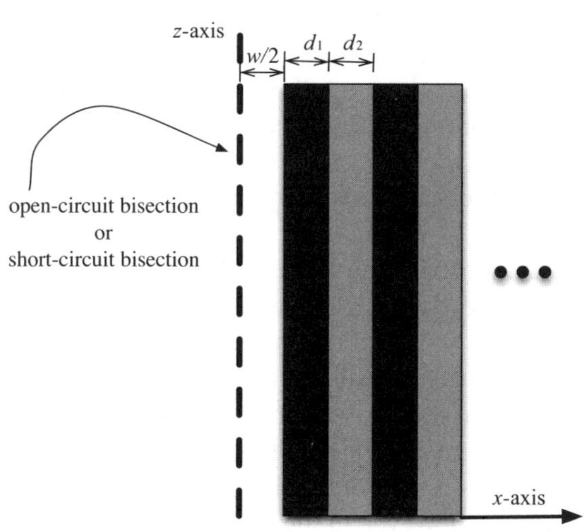

Figure 5.12 OCB and SCB of the wave-guiding structure shown in Figure 5.1

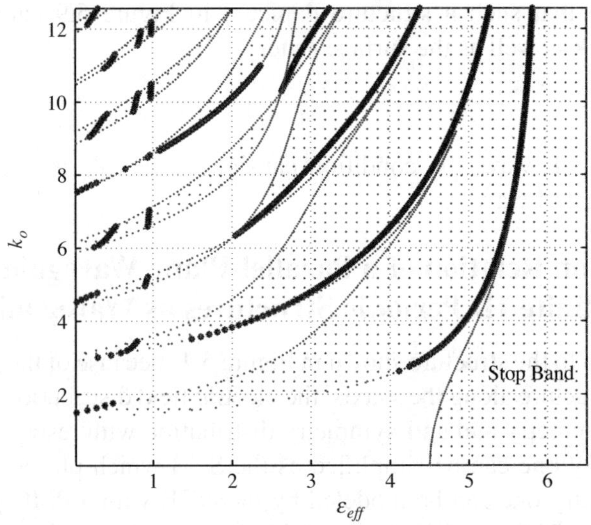

Figure 5.13 Dispersion diagram (TE polarization) of the OCB of the structure shown in Figure 5.12; the heavy dotted line is the dispersion relation of the waveguide; the half-width of the waveguide is 1

5.2.1 Numerical Example

In this example, we employ the same semi-infinite 1D periodic medium discussed in the previous examples as the waveguide side wall for designing the parallel-plane waveguide shown in Figure 5.1. The waveguide is filled with air with the width 2 (arbitrary units) along the x-axis.

Figure 5.13 is the dispersion diagram of the waveguide with the OCB termination condition. The heavy dotted line is the dispersion relation of the waveguide; the lines in the region $\varepsilon_{\text{eff}} < 1.0$ are the *parallel-plate-waveguide-like* modes,while the others ($\varepsilon_{\text{eff}} > 1$) are the surface state caused by the semi-infinite 1D periodic stratified medium.

In Figure 5.14 we demonstrate the tangential electric field strength of the mode supported in the waveguide channel. Apparently, such a mode has the maximum electric field strength at $x = 0$, which resembles the fundamental mode in a metallic ppwg. Owing to the stop-band condition of the semi-infinite 1D periodic medium, the field decays exponentially in the periodic medium, which is very different from the metallic ppwg with zero tangential electric field on its boundaries. Notably, because of $\varepsilon_{\text{eff}} < 1$, the field exhibits a sinusoidal distribution in the two dielectric mediums of the 1D periodic structure, however, gradually decaying due to the attenuation in the stop-band operation. From the field distribution shown in this figure, in general, the *effective* width of the waveguide is larger than the actual channel width w.

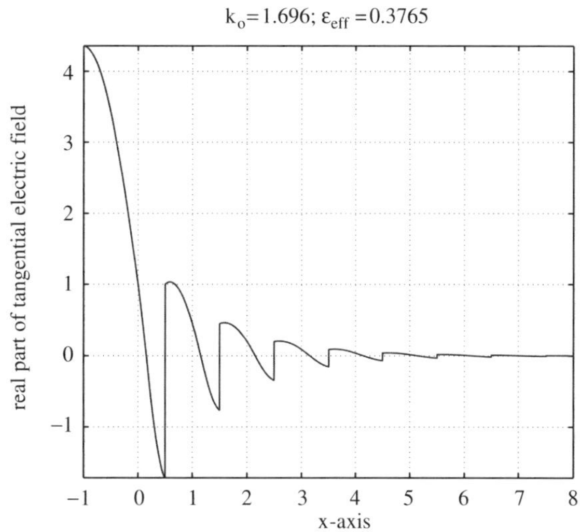

Figure 5.14 Tangential electric field strength of the mode supported in the waveguide channel: fundamental TE mode for the OCB condition

In addition to the fundamental TE mode, we also plot the electric field strength of the higher order mode in Figure 5.15. Similarly, the higher order mode experiences an exponential decay in the 1D periodic medium due to the stop-band condition.

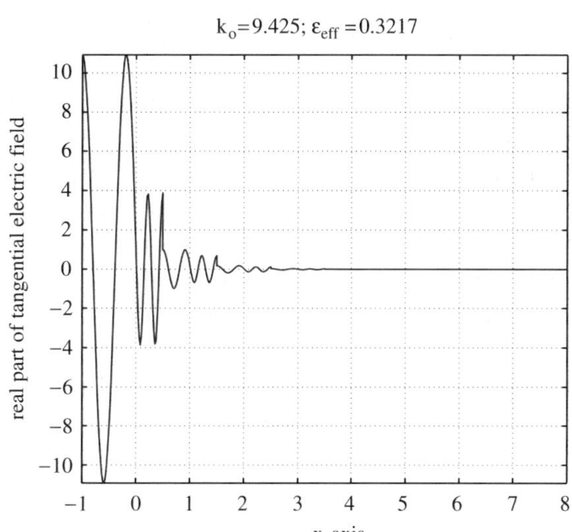

Figure 5.15 Tangential electric field strength of the mode supported in the waveguide channel: higher order TE mode for the OCB condition

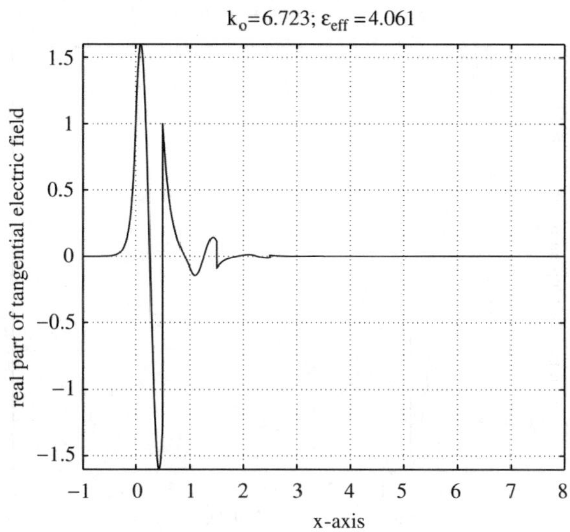

Figure 5.16 Tangential electric field strength of the surface state (TE mode): OCB condition

The other mode supported in such a waveguide is also drawn, which is the surface state already elucidated previously. Since the effective dielectric constant is greater than unity, it exhibits evanescence in the waveguide channel and decays exponentially in the 1D periodic medium due to stop-band operation; see Figure 5.16. Notably, due to $\varepsilon_{\text{eff}} < \varepsilon_1$ and $\varepsilon_{\text{eff}} > \varepsilon_2$, the field distributes sinusoidally along the x-axis in ε_1 and decays exponentially in ε_2.

Additionally, without loss of generality, the waveguide with the SCB termination condition is also demonstrated in Figure 5.17. The curves with $\varepsilon_{\text{eff}} < 1$ represent the waveguide mode concentrating most of its field strength in the channel, while those with $\varepsilon_{\text{eff}} > 1$ are the surface state surviving at the interface, at $x = 0$, but decaying in the $\pm x$-axis. Furthermore, the field profile of the first higher order mode in the channel is plotted in Figure 5.18, which is similar to the TE_2 of an ideal ppwg. Owing to the SCB, the tangential electric field returns to zero at the termination plane at $x = -1$.

The surface state supported by the same structure is also plotted in Figure 5.19. Notably, the decay of the wave in the second medium is obvious due to the condition that $\varepsilon_{\text{eff}} > \varepsilon_2$.

Problem 5.5

Consider a finite-thickness 1D periodic structure instead of the semi-infinite one presented in the previous example; however, the number of periods is not large enough to

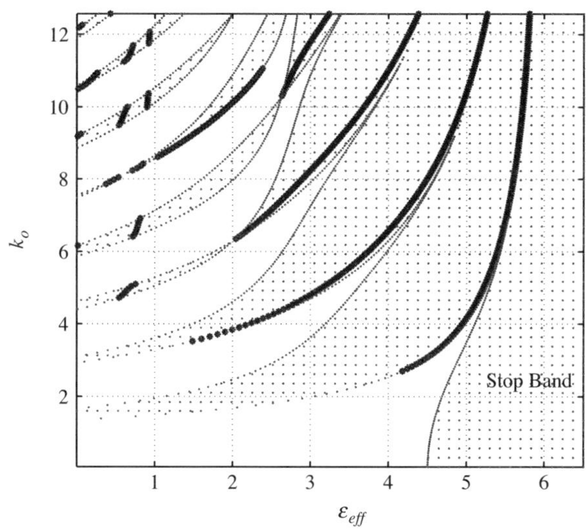

Figure 5.17 Dispersion diagram (TE polarization) of the SCB of the structure shown in Figure 5.12; the heavy dotted line is the dispersion relation of the waveguide; the half-width of the waveguide is 1

regard it as a periodic structure; for example, two periods only. Although the *surface-state-like* mode may be found, the band-gap map seems not to be suitable for us to assess the type of mode. How do you explain the physical meaning of the *surface-state-like* mode based on the *perturbation theory* of coupled waveguides?

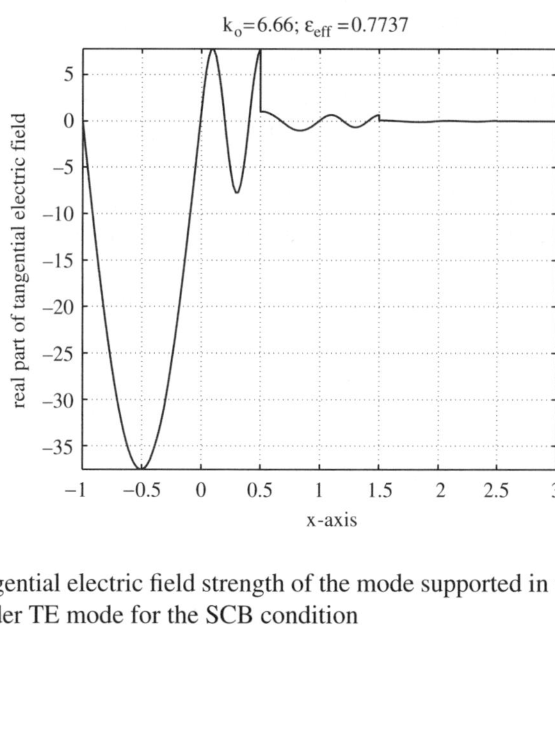

Figure 5.18 Tangential electric field strength of the mode supported in the waveguide channel: first higher order TE mode for the SCB condition

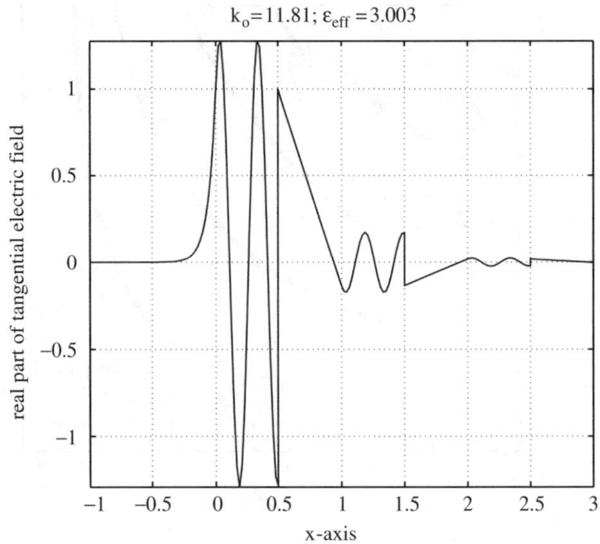

Figure 5.19 Tangential electric field strength of the surface state (TE mode): SCB condition

5.3 A Parallel-Plane Waveguide with 2D Dielectric Periodic Structures as its Side Walls

In addition to the 1D periodic stratified medium, the 2D periodic structure can also serve as the waveguide side wall. Figure 5.20 shows a typical 2D phtonic crystal waveguide. The basic idea is to carve a channel out of a perfect 2D photonic crystal. Light is able to bounce back and forth between the side walls and propagate along the waveguide axis, when the periodic side walls are operated in the stop-band condition or below cutoff. As is well known, an optical fiber can guide visible light except for tight corners. However, a photonic crystal waveguide does not rely on total internal reflection, so that it can confine light even around tight corners. Ideally, the waves will propagate in a zigzag pattern within the channel without any loss if the guiding channel is filled with air and the field penetrating into the periodic medium is negligible. However, owing to the finite thickness of the waveguide, side walls made of periodic structure definitely cause the power leakage from the waveguide into the surrounding medium. Although the guiding phenomenon was observed in experimental and numerical studies, a systematic investigation of the dispersion characteristics, including the variation of phase and attenuation constants against the operational wavelength, was rarely conducted.

In fact, the MMA described in the previous chapters is very suitable for analyzing the dispersion relation of the waveguide. Although the structure under consideration must be infinite in extent along the wave-guiding axis, the guiding characteristics of each

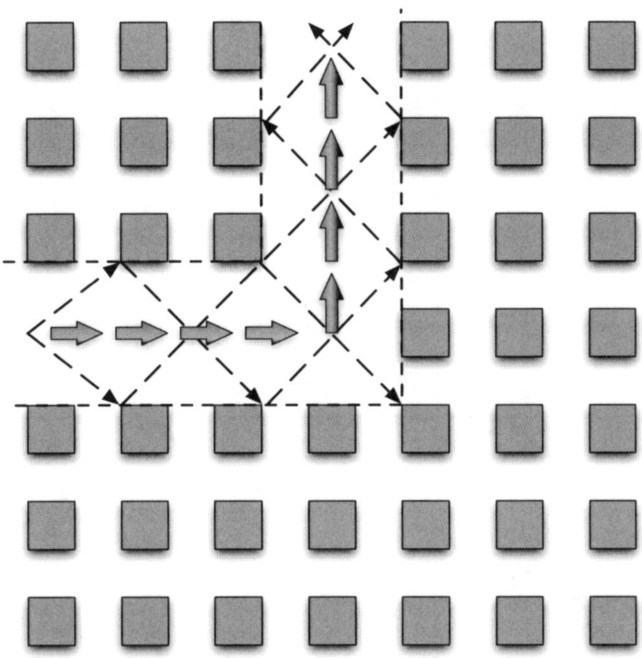

Figure 5.20 A 2D photonic crystal waveguide with a tight corner: the guiding channel with 2D periodic structures as its side walls; each block represents a dielectric rod with infinite length

mode, including both the propagating and nonpropagating ones, can be completely understood by solving the transverse resonance equation. Specifically, the dispersion characteristic consisting of the phase and attenuation constants can be readily determined once the structure configuration of the waveguide is specified. On the other hand, other numerical methods, such as finite-difference time (or frequency) domain, method of moments, and the finite-element method, can get the electromagnetic field information over the grids of the finite structure; nevertheless, the modal profile and amplitude have to resort to the numerical extraction method, which may cause significant numerical errors.

Incidentally, as far as the photonic-crystal-based waveguide is concerned, the unit-cell pattern, periodicity, and the lattice pattern are the important factors affecting the dispersion characteristics of such a class of waveguide. I have published a series of papers discussing the above-mentioned problems (Hwang 2007a, 2009; Hwang and Hsiao 2006; Hwang and Peng 2003); the numerical examples were demonstrated there; the mathematical details will be elucidated in the following sections. In this section I will focus on the guided-wave properties of the waveguide modes supported in a channel sandwiched by a pair of 2D periodic structures (or photonic crystal). Through

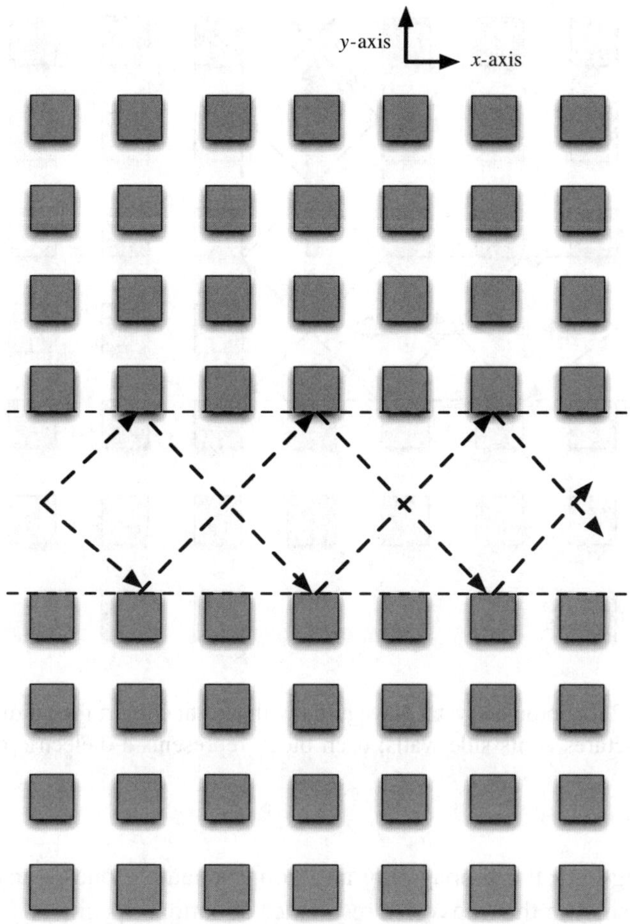

Figure 5.21 A channel sandwiched by a pairs of 2D periodic structures as its side walls; each block represents a dielectric rod with infinite length

the numerical results and physical interpretation, I intend to let you know that the wave propagation in a photonic-crystal waveguide is absolutely not an exotic phenomenon. The electromagnetic guided-wave theory has already revealed the secret.

Figure 5.21 shows the channel waveguide sandwiched by a pair of 2D periodic structures serving as waveguide side walls. The dielectric square blocks are made of dielectric material, and they are assumed to be infinite in extent along their length direction, the z-axis. The fields along the length direction are assumed to have no variation. Therefore, the problem can be regarded as a 2D problem (in-plane propagation), and the E_z and H_z polarizations can be treated independently. The zigzag pattern represents that the waveguide mode is bouncing back and forth between the

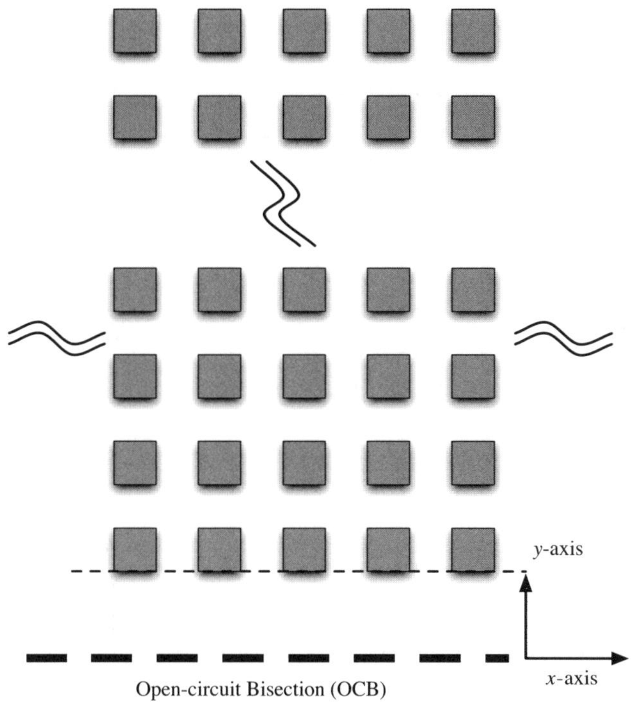

Figure 5.22 OCB of the original waveguide shown in Figure 5.21

two side walls. Notably, in general, the dielectric mediums in the channel and that surrounding the 2D dielectric columns array can be different.

Owing to the symmetry of the structure along the y-axis, the electromagnetic field along that direction contains symmetry and anti-symmetry patterns. Consequently, the original structure can be handled as two independent schemes, which are the OCB and SCB shown in Figures 5.22 and 5.23. With the short-circuit termination (terminated by a PEC at the bisection plane), the electric field distribution is anti-symmetric with respect to the plane at $y = 0$. Contrarily, the PMC is placed at the bisection plane to obtain a symmetric distribution of the electric field with respect to $y = 0$ for the OCB condition. It is noted that the period along the x-axis is assumed to be infinite in extent, while the period in the y-direction can be a finite one. The width of the guiding channel in Figures 5.22 and 5.23 is half of the original structure shown in 5.21.

5.3.1 Method of Mathematical Analysis

In light of the assumption that the 2D periodic structure is finite along the y-axis but is infinite along the x-axis, the 2D periodic structure can be regarded as a finite stack of 1D periodic dielectric layers along the y-direction. In Section 4.2.1 we introduced the

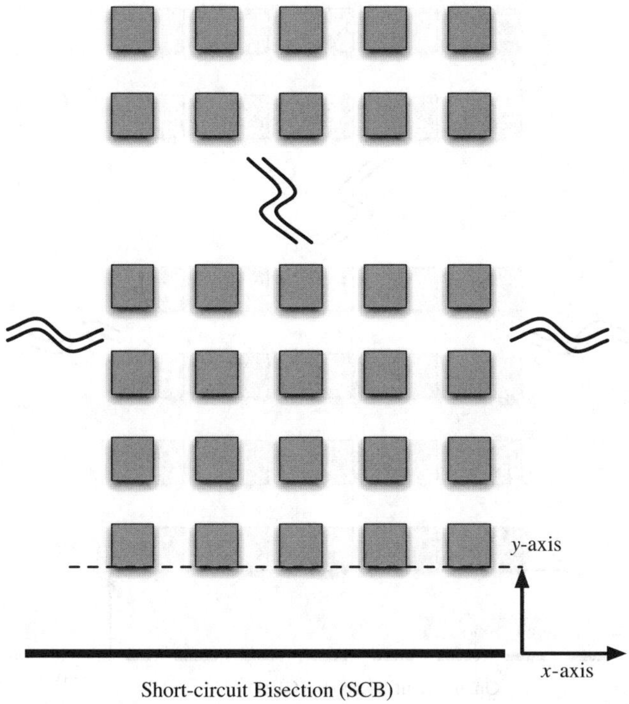

Figure 5.23 SCB of the original waveguide shown in Figure 5.21

generalized scattering matrix approach for determining the scattering characteristics of a finite stack of 1D dielectric periodic layers by cascading each scattering matrix of the 1D periodic layer. Here, we will not repeat the process; the detailed mathematical procedures can be found in Chapter 4. We assume the generalized scattering matrix of the 2D periodic structure composed of N_y 1D periodic layers is written as

$$\begin{pmatrix} \underline{b}_1 \\ \underline{b}_2 \end{pmatrix} = \begin{pmatrix} \mathbb{S}_{11} & \mathbb{S}_{12} \\ \mathbb{S}_{21} & \mathbb{S}_{22} \end{pmatrix} \begin{pmatrix} \underline{a}_1 \\ \underline{a}_2 \end{pmatrix} \tag{5.21}$$

Owing to the termination condition at the bisection plane, the incident and reflected waves at port 2 of the generalized scattering matrix $\mathbb{S}$ satisfy the condition

$$\underline{a}_2 = \Gamma_l \, \mathrm{diag}[\mathrm{e}^{-\mathrm{j}2\underline{\kappa}(0.5w)}]\underline{b}_2 \tag{5.22}$$

where $\underline{\kappa}$ is a row vector with its elements representing the propagation constant of each space harmonic along the y-axis; for example, the nth element is characterized as $\kappa_n = \sqrt{k_0^2\varepsilon_a - (k_x + n2\pi/d_x)^2}$, where k_x is the propagation constant of the channel waveguide to be determined and d_x is the period along x-axis. Parameter Γ_l is

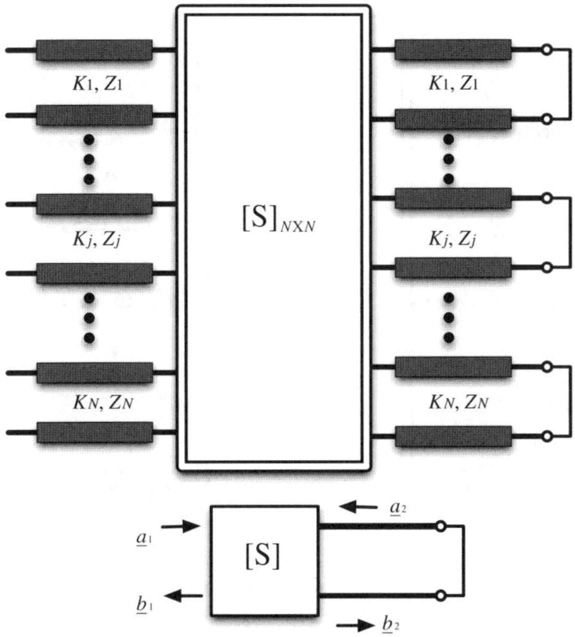

Figure 5.24 Transmission-line network representation of the generalized scattering matrix terminated by the short-circuit condition

the reflection coefficient at the termination plane, which is +1 for OCB and −1 for SCB; "diag" stands for the diagonal matrix with the main diagonal vector given in the brackets. The channel width is designated as $0.5w$. The transmission-line network representations of the short-circuit and open-circuit terminations are shown in Figures 5.24 and 5.25, respectively.

For the guided-wave condition, the incident wave vector at port 1, $\underline{a}_1$, vanishes. Therefore, from Equations (5.21) and (5.22), we have the following:

$$[\underline{\underline{I}} - \mathbb{S}_{22}\Gamma_l \,\mathrm{diag}(\mathrm{e}^{-\mathrm{j}\underline{\kappa} w})]\underline{b}_2 = 0 \tag{5.23}$$

The existence of the nontrivial solution, $\underline{b}_2$, leads to the dispersion relation written as

$$\det[\underline{\underline{I}} - \mathbb{S}_{22}\Gamma_l \,\mathrm{diag}(\mathrm{e}^{-\mathrm{j}\underline{\kappa} w})] = 0 \tag{5.24}$$

Problem 5.6

Prove Equation (5.23).

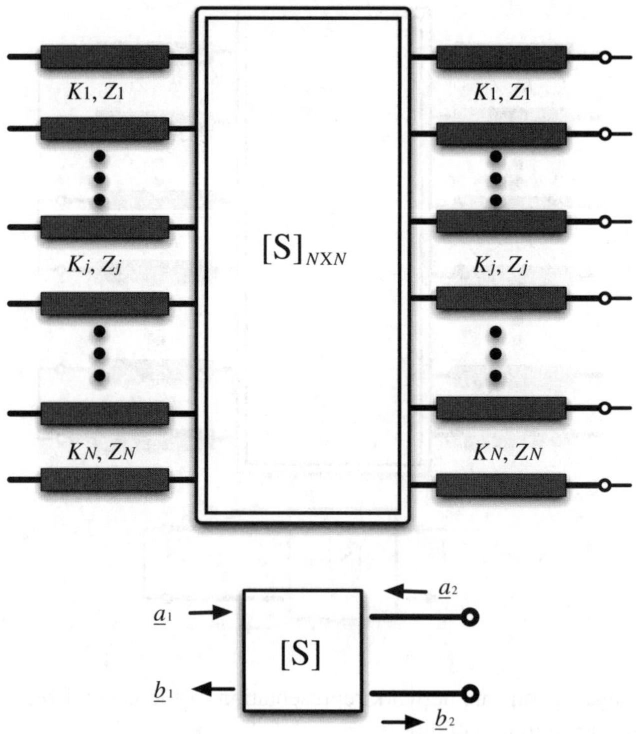

Figure 5.25 Transmission-line network representation of the generalized scattering matrix terminated by the open-circuit condition

5.3.2 Dispersion Relation of a Channel with a Pair of 2D Periodic Structures as its Waveguide Side Walls

Based on the method of analysis, a computer code is written to calculate the guiding characteristics of the waveguides;and some numerical examples will be demonstrated in this section. The problem to be dealt with is an air channel sandwiched by a pair of 2D periodic structures. Please refer to the similar figure shown in Figure 5.21. The 2D square columns array is immersed in a uniform dielectric medium with relative dielectric constant $\varepsilon_i = 3.0$. The dimension and relative dielectric constant of the dielectric column are respectively 0.5 (in arbitrary units) and 6.0. The periods along the x- and y-axes are both $d_x = d_y = 1$. The structure is infinite in extent along the x-axis but finite along the y-axis. The channel width is $w = 2.0$. In fact, the 2D periodic structure can be considered as a finite stack of 1D periodic layers in the y-axis; for example, in this figure it consists of four 1D periodic layers. Now, once the geometric parameters are prescribed, we may start to carry out the numerical calculation.

Recalling the channel waveguide equipped with 1D periodic medium side walls discussed in Section 5.1, the field penetrating into the 1D periodic medium decays rapidly through several periods when the 1D periodic medium is operated in the stop-band condition; see Figures 5.14 and 5.15. However, for the finite-thickness periodic structure, we may conjecture that it cannot completely forbid the propagation of the electromagnetic field even in the stop-band condition. In the following examples, the 2D periodic structures are designed to be operated in the stop-band condition and to evaluate the guiding characteristics of the waveguide.

5.3.2.1 Short-Circuit Bisection

The first example is the SCB of the channel waveguide shown in Figure 5.22. The dispersion relation, which includes the normalized phase (β/k_o) and attenuation (α/k_o) constants, against the normalized frequency d_x/λ for various numbers of 1D periodic layers stacked up along the y-axis is shown in Figure 5.26. The region with $\beta/k_o < 1$ is the fast wave. However, since its α/k_o is small enough, it can also be regarded as a guided wave. This is different from the metallic waveguide; for instance, the ppwg; the attenuation constant vanishes above and is significant before the cutoff

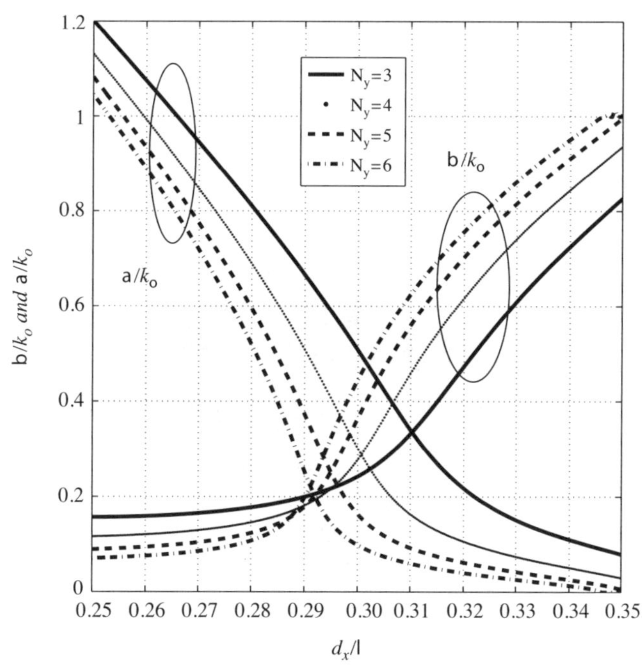

Figure 5.26 Dispersion relation of the channel waveguide with short-circuit termination at its bisection plane; the structure configuration is shown in Figure 5.22. Parameter N_y represents the number of 1D periodic layers stacked up along the y-axis

frequency. In this example, the attenuation constant does not return to zero, so that the cutoff frequency is hard to define clearly. For ease of reference, the normalized frequency satisfying the condition $\beta = \alpha$ (the intersection point between the phase and attenuation constants curves) is defined, throughout this book, as the cutoff frequency of the dispersion relation of a leaky mode. It is obvious that the dispersion curve has a significant attenuation constant before the cutoff frequency; this means that the mode is evanescent before the cutoff frequency. As I have mentioned previously, the 2D periodic structure is designed at its stop-band condition. Consequently, the increase in the number of 1D periodic layers along the y-axis, N_y (or thickness of 2D periodic structure), is indeed able to significantly reduce the normalized attenuation constant. Owing to the similar dispersion diagram compared with that of the metallic ppwg, such a guided wave is termed the *ppwg-like mode* hereafter.

The tangential electric field distribution in between the 2D periodic structure and SCB is shown in Figure 5.27 for the case $N_y = 3$. The condition of this plot is attached in the legend of this figure. The axis ranging from $y = 0$ to $y = 1$ is along the channel width direction, where $y = 1$ is at the short-circuit termination plane. Obviously, the field vanishes at the SCB plane ($y = 1$) because of the electromagnetic boundary condition.

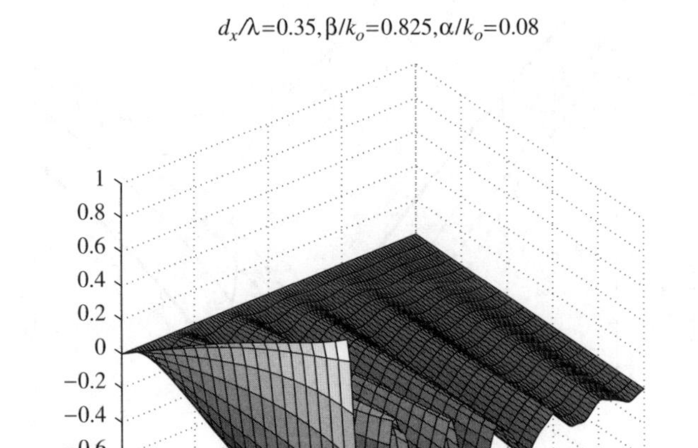

Figure 5.27 Tangential electric field distribution in the channel with short-circuit termination at its bisection plane; the structure configuration is shown in Figure 5.22 for the case of $N_y = 3$

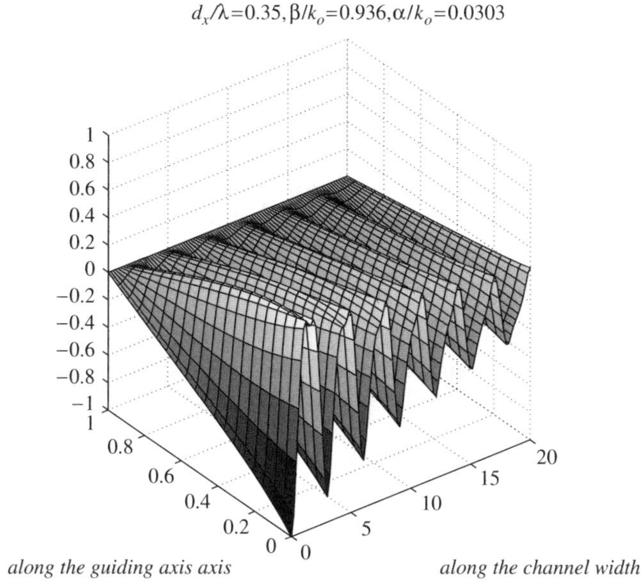

Figure 5.28 Tangential electric field distribution in the channel with short-circuit termination at its bisection plane; the structure configuration is shown in Figure 5.22; the case of $N_y = 4$

Owing to the significant attenuation constant, the amplitude of the electric field decays along the waveguide axis, the x-axis, ranging from 0 to $20d_x$. For an ideal ppwg, the field strength must vanish at the position of the side wall ($y = 0$). Nevertheless, the existence of the attenuation constant causes the power leakage and nonzero electric field over the side wall. From the power-conservation law, for a lossless structure, the decaying power will be converted into that radiating into the surrounding medium.

Additionally, we increase N_y to plot the tangential electric field at the same normalized frequency, $d_x/\lambda = 0.35$. For the case of $N_y = 4$, the normalized attenuation constant is reduced from $\alpha/k_o = 0.08$ to $\alpha/k_o = 0.0303$. It is conjectured that the decrease in the attenuation constant may reduce the decay of the wave along the guiding axis. As shown in Figure 5.28, the decay rate is apparently smaller than that in Figure 5.27. Moreover, at $d_x/\lambda = 0.35$, when N_y is increasing to 5, the normalized attenuation constant is decreasing to 0.006 92. We can see in Figure 5.29 that the attenuation of the field strength is negligible; the amplitude level can almost remain along the waveguide axis.

5.3.2.2 Leaky-Wave Phenomenon

Recalling the leaky surface-wave supported by the 1D waveguide grating described in Section 3.8.2, owing to the periodic variation, the higher order space harmonic of the bound wave (surface wave in a dielectric waveguide) has a phase constant smaller than

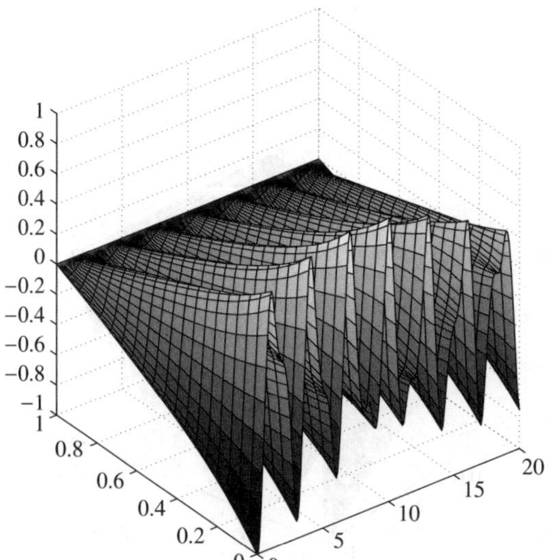

Figure 5.29 Tangential electric field distribution in the channel with short-circuit termination at its bisection plane; the structure configuration is shown in Figure 5.22; the case of $N_y = 5$

that of the free-space wavenumber k_o, enabling the occurrence of power leakage. For example, the dispersion relation of the -1th-order space harmonic of the surface wave leaves the bound-wave triangle and enters into the leaky-wave region; see Figure 3.45.

Different from the leaky wave due to the -1th-order space harmonic of surface wave, the leaky-wave phenomenon is caused by the *leaky ppwg-like* mode. It is almost similar to a metallic waveguide equipped with radiating slit cut along its side wall, as shown in Figure 5.30. Since leakage is over the length of the 2D periodic structure, the whole length determines the effective aperture of the leaky-wave structure unless the attenuation constant is so large that the input power has completely leaked into the air before reaching the end of the 2D periodic structure along the x-axis. Owing to the leakage along the waveguide axis, the waveguide mode has a complex propagation constant $(\beta - j\alpha)$ with phase constant β and attenuation (leaky) constant α. The phase constant satisfies the fast-wave condition $\beta < k_o$. The leaky constant α relates to the beam width of the far-field radiation pattern. A larger α implies that the large leakage rate causes a short effective aperture, so that the radiation pattern has a wide beam width. Conversely, as long as the physical aperture is sufficiently long, the small α results in a long effective aperture and a narrow beam width in the far field.

We have mentioned that the attenuation of the electric field along the waveguide axis results in a leaky-wave in the air. In fact, such a leaky-wave phenomenon can be verified by computing the radiation pattern of the leaky mode. Once the eigenvalue

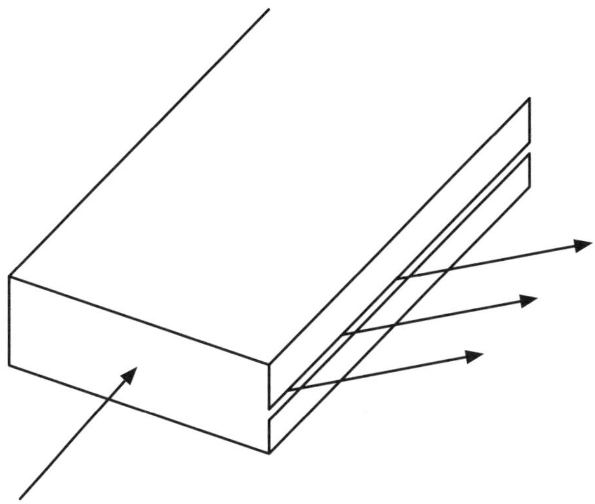

Figure 5.30 The earliest example of a leaky-wave antenna: a rectangular waveguide with a continuous slit cut along one of its side walls

in Equation (5.24) is solved, its associated eigenvector in Equation (5.23) can be readily determined. Thus, the electromagnetic field in the air region can be obtained by summing up all the contribution from the space harmonics in free space. Moreover, the radiation far-field pattern is thus determined. In the following cases, I calculate the far-field radiation pattern corresponding to the examples shown in Figures 5.27 and 5.28 for demonstrating the leaky-wave phenomenon of the *leaky ppwg-like* mode. Figure 5.31 is the radiation far-field pattern of the example in Figure 5.27. The far field is calculated for the radiation aperture truncated to $l = 20d_x$. In this figure, the main beam exhibits a commonly seen leaky-wave pattern with high directivity accompanied by significant sidelobes clearly present. As the length of the radiating aperture increases to $50d_x$, apparently, the sidelobes are absent, as shown in Figure 5.32; evidently, the antenna length is effectively infinite.

Moreover, the radiation patterns of the same waveguide but with $N_y = 4$ at the same normalized frequency ($d_x/\lambda = 0.35$) are demonstrated in the following two figures. In this case, because of large N_y the leaky constant can be further reduced; the structure needs a long aperture to dissipate the input power. Figure 5.33 and 5.34 show the radiation far-field patterns for $l = 20d_x$ and $l = 50d_x$, respectively. In comparison with the two cases, the pattern in Figure 5.34 with lower sidelobes since it has longer radiating aperture. However, since the aperture is not effectively long enough, the pattern is not so smooth as that shown in in Figure 5.32. Additionally, the beam in long-length case has a narrow width since it has an long effective aperture. Although the case of $N_y = 5$ is calculated for its electric-field pattern in the channel, we neglect its far-field radiation since the leaky constant is negligible.

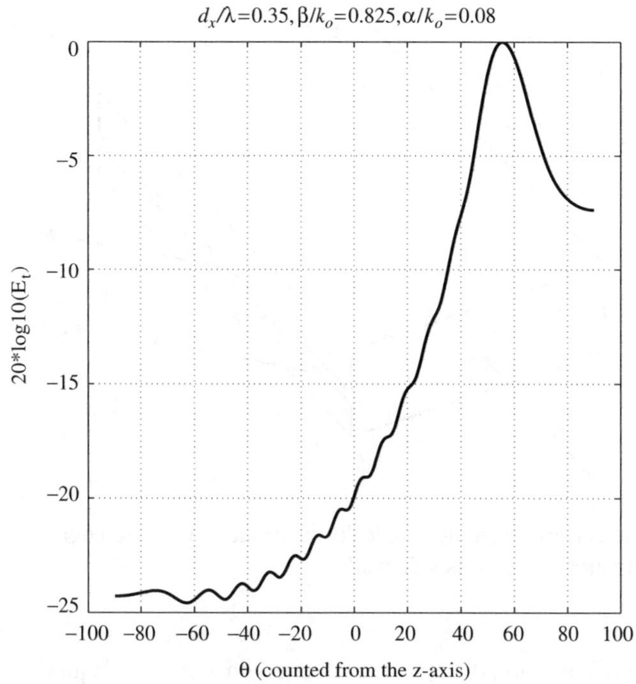

Figure 5.31 Radiation pattern of the leaky wave supported in the waveguide shown in Figure 5.23; the case of $N_y = 3$; the length of the waveguide for calculating the leaky-wave pattern is designated as $l = 20d_x$

5.3.2.3 Determination of Main Beam Angle

The radiation main beam angle of the leaky-wave structure can, in fact, be determined by the phase constant of the leaky mode. From Figure 5.35, the leaky-wave main-beam angle, which is counted from the y-axis, can be determined by

$$\theta = \sin^{-1} \beta / k_o \tag{5.25}$$

Thus, the main-beam angle in Figures 5.31 and 5.32 can be obtained by calculating $\sin^{-1} \beta/k_o$, which is $\sin^{-1} 0.825 = 55.59°$ corresponding to the ones shown in the figure. Moreover, the main-beam angles in Figures 5.33 and 5.34 are both around $\sin^{-1} 0.936 = 69.4°$, close to those in the two figures.

5.3.2.4 Open-Circuit Bisection

In addition to the SCB condition, I also calculate the case of OCB. Without loss of generality, the dispersion diagrams for various 2D periodic structure thicknesses are

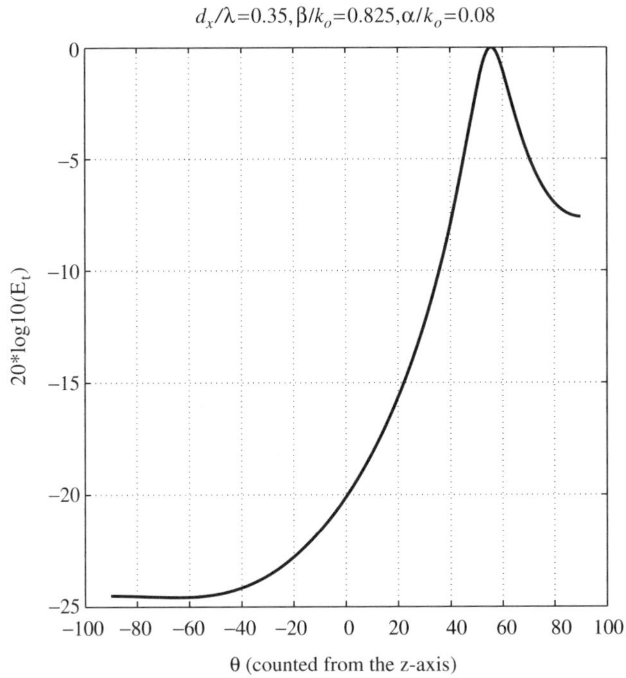

Figure 5.32 Radiation pattern of the leaky wave supported in the waveguide shown in Figure 5.23; the case of $N_y = 3$; the length of the waveguide for calculating the leaky-wave pattern is designated as $l = 50d_x$

calculated and shown in Figure 5.36. As was known in SCB case, the increase in N_y causes a decrease in the attenuation constant. Additionally, the hyperbolic shape of the attenuation constant around $d_x/\lambda = 0.375$ is the stop-band due to the contraflow interaction between the fundamental *ppwg-like* mode and its -1th-order space harmonics. However, since no new physical phenomenon was found, I turn to discuss the beam-steering (or scanning) mechanism for the same structure operating at different normalized frequencies.

Before investigating the leaky-wave phenomenon, I will present the tangential electric-field strength distribution in the waveguide channel for the normalized frequencies designated at $d_x/\lambda = 0.32$, 0.33, 0.34, and 0.35, which are individually shown in Figures 5.37–5.40. Owing to the open-circuit termination at $y = 1$, the tangential electric field has a maximum value at the bisection plane for the fundamental *ppwg-like* mode, which can be confirmed from these figures. Additionally, among those figures, the case in Figure 5.37 has the largest decay constant, so that the field is rapidly attenuating to zero. The case in Figure 5.40 has the smallest

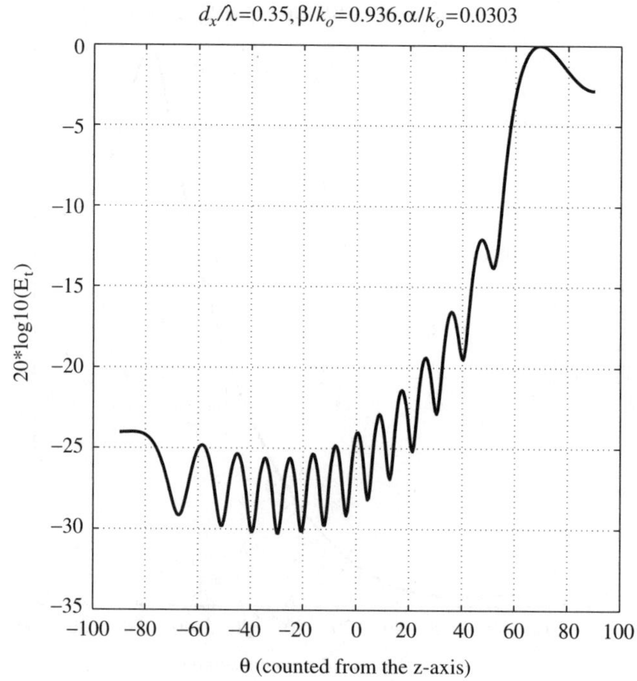

Figure 5.33 Radiation pattern of the leaky wave supported in the waveguide shown in Figure 5.23; the case of $N_y = 4$; the length of the waveguide for calculating the leaky-wave pattern is designated as $l = 20d_x$

attenuation constant; therefore, the field can almost survive after the same traveling distance ($20d_x$).

5.3.2.5 Beam-Steering Phenomenon

With the evidence of wave decay along the waveguide axis, we know that the waveguide mode, fundamental *ppwg-like* mode, leaks its energy into the air. The above four cases belong to the same mode, but operate at different normalized frequencies. The far-field radiation patterns of the four cases are plotted in the same diagram, as shown in Figure 5.41. Interestingly, each of the operation frequencies has its radiation main beam angle. Alternatively, if the operational frequency is changing, the main-beam angle is steering accordingly. Such a phenomenon can be applied to design a beam-steering antenna in a radar system. Again, the main beam angles for the four cases are figured out by the formula given in Equation (5.25), which are 30.731°, 39.494°, 45.972°, and 51.169° for the four cases with the normalized frequencies d_x/λ ranging from 0.32 to 0.35 with step 0.01, respectively.The estimated angles agree perfectly with those shown in this figure. Additionally, since the length of the radiating

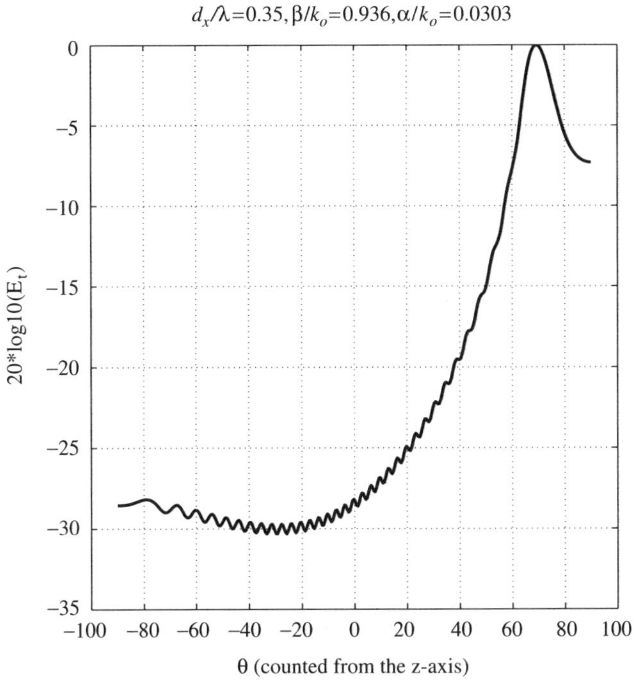

Figure 5.34 Radiation pattern of the leaky wave supported in the waveguide shown in Figure 5.23; the case of $N_y = 4$; the length of the waveguide for calculating the leaky-wave pattern is designated as $l = 50d_x$

structure is assumed to be $50d_x$, it can almost radiate its power before reaching its end. Therefore, the sidelobes are insignificant in this figure. Moreover, since the smallest attenuation case has the longest effective radiating aperture, it possesses the narrowest beam width. Conversely, the largest attenuation constant has the shortest effective aperture; and it has the widest beam width. The above-mentioned phenomena can be perfectly confirmed from the figure.

5.4 Scattering Characteristics of a Periodic Structure with Defects

In addition to the guiding properties of the mode supported in the channel with a 2D periodic structure as its side walls, the scattering characteristics of a plane wave by the same structure are also a research topic intensively studied in the optics and microwave community, and some important applications have been reported in the literature. Figure 5.42 shows a schematic diagram of the waveguide structure similar to that in Figure 5.21. Actually, the figure shown here can also be regarded as a 2D periodic structure with several rows of 1D periodic layers carved out. The region

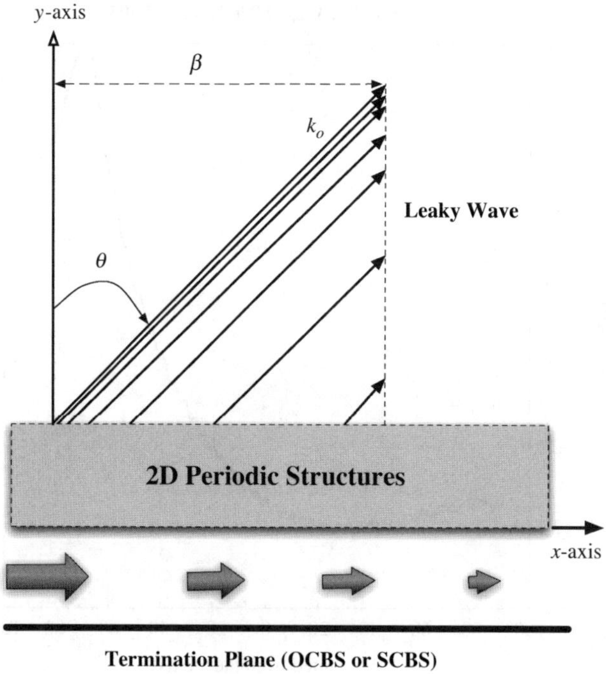

Figure 5.35 Schematic diagram of a leaky-wave waveguide consisting of 2D periodic structure as its side wall

denoted *defect* is the location of the waveguide channel in Figure 5.21. Such a structure was found to have interesting scattering properties, such as creating a narrow pass-band in the forbidden condition of the complete 2D periodic structure. In this section, not only the above-mentioned phenomenon will be observed, but also the underlying physics will be systematically interpreted.

The mathematical analysis for the structure shown in Figure 5.42 has been introduced previously. The generalized scattering matrix of the overall structure can be considered as a finite stack of 1D periodic layers and uniform separators. Notably, in general, a position shift of the distance s in the lateral direction between two adjacent 1D periodic layers is allowed, so that we can study the effect of lattice patterns on the scattering characteristics of a 2D periodic structure by adjusting the parameter s in our analysis. For example, we have a square lattice for $s = 0$ and a triangular lattice for $s = 0.5a$. Notably, for an arbitrary value of s, $b = b_1 + b_2$ is not necessarily the period along the y-axis, as shown in Figure 5.43. In fact, the structure has a period $\sqrt{s^2 + b^2}$ along the direction at the angle $\sin^{-1}(s/b)$ counted from the y-axis. The period along the x-axis is $a = a_1 + a_2$, while the ratio of b_1/b will be referred to as the aspect ratio in the y-axis.

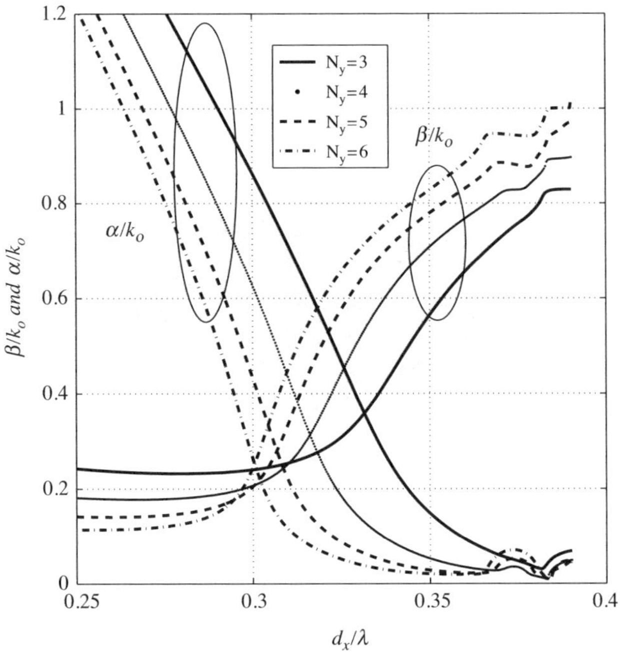

Figure 5.36 Dispersion relation of the channel waveguide with open-circuit termination at its bisection plane; the structure configuration is shown in Figure 5.22. Parameter N_y represents the number of 1D periodic layers stacked up along the y-axis

The input–output relation of the overall structure can be obtained by cascading the generalized scattering matrix of each layer, which is a 1D periodic or uniform one. Therefore, the scattering characteristic including the reflectance and transmittance of each space harmonic can be readily determined.

Problem 5.7

Derive the input–output of a 1D dielectric grating of finite thickness if a lateral shift distance s is present.

In the following example, we will demonstrate the numerical results concerning the scattering analysis of plane wave by a 2D periodic structure with a defect region. The structure configuration is shown in Figure 5.42. The square dielectric columns are arranged in a square lattice ($s = 0$) and immersed in air. The width and period are respectively $a_1 = b_1 = 0.5$ and $d_x(a) = d_y(b) = 1.0$. The relative dielectric constant of the dielectric block is $\varepsilon_s = 8.9$. The width of the channel (defect region) is denoted w. Because of the symmetry of the structure with respect to the bisection plane ($y = 0$),

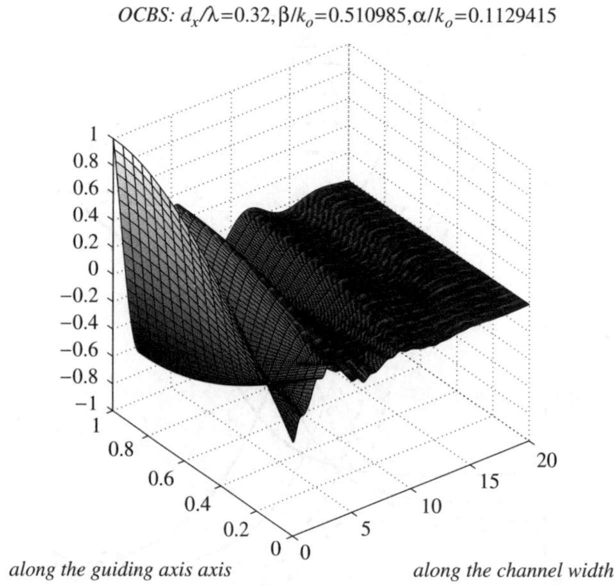

Figure 5.37 Real part of the tangential electric-field strength distribution in the channel waveguide with open-circuit termination at its bisection plane; the structure configuration is shown in Figure 5.22; $N_y = 5$ and $d_x/\lambda = 0.32$

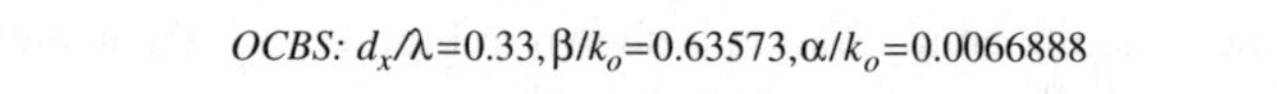

Figure 5.38 Real part of the tangential electric-field strength distribution in the channel waveguide with open-circuit termination at its bisection plane; the structure configuration is shown in Figure 5.22; $N_y = 5$ and $d_x/\lambda = 0.33$

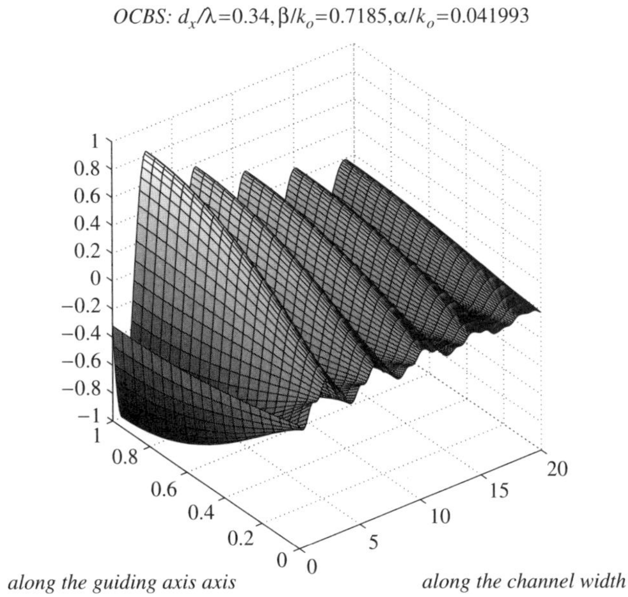

Figure 5.39 Real part of the tangential electric-field strength distribution in the channel waveguide with open-circuit termination at its bisection plane; the structure configuration is shown in Figure 5.22; $N_y = 5$ and $d_x/\lambda = 0.34$

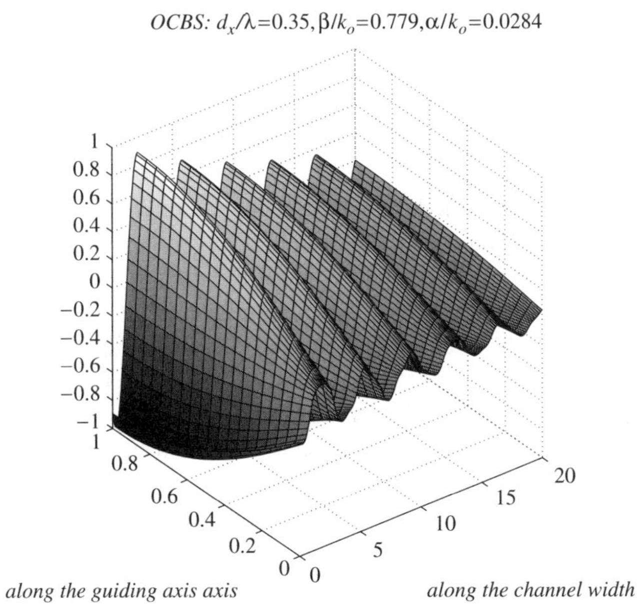

Figure 5.40 Real part of the tangential electric-field strength distribution in the channel waveguide with open-circuit termination at its bisection plane; the structure configuration is shown in Figure 5.22; $N_y = 5$ and $d_x/\lambda = 0.35$

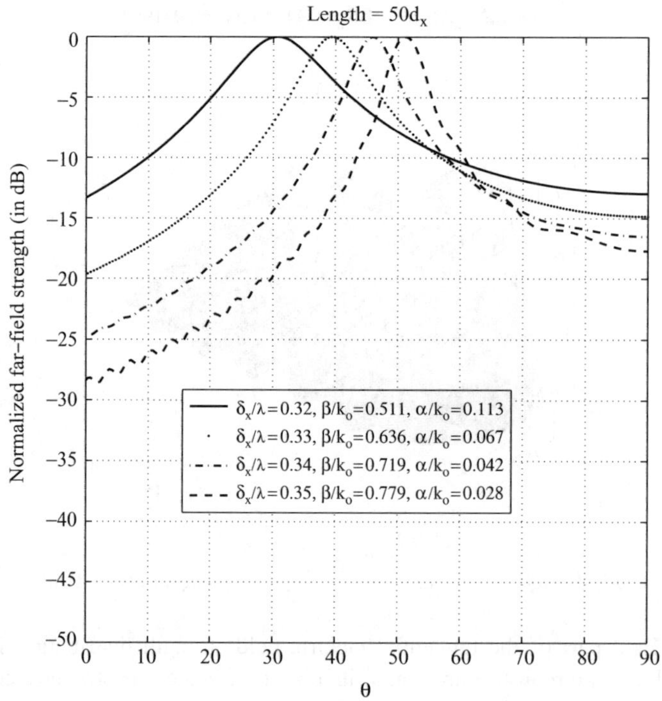

Figure 5.41 Radiation far-field pattern of the leaky-wave structure operated at different normalized frequencies; the structure configuration is shown in Figure 5.22; $N_y = 5$

we have $w = 2h$, where h is the half-width of the channel. The plane wave with TE polarization (E_z wave) is considered hereafter.

Figure 5.44 shows the transmittance of the fundamental space harmonic ($n = 0$) against the normalized frequency for various defect widths. The curve with circles is the result without a defect region (a complete 2D periodic structure consisting of six 1D periodic layers). Obviously, there is a stop-band, where the wave propagation is forbidden. However, when the defect region is created in the central region of the structure, a pass-band with a very narrow bandwidth is seen in the stop-band region. The three peaks (from left to right) correspond to the defect widths, which are 1.05, 1.0 and 0.95, respectively. From the spectrum behavior of the three peaks, we may conjecture that they are due to the resonant effect because of their considerable quality factor Q. Although not shown here, the peak value decreases as the number of 1D periodic layers N_y increases, indicating that the reflection property of the 2D periodic structure operating in the stop-band region affects the resonant transmission of the structure under consideration.

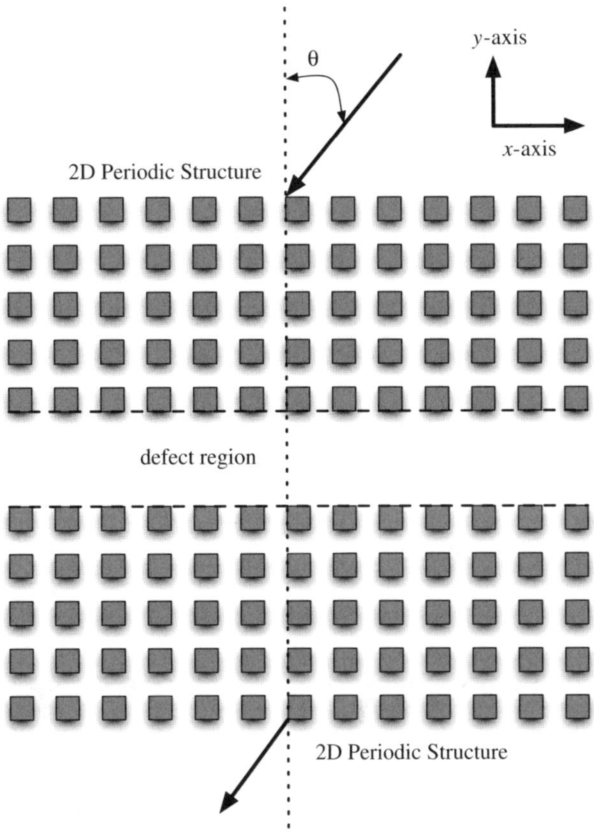

Figure 5.42 Schematic diagram of a 2D periodic structure with defect region

5.4.1 Fabry–Perot Etalon

In optics, a Fabry–Perot etalon or interferometer is typically made of a transparent plate with two parallel highly reflecting surfaces. Its transmission spectrum is a function of wavelength, exhibiting peaks of considerable transmission, which correspond to the resonance of the etalon. The wavelength-selective transmission is caused by interference between the bouncing rays in between the two reflection mirrors. If the transmitted beams are in phase, constructive interference takes place and a transmission maximum is exhibited, whereas destructive interference occurs (transmission minimum) if the beams are out of phase. The transmission function of the etalon relates to the following factors: reflectivity of the mirrors, distance between two reflecting plates, incident angle, and loss coefficient of the etalon.

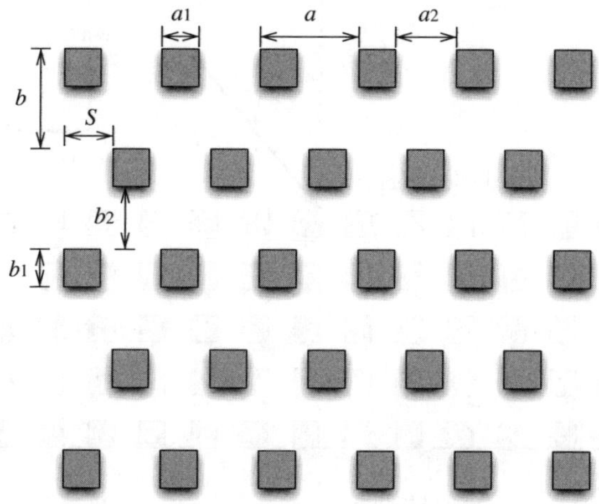

Figure 5.43 Schematic diagram of a 2D periodic structure with an arbitrary lattice pattern determined by a lateral shift distance s

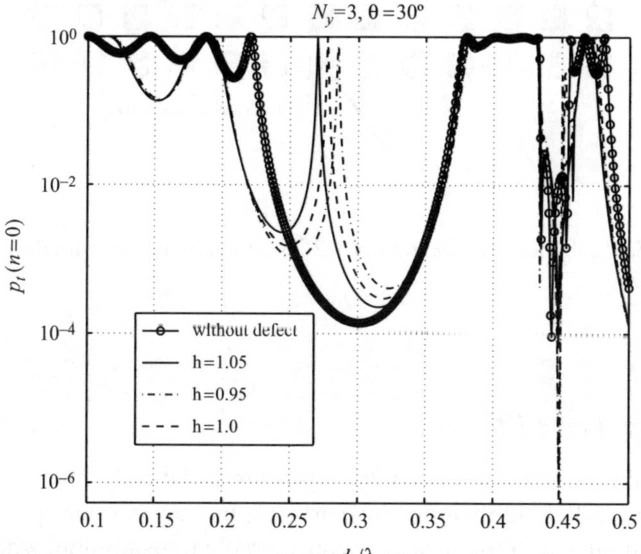

Figure 5.44 Transmittance of the zeroth-order space harmonic, which is defined as the ratio of the zeroth-order transmitted power to the incident power, through the 2D periodic structure with a defect region; the defect widths are 0.95, 1.0, and 1.05, and the case without a defect region ($h = 0.25$); the incident angle is 30°; the number of 1D periodic layers for each waveguide side wall is $N_y = 3$

Conceptually, the structure under consideration can be regarded as a Fabry–Perot etalon with a 2D periodic structure acting as a partially reflecting mirror. The transmission behavior may be roughly estimated by the formula, which can be easily found in a textbook. However, as far as a periodic structure is concerned, an infinite number of space harmonics must be taken into account for the plane-wave excitation. The approach based on the ray tracing is not easy to implement.

5.4.2 *The Correlation between the Scattering and Guiding Characteristics*

In consideration of the disadvantage of using the Fabry–Perot resonance to explain the behavior shown in Figure 5.44, I will try to develop a rigorous and concise theory, which is on the basis of guided-wave theory described in the previous section, for exploring the physical picture of the wave process involved in the 2D periodic structure composed of defects. Before tackling the problem, the effect of the incident angle on the wavelength-selective transmission is shown through the following figures.

Figures 5.45–5.47 show the transmittance of $n = 0$ against the normalized frequency for three different incident angles. The structure parameters and incident

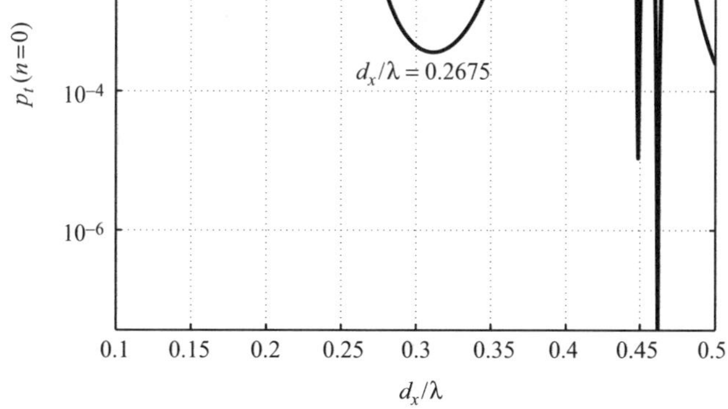

Figure 5.45 Transmittance of the zeroth-order space harmonic through a 2D periodic structure with defect widths $w = 2.0$; the number of 1D periodic layers of the waveguide side wall is $N_y = 3$; incident angle $\theta = 23.58^\circ$

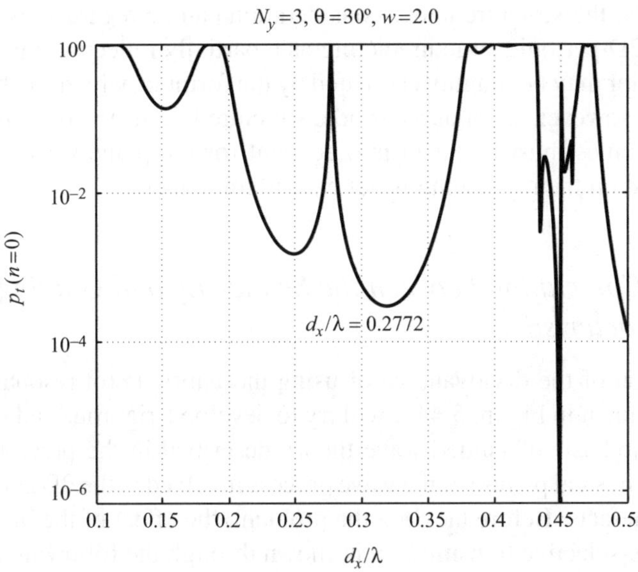

Figure 5.46 Transmittance of the zeroth-order space harmonic through a 2D periodic structure with defect widths $w = 2.0$; the number of 1D periodic layers of the waveguide side wall is $N_y = 3$; incident angle $\theta = 30°$

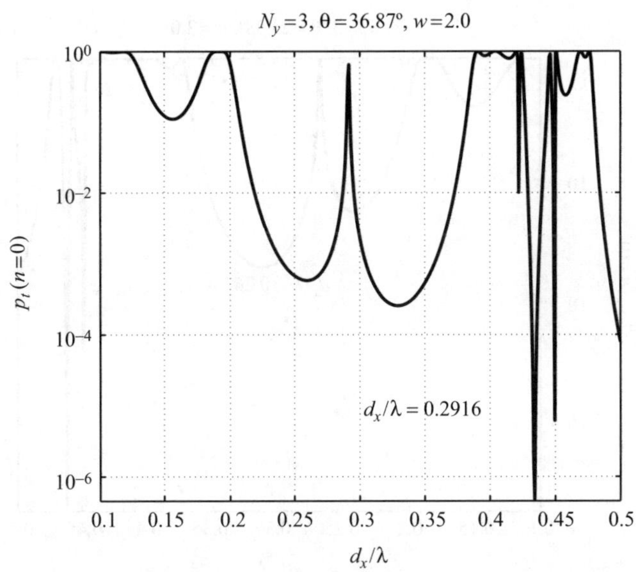

Figure 5.47 Transmittance of the zeroth-order space harmonic through a 2D periodic structure with defect widths $w = 2.0$; the number of 1D periodic layers for each waveguide side wall is $N_y = 3$; incident angle $\theta = 36.87°$

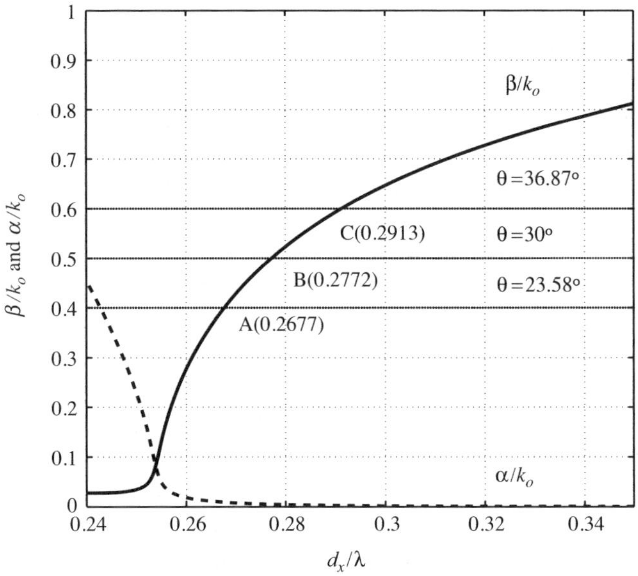

Figure 5.48 Dispersion diagram of the OCB of the structure shown in Figure 5.22; the defect widths are $h = 1.0$; the number of 1D periodic layers of the waveguide side wall is $N_y = 3$; the duty cycle is 50% along both the x- and y-axes; $d_x = d_y = 1$; $\varepsilon_s = 8.9$

condition are described in each caption. It is apparent that the normalized frequency of the resonant peak changes as the incident angle varies. For ease of reference, the peak frequencies are denoted in each figure.

The dispersion diagram of the OCB of a periodic structure with a defect region (see Figure 5.42) is shown in Figure 5.48. The solid and dashed curves are the phase and attenuation constants, respectively. Below the interception point (defined as the cutoff condition) between the above two curves, the evanescent wave has a considerable attenuation constant, while the attenuation constant is decreasing once the operation frequency is above that. The wave with insignificant attenuation constant and $\beta/k_o < 1$ represents the guided mode, which is termed a *ppwg-like mode*, behaving similar to the TE_1 mode in a ppwg. Since the attenuation constant is nonzero, the decaying wave along the guiding axis enables power leakage into the air, which has been clearly demonstrated previously. Such a waveguide mode can be excited by placing a line source (along the z-axis) in the channel at the position where the TE_1 mode has the maximum tangential electric field. Alternatively, the plane-wave excitation is also an approach for excitation. As is well known from Chapter 4, the plane wave can be coupled into the dielectric waveguide via the grating. By matching the tangential component of the phase constant along the interface direction and that in the grating waveguide – that is, $k_x = \beta_{sw} + n2\pi/d_x$, the plane wave can be coupled into the grating waveguide, where $k_x = k_o \sin\theta_{\text{inc}}$. From the reciprocal theory, the periodic perturbation of the waveguide can convert the guided mode into a leaky-wave radiating into the air.

Different from the scenario mentioned previously, we consider here the coupling of a plane wave in air into the fast wave, *ppwg-like* mode, in the defect channel. Because of the small attenuation constant, the phase-match condition between the incident plane wave and the guided mode certainly can occur; that is, $\beta_x = k_o \sin\theta$, where θ is the incident angle of the plane wave. Consequently, the normalized phase constant $k_x/k_o = \sin\theta$. The dispersion diagram of the channel waveguide is plotted in Figure 5.48 together with that of the incident plane wave. The incident angle determines the value of k_x/k_o, shown in the figure. By the phase-match condition, the intersection between the dispersion curves of the guided-mode and that of the incident plane wave determines the resonant-coupling frequency. The frequencies denoted by A, B, and C respectively correspond to the cases of $\theta = 23.58°$, $30°$, and $36.87°$. This means that, through the phase-match condition, the incident plane wave can couple into the TE_1 of the defect waveguide. Moreover, owing to the nonzero attenuation constant, the leaky-mode can again radiate its energy back to the air. However, the backward beam is partially cancelled with the incident beam (minimum reflection), while the maximum transmission is contributed by the forward radiation beam.

Problem 5.8

Can the phase-match condition between the incident plane wave and the phase constant of the guided mode occur if the attenuation of a *ppwg-like mode* is considerable?

In the previous analysis, the rigorous MMA is exploited for determining the dispersion relation of the waveguide mode; i.e., *ppwg-like* mode. Recalling the tangential electric field pattern in the defect region, which we demonstrated in Section 5.3.2, the pattern exhibits the maximum field strength on the OCB plane. It is almost similar to that of the TE-polarized fundamental mode of the ppwg; that is the reason why this mode is named as the *ppwg-like* mode. In addition to the field pattern, the dispersion diagram has to also be examined. For easy reference, the dispersion diagram of the ideal ppwg, which is composed of two parallel PEC plates separated by a distance w, is written as

$$k_x/k_o = \sqrt{\varepsilon_a - (\lambda/2w)^2} \tag{5.26}$$

This dispersion diagram of the structure under consideration using the MMA together with the ideal ppwg approximation is sketched in Figure 5.49. In this figure, the dashed line and dot-dashed line respectively represent the cases of 2D periodic structure with three and six 1D periodic layers; they coincide with each other after $d_x/\lambda > 0.26$. The curve with a solid line is the dispersion curve of the ideal ppwg.

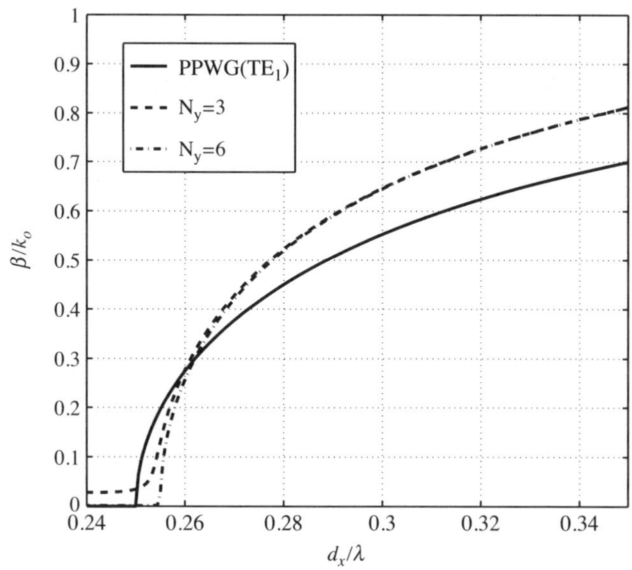

Figure 5.49 Dispersion diagram of the OCB of the structure shown in Figure 5.22 by the rigorous MMA and that based on the ideal ppwg model; the defect widths are $h = 1.0$; the number of 1D periodic layers for the waveguide side wall is $N_y = 3$; the duty cycle is 50% along both the x- and y-axes; $d_x = d_y = 1$; $\varepsilon_s = 8.9$

Although the case of $N_y = 6$ has almost insignificant attenuation constant below the cutoff frequency, it is very different from that of the ideal ppwg with a PEC as its side wall. As a consequence, if the dispersion relation of the ideal ppwg is employed to predict the resonant transmission frequency, there exists an obvious discrepancy. From the above-mentioned results, we find that the reflection property of the stop-band is not the same with that of the PEC plate. The reflection plane is not exactly on the outer surface of the 2D periodic structure, producing a lateral shift when total internal reflection takes place (Goos–Hanchen effect).

Problem 5.9

Calculate the Goos–Hanchen shift distance along the lateral direction in the waveguide.

The quality factor of the channel waveguide depicted in this section was investigated and reported by Hwang and Hsiao (2006). Additionally, the electric and magnetic field patterns together with the Poynting-vector distribution in the whole structure, including the waveguide channel and the 2D periodic structures, were presented in Hwang and Peng (2003).

Regarding the *surface-state* described in the asymmetric structure consisting of semi-infinite 1D periodic layers, this can also be observed in the structure shown in this section that consists of a 2D periodic structure of finite thickness. The *volume wave* concentrated in the 1D periodic layers and the *surface state* supported at the interface between the uniform and 2D periodic layers were reported by Hwang (2007a).

5.5 A Parallel-Plane Waveguide with 2D Metallic Periodic Structures as its Side Walls

The ppwg, which consists of a uniform medium sandwiched by two parallel metallic plates, is commonly used in microwave engineering since it is easy to manufacture and to predict its dispersion characteristics of guided modes. The 2D dielectric periodic structures employed to serve as the side wall have been discussed systematically, particularly with regard to the dispersion characteristics of the guided modes (Hwang 2007b). Recently, a similar structure but using a metallic rods array as the waveguide side walls (see Figure 5.50) was extensively studied and has found potential applications in microwave and millimeter- wave engineering. Such a structure is termed a *substrate integrated waveguide*. The structure contains a dielectric substrate sandwiched by two parallel metal plates; the metallic rods array electrically connects the two metal plates. The region within the metallic rods arrays is the channel where the waveguide mode is guided. Therefore, the waveguide is a *quasi*-closed structure

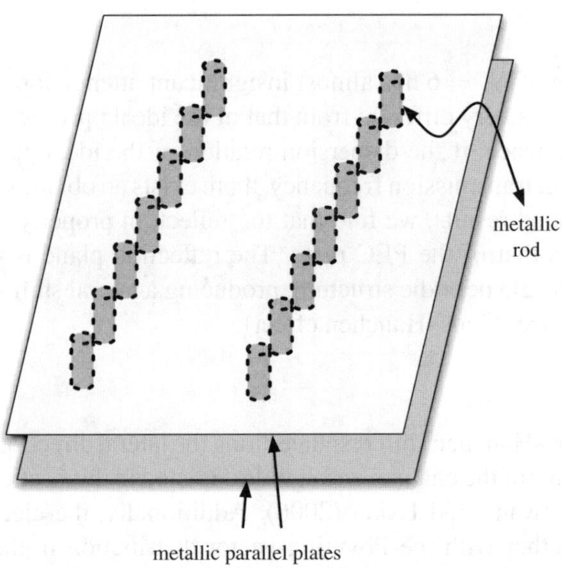

Figure 5.50 Schematic diagram of a quasi-rectangular waveguide with a metallic rods array as its side wall (also termed a substrate integrated waveguide); the region sandwiched by the two metal plates is a dielectric layer

bounded by the parallel plates and the two metallic rods arrays. Additionally, the structure is usually implemented by the printed-circuit board process; the via (through)-hole is made conductive by electroplating to replace the metal rod. Some researchers claim that it can retain the advantages of the traditionally used rectangular metallic waveguide. Specifically, the compact size and high level of integration with passive and active components make it suitable as a candidate for planar transmission lines. In addition to the traditionally used structure, a novel structure equipped with additional slits outside the via-hole arrays was reported to prevent the power leakage (Hwang and Chin 2009).

Because the thickness of the dielectric layer between the parallel metal plates is very thin (much smaller than the channel width), the electric field strength of the fundamental mode is assumed to be invariant along the thickness direction (the direction perpendicular to the metal plate). Therefore, the problem can be simplified as a 2D problem, as shown in Figure 5.51. The metal rod is infinite in extent along the length direction, the z-axis, and the field has no variation along that direction. In doing so, the TE and TM polarizations of the waveguide mode can be individually handled. The background medium is homogeneous dielectric material of infinite extent. As for the

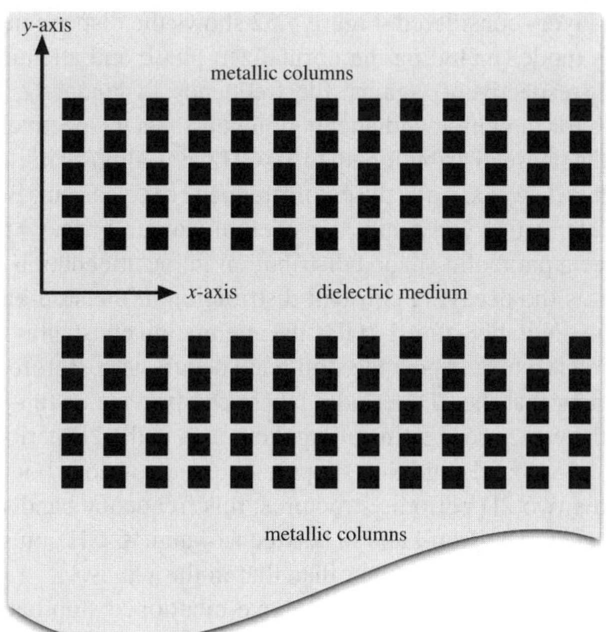

Figure 5.51 The 2D model of the waveguide structure shown in Figure 5.50

mathematical formulation, as usual, the number of periods along the x-axis is assumed to be infinity, while the unit-cell number along the y-axis is a finite number. Moreover, the mathematical procedures for analyzing the dispersion relation is almost similar to the structure with 2D dielectric periodic structures. The only difference is at the input–output relation of the 1D periodic layer. Regarding the mathematical formulation, the reader can go through the details in Section 4.4.1.

Example 5.2

Consider the 2D model of the waveguide in Figure 5.51 with square rods of width of 1.0 mm and the pitch between two successive rods (along the x-axis) of 5.0 mm. The channel width, which is defined as the separation distance between two 2D periodic structures, is 9.15 mm. The distance between two adjacent 1D periodic layers (along the x-direction) is 1.0 mm. The relative dielectric constant of the background dielectric medium is $\varepsilon_s = 3.55$.

Solution

By solving the transverse resonance equation, the dispersion diagram of the fundamental mode (E_z polarization) is determined for the given structure parameters and 10 1D periodic layers considered. Figure 5.52 shows the dispersion relation of the fundamental E_z mode (including the normalized phase and attenuation constants β/k_o and α/k_o, respectively) against the frequency in gigahertz. The solid and dashed lines are for the phase and attenuation constants, respectively. Apparently, if we focus on the phase distribution up to 14 GHz, it is almost similar to that of the ideal ppwg with PEC walls, since the cutoff frequency is clear-cut. Below the cutoff frequency, the attenuation constant is considerably larger. From 14 to 16 GHz, it is interesting to see a parabolic-shaped distribution in the attenuation constant. Such a bump, in fact, is the decaying constant distribution in the stop-band region of a periodic structure. Notably, after 14 GHz the attenuation constant is no longer zero. The next question is why and how the stop-band occurs here. Before answering this question, we know that the 2D periodic layers can provide a stop-band along the y-axis; that is, the wave propagating along the y-axis in the 2D periodic structure is evanescent. The stop-band condition supports the guided-mode bouncing back and forth between the two 2D periodic structures; this frequency band shall start from the cutoff frequency. The stop-band in between 14 and 16 GHz must be caused by the periodicity along the x-axis rather than that in the y-axis.

To explore the physical insight of the above-mentioned stop-band, Figure 5.52 is redrawn in the format of a Brillouin diagram in Figure 5.53. The horizontal and vertical axes represent the normalized phase constant along the x-axis and the free-space wavenumber; the normalization is obtained by multiplying those by $d_x/2\pi$.

The two dashed lines are the light lines with wavenumber $k = k_0\sqrt{\varepsilon_s}$, which are the boundary between the fast- and bound-wave regions. From this figure, we know that the region bound by the two light lines is the bound-wave triangle. Outside the triangle is the fast-wave region where the leaky wave exists. The dispersion curve of the waveguide mode belongs to the fast-wave; however, owing to the insignificant attenuation constant, it still is a guided wave.[1] As is well known, the stop-band satisfies the *Bragg condition* with $\beta_x d_x/2\pi = 0.5$. The stop-band is caused by the contraflow interaction between the *ppwg-like* TE_1 mode and its -1th-order space harmonic. Within the stop-band region, the forward propagation mode transfers its power into the backward-propagating one (-1th-order space harmonic), becoming the reflection wave. For frequencies above 16 GHz, the attenuation constant does not vanish because the 2D periodic structure enters the pass-band region; the power leakage into the surrounding medium is obvious.

Effect of the Number of 1D Metallic Periodic Layers on the Propagation Constant Intuitively, the propagation constant will be changed as the number of 1D periodic layers is altered. In the following example, the normalized phase constant distribution for different numbers of 1D periodic layers is plotted in Figure 5.54. We concentrate on the region around the stop-band condition. From this figure, we find that if the number of 1D periodic layers is small, the leaky constant of the mode is significant; the contraflow interaction of the leaky mode and its -1th-order space harmonic results in the different phase constants during coupling; otherwise, they share the same phase constant if their leakage is negligible. It is noted that the phase relation of the -1th-order space harmonic is symmetric with respect to the line $\beta_x d_x/2\pi = 0.5$.

Additionally, observing the attenuation constant distribution depicted in Figure 5.55 for the case with 10 1D periodic layers, the stop-band is obvious; this means that the mode experiences strong reflection in the stop-band region; the radiation loss is insignificant. Contrarily, for the case with a thin 2D periodic structure, the mode continuously leaks its power into the surrounding medium; inside the stop-band region, the attenuation is due to reflection as well as radiation loss. As we mentioned in the previous paragraph, above the upper band-edge the 2D periodic structure enters the pass-band region; the mode leaks its energy into the surrounding

[1] Here, because 10 1D metal gratings are used, the leakage is insignificant. For the case with one or two 1D grating layers (with the same pitch), the coupling due to the two radiation space harmonics, ($n = 0$) and ($n = -1$), does *not* result in a band effect. During the coupling, the two oppositely directed space harmonics have equal amplitudes, causing a standing wave. The amplitude of the radiating space harmonic drops drastically because of this coupling, and the radiation power is reduced significantly at theses coupling points accordingly.

medium. This can explain why the increase in the number of 1D periodic layers cannot effectively reduce the attenuation constant, as shown in this figure.

Effect of Pitch Width on the Propagation Constant

Finally, the effect of the pitch width on the phase and attenuation constants is investigated. Intuitively, a decrease in pitch width (the distance between two rods along the x-axis) will reduce the leakage. This can be confirmed from Figure 5.56. Additionally, owing to the period along the x-axis being larger than that in the previous examples, the stop-band moves to higher frequency and beyond the scope of this figure.

5.6 Other Applications in Microwave Engineering

In addition to the waveguide application, a ppwg-like structure composed of one metal plate and a 1D metal strips array coated on a dielectric substrate for replacing the solid metal plate was developed (Hwang and Pu 2007). Such a structure is excited by a line source within the guiding channel and found to be able to radiate a shaped beam

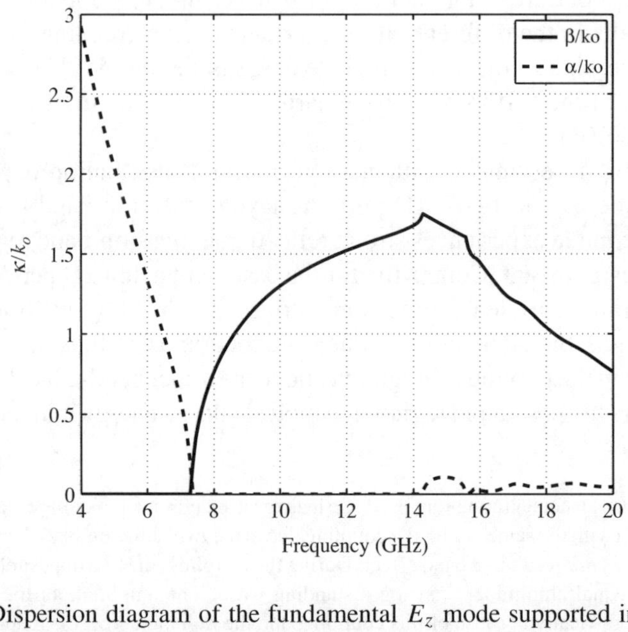

Figure 5.52 Dispersion diagram of the fundamental E_z mode supported in the waveguide structure shown in Figure 5.51; the solid and dashed curves are respectively the normalized phase and attenuation constants. There are 10 1D periodic layers considered

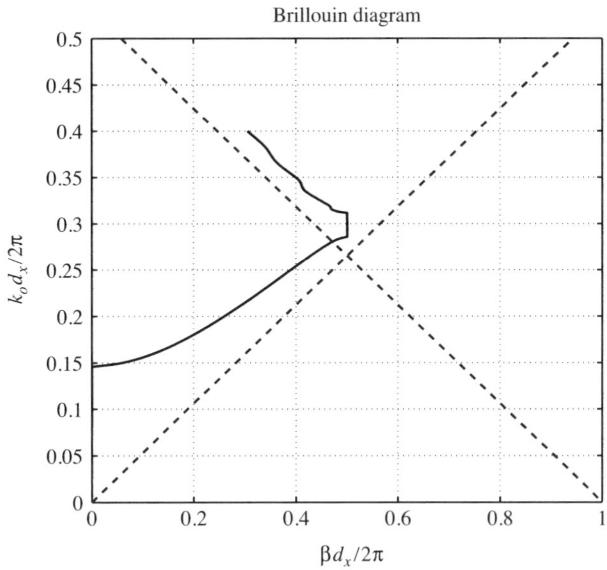

Figure 5.53 Brillouin diagram representation of Figure 5.52

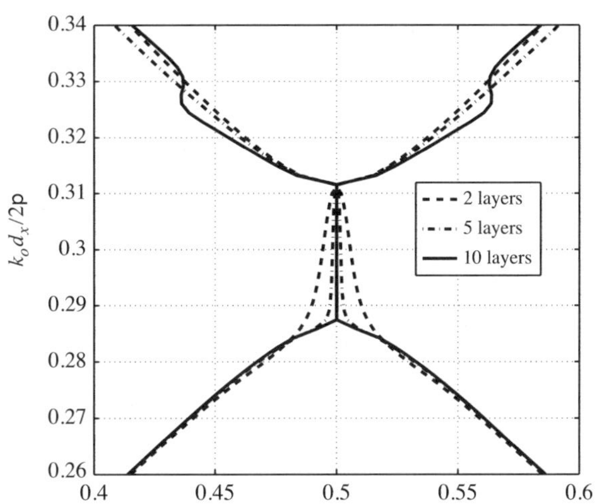

Figure 5.54 Normalized phase constant against the normalized frequency for various numbers of 1D periodic layers plotted in the format of a Brillouin diagram; the coupling effect is caused by the fundamental and -1th-order space harmonic outside the bound-wave triangle. Regarding the case with two 1D periodic layers, the leaky phenomenon is obvious for the two above-mentioned space harmonics. The coupling between them does not result in a band effect depicted in this figure. It is seen from the case with 10 1D periodic layers; the apparent stop-band is produced by the two space harmonics with negligible attenuation constants, which can be clearly seen in Figure 5.55 concerning the attenuation constant distribution

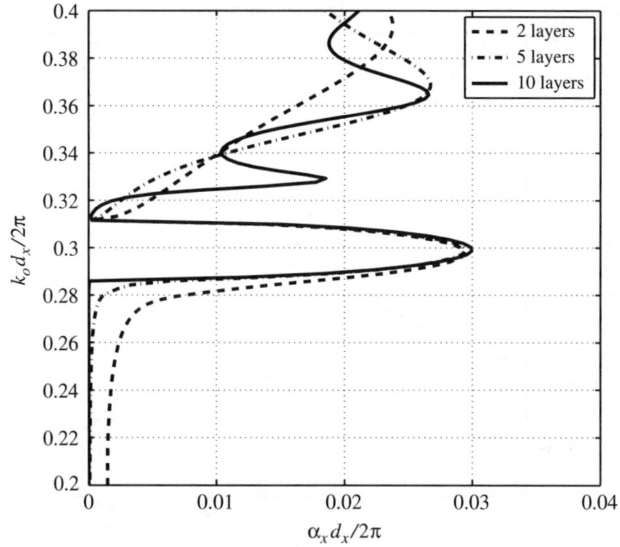

Figure 5.55 Normalized attenuation constant against the normalized frequency for various numbers of 1D periodic layers plotted in the format of a Brillouin diagram. Notably, the case with 10 1D metallic gratings exhibits a insignificant attenuation constant even though it is outside the bound-wave triangle. Therefore, the coupling of the two space harmonics with negligible attenuation constants can produce an *open stop-band*, which is similar to that inside the bound-wave region

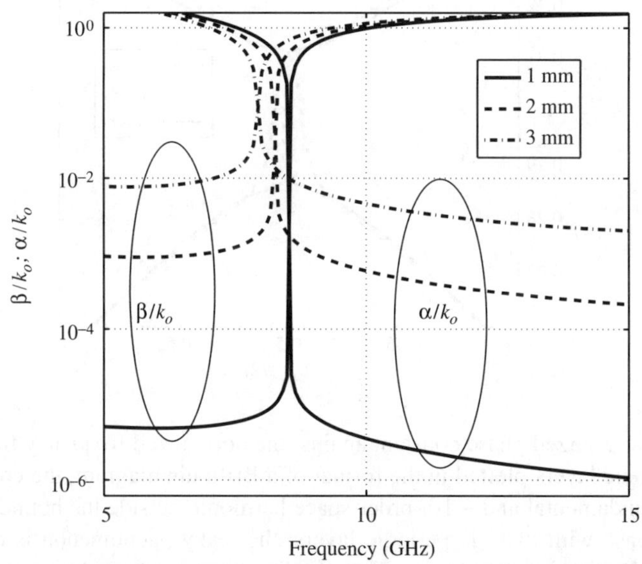

Figure 5.56 Effect of pitch width on the propagation constant of the waveguide mode; the number of 1D periodic layers is one; the other parameters remain the same as in the previous example

caused by the superposition of two leaky waves together with the direct radiation from the line source. It has potential application in indoor hyper-LAN antenna systems for generating a uniform coverage over a plane.

Incidentally, the investigation on the resonant transmission properties of the parallel-plane waveguide consisting of 2D metallic periodic structures was reported by Hwang (2009). The 2D periodic structure is implemented by stacking the 1D periodic layers, which is made of metal strips coated on a foam with relative dielectric constant near unity. The measurement is conducted using microwave horn antennas and a vector network analyzer. The detailed procedures can be found in the reference.

References

Hwang, R. (2007a) Surface waves and volume waves in a photonic crystal slab. *Optics Express* **15**, 7913–7921.

Hwang, R.-B. (2007b) Leaky wave phenomena of a conductor-back dielectric slab covered with a two-dimensionally electromagnetic band gap superstrate. *Radio Science* **42**, RS4010. DOI: 10.1029/2005RS003445.

Hwang, R.-B. (2009) Resonance transmission through electromagnetic crystals consisting of metal strips. *Microwave and Optical Technology Letters* **51**, 1209–1212.

Hwang, R.-B. and Chin, C.-Y. (2009) Substrate integrated waveguide with moats. *Journal of Electromagnetic Waves and Applications* **23**, 1101–1112.

Hwang, R.-B. and Hsiao, C. C. (2006) Frequency-selective transmission by a leaky parallel-plate-like waveguide. *IEEE Transactions on Antennas and Propagation* **54**, 121–129.

Hwang, R.-B. and Peng, S.-T. (2003) Scattering and guiding characteristics of waveguides with two-dimensionally periodic walls of finite thickness. *Radio Science* **38**, 1091. DOI: 10.1029/2002RS002847.

Hwang, R.-B. and Pu, T.-C. (2007) A planar shaped-beam antenna for indoor wireless LAN access points. *IEEE Transactions on Antennas and Propagation* **55**, 1871–1879.

Larsen, R. P. and Oliner, A. A. (1967) A new class of low loss recative wall waveguides. In *G-MTT International Microwave Symposium Digest*, pp. 17–22.

Yeh, P., Yariv, A., and Hong, C.-S. (1977) Electromagnetic propagation in periodic stratified medium. I. General theory. *Journal of the Optical Society of America* **67**, 423–437.

6

Periodic Impedance Surface

In the previous chapters, the rigorous MMA was developed for dealing with the scattering and guiding characteristics of a dielectric waveguide incorporating periodic structures and numerous numerical examples were demonstrated for exploring the unique physical picture of the wave process therein. In addition to the above-mentioned approach, in this chapter we will introduce an alternative approach, the periodic IBC, for analyzing the scattering properties of a plane wave by a periodic structure. As is well known, the IBC was first developed (Leontovich 1948) in the early 1940s for solving the problem of radio waves propagating over the Earth. Leontovich showed that, on the surface of a nearly perfect conductor, the IBC related the tangential electric and magnetic fields via a surface impedance defined by the electromagnetic properties of the scatterer. Because the approximate IBC relates only the electromagnetic fields outside the scatterer, the scattering fields can be assessed without considering the internal fields; therefore, such an approach can further simplify the mathematical analysis. Furthermore, a 1D periodic impedance surface was employed to successfully explain the anomalous absorption of a lossy optical grating (Hessel and Oliner 1965), which is the so-called Wood's anomaly.

In this chapter, the rigorous[1] mathematical formulation for formulating the scattering of a plane wave by 1D scalar periodic impedance surfaces will be carried out. Specifically, the impedance surface approximated by the input impedance surface of the open parallel-plate cavity will be employed to calculate the reflection properties of a composite periodic structure composed of alternating PEC and AMC surfaces. Additionally, the rigorous MMA is also exploited to validate such a periodic impedance model.

[1] The scattering analysis using periodic IBC, in fact, is rigorous as long as the periodic IBC is perfectly correct; that is, the error arises from the error of the approximated periodic IBC.

Periodic Structures: Mode-Matching Approach and Applications in Electromagnetic Engineering, First Edition.
Ruey-Bing (Raybeam) Hwang.

Finally, the mathematical formulation will be extended to be able to rigorously deal with the scattering of a plane wave by a general 2D dyadic periodic impedance surface.

6.1 Scattering Characteristics of Plane Wave by a 1D Periodic Structure Consisting of a Cavities Array

Before introducing the periodic impedance surface approach (PISA) for analyzing the scattering properties of a 1D periodic structure, the rigorous mode-matching analysis for a 1D periodic structure consisting of multiple open cavities in a period will first be performed for determining the *exact* solutions. With the exact solution, we may understand the viability and accuracy of the approximate solutions obtained using the PISA, which will be discussed in the next section.

As shown in Figure 6.1, a uniform plane wave of E_y or H_y modes is obliquely incident on the structure with incident angle θ. Here, we assume that the incident wave is on the x–z plane, while the propagation constant along the y-axis, k_y, vanishes. Actually, each open cavity can be regarded as a ppwg with short-circuit termination at its one end and open at the other end. The number of cavities in a unit cell, which is characterized by d_x, can be an arbitrary number. The surrounding medium outside the periodic structure is assumed to be the air.

Regarding the mode-matching analysis, first of all we partition the structure into several constituent regions where the modal solutions can be determined individually. Here, we have p open cavities consisting of uniform medium. In each open cavity (or ppwg), the tangential electric and magnetic fields of the H_y (or TM) mode are respectively written as

$$E_x^{(i)}(x,z) = \sum_{n=0}^{\infty} V_n^{(i)}(z)\phi_n^{(i)}(x) \tag{6.1}$$

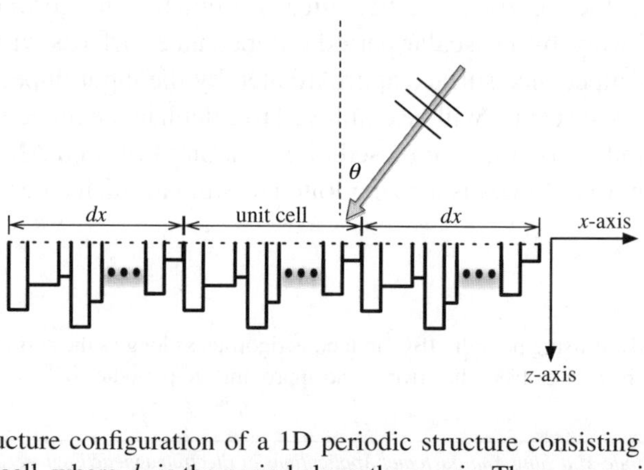

Figure 6.1 Structure configuration of a 1D periodic structure consisting of multiple open cavities in a unit cell, where d_x is the period along the x-axis. There are an infinite number of periods along the x-axis

$$H_y^{(i)}(x,z) = \sum_{n=0}^{\infty} I_n^{(i)}(z)\phi_n^{(i)}(x) \tag{6.2}$$

$$\phi_n^{(i)}(x) = \sqrt{\frac{\gamma_n}{w_i}} \cos\frac{n\pi(x-x_i)}{w_i}; \quad n = 0, 1, 2, \ldots, \infty \tag{6.3}$$

where $\gamma_n = 1.0$ for $n = 0$ and 2.0 otherwise. Parameters x_i and w_i are the starting position and width of the ith cavity.

The tangential electric and magnetic fields in the air region are written as

$$E_x^{\mathrm{a}}(x,z) = \sum_{n=-\infty}^{\infty} V_n(z)\psi_n(x) \tag{6.4}$$

$$H_y^{\mathrm{a}}(x,z) = \sum_{n=-\infty}^{\infty} I_n(z)\psi_n(x) \tag{6.5}$$

$$\psi_n(x) = \frac{1}{\sqrt{d_x}} \exp(-\mathrm{j}k_{xn}x); \quad n = +\infty, \ldots, -\infty \tag{6.6}$$

$$k_{xn} = k_{xo} + n\frac{2\pi}{d_x} \tag{6.7}$$

By the electromagnetic boundary conditions, the tangential electric and magnetic fields must be continuous at the interface (or discontinuity) between the air and open cavities; that is

$$E_x^{\mathrm{a}}(x, z=0^-) = \begin{cases} E_x^{(1)}(x, z=0^+); & x \in \Omega_1 \\ E_x^{(2)}(x, z=0^+); & x \in \Omega_2 \\ \vdots & \\ E_x^{(i)}(x, z=0^+); & x \in \Omega_i \\ \vdots & \\ E_x^{(p)}(x, z=0^+); & x \in \Omega_p \end{cases} \tag{6.8}$$

$$H_y^{\mathrm{a}}(x, z=0^-) = \begin{cases} H_y^{(1)}(x, z=0^+); & x \in \Omega_1 \\ H_y^{(2)}(x, z=0^+); & x \in \Omega_2 \\ \vdots & \\ H_y^{(i)}(x, z=0^+); & x \in \Omega_i \\ \vdots & \\ H_y^{(p)}(x, z=0^+); & x \in \Omega_p \end{cases} \tag{6.9}$$

where Ω_i denotes the aperture on the ith open cavity.

By multiplying the complex conjugate of the modal function in the air region on both sides of Equation (6.8) and taking the integration over the period along the x-axis, we obtain

$$\left\langle \psi_m^*(x) \,\middle|\, \sum_{n=-\infty}^{\infty} V_n(0^-)\psi_n(x) \right\rangle = \sum_{i=1}^{p} \left\langle \psi_m^*(x) \,\middle|\, \sum_{n=0}^{\infty} V_n^i(0^+)\phi_n^{(i)}(x) \right\rangle \tag{6.10}$$

$$\sum_{n=-\infty}^{\infty} V_n(0^-) \langle \psi_m^*(x) \mid \psi_n(x) \rangle = \sum_{i=1}^{p} \sum_{n=0}^{\infty} V_n^{(i)}(0^+) \left\langle \psi_m^*(x) \,\middle|\, \phi_n^{(i)}(x) \right\rangle \tag{6.11}$$

The orthonormal of the modal functions in the air region results in the completeness relation given below:

$$\langle \psi_m^*(x) \mid \psi_n(x) \rangle = \delta_{mn} \tag{6.12}$$

The overlap integral on the right-hand side of the Equation (6.11) is defined as

$$r_{mn}^{(i)} = \langle \psi_m^*(x) \mid \phi_n^{(i)}(x) \rangle \tag{6.13}$$

which can be determined by evaluating the following integration:

$$r_{mn}^{(i)} = \sqrt{\frac{1}{d_x}} \sqrt{\frac{\gamma_n}{w_i}} \int_{x_i}^{x_i + w_i} \exp(\mathrm{j}k_{xm}x) \cos \frac{n\pi(x - x_i)}{w_i} \,\mathrm{d}x \tag{6.14}$$

Thus, Equation (6.11) can be written as

$$V_m(0^-) = \sum_{i=1}^{P} \sum_{n=0}^{\infty} V_n^{(i)}(0^+) r_{mn}^{(i)} \tag{6.15}$$

By collecting the space harmonics from negative to positive infinity together with the waveguide-mode index from zero to infinity, we obtain the system of homogeneous equations, which are expressed in terms of the matrix equation given below:

$$\underline{V}(0^-) = \sum_{i=1}^{P} \mathbf{R}^{(i)} \underline{V}^{(i)}(0^+) \tag{6.16}$$

where $\underline{V}(0^-)$ is a vector with $V_m(0^-)$ at its mth element; $V_n^{(i)}(0^+)$ is the nth element in vector $\underline{V}^{(i)}(0^+)$; matrix $\mathbf{R}^{(i)}$ contains $r_{mn}^{(i)}$ at the mnth element, where the index m runs from negative to positive infinity while the index n ranges from zero to infinity.

The dimensions of the matrix and vector, in fact, are infinity; however, being limited to the finite computational resources available, they have to be truncated into finite dimensions. We consider that the number of space harmonics in the air and the total number of waveguide modes in the unit cell of a cavities array are the same, denoted N; that is

$$N = \sum_{i=1}^{P} N_i \tag{6.17}$$

where N_i is the number of the ppwg in the ith cavity. Equation (6.16) can be rewritten as

$$\underline{V}(0^-) = \mathbf{R}\bar{\underline{V}}(0^+) \tag{6.18}$$

$$\mathbf{R} = \begin{bmatrix} \mathbf{R}^{(1)} & \dots & \mathbf{R}^{(P)} \end{bmatrix} \tag{6.19}$$

$$\bar{\underline{V}}(0^+) = \begin{pmatrix} \underline{V}^{(1)}(0^+) \\ \vdots \\ \underline{V}^{(P)}(0^+) \end{pmatrix} \tag{6.20}$$

The dimension of $\mathbf{R}^{(i)}$ is $N \times N_i$, where i runs from 1 to P. We collect the sub-matrix $\mathbf{R}^{(i)}$ and put this into a super-matrix $\mathbf{R}$ denoted in Equation (6.19). Additionally, the super-vector is composed of the sub-vectors, each of which contains the modal amplitudes in the respective ppwg.

Returning to Equation (6.8), if we multiply both sides by the mth modal function in ith waveguide and perform the integration over the waveguide aperture, we obtain

$$\left\langle \phi_m^{(i)}(x) \,\middle|\, \sum_{n=-\infty}^{\infty} V_n(0^-)\psi_n(x) \right\rangle = \left\langle \phi_m^{(i)}(x) \,\middle|\, \sum_{n=0}^{\infty} V_n^{(i)}(0^+)\phi_n^{(i)}(x) \right\rangle \tag{6.21}$$

By the completeness relation of the ppwg mode, which is written below,

$$\langle \phi_m^{(i)}(x) \,|\, \phi_n^{(i)}(x) \rangle = \delta_{mn} \tag{6.22}$$

the above equation can be written as

$$\sum_{n=-\infty}^{\infty} V_n(0^-)\langle \phi_m^{(i)}(x) \,|\, \psi_n(x) \rangle = V_m^{(i)}(0^+) \tag{6.23}$$

From the parameter defined in Equation (6.13), the inner product in Equation (6.23) can be written as

$$\langle \phi_m^{(i)}(x) \,|\, \psi_n(x) \rangle = [r_{nm}^{(i)}]^*$$

By collecting the index of the waveguide mode and the index of the space harmonic, the system of homogeneous equations represented by a matrix equation can be written as

$$\underline{V}^{(i)}(0^+) = \mathbf{R}_i^{\mathrm{H}}\underline{V}(0^-) \tag{6.24}$$

where $\mathbf{R}_i^{\mathrm{H}}$ represents the Hermitian matrix of $\mathbf{R}_i$.

Thus, after collecting each of the voltage vectors in the respective ppwg and putting into a super-vector, the above equation can be rewritten as:

$$\underline{\bar{V}}(0^+) = \mathbf{R}^{\mathrm{H}} \underline{V}(0^-) \tag{6.25}$$

$$\mathbf{R}^{\mathrm{H}} = \begin{bmatrix} \mathbf{R}_1^{\mathrm{H}} & \ldots & \mathbf{R}_P^{\mathrm{H}} \end{bmatrix}^{\mathrm{T}} \tag{6.26}$$

Here, by truncating the matrix and vector based on the same rules described previously, the matrix $\mathbf{R}$ is a square matrix with dimension $N \times N$. Notably, $\mathbf{R}^{\mathrm{H}}$ represents the Hermitian of the square matrix $\mathbf{R}$.

Now, we have to construct the coupling matrix for the voltage amplitudes at the discontinuity between the cavities array and uniform medium in Equations (6.18) and (6.25). Substituting Equation (6.25) into Equation (6.18), we obtain

$$\mathbf{R}\mathbf{R}^{\mathrm{H}} = \mathbf{I} \tag{6.27}$$

Contrarily, if we substitute Equation (6.18) into Equation (6.25), we have

$$\mathbf{R}^{\mathrm{H}}\mathbf{R} = \mathbf{I} \tag{6.28}$$

The matrix $\mathbf{I}$ is the identity matrix, with the main diagonal terms equal to one and zero otherwise. The above two equations indicate that the matrix $\mathbf{R}$ satisfies the unitary condition.

Repeating the above-mentioned mathematical procedures for matching the tangential magnetic fields over the discontinuity at $z = 0$, shown in Equation (6.9), we obtain the input–output relation of the current waves between the air and waveguide cavities array, which are written below:

$$\underline{I}(0^-) = \mathbf{R}\underline{\bar{I}}(0^+) \tag{6.29}$$

$$\underline{\bar{I}}(0^+) = \mathbf{R}^{H} \underline{I}(0^-) \tag{6.30}$$

In the ith parallel waveguide (or cavity), the voltage and current waves of each waveguide mode satisfy the following relationship:

$$V_n^{(i)}(0^+) = Z_{\mathrm{in},n}^{(i)} I_n^{(i)}(0^+) \tag{6.31}$$

$$Z_{\mathrm{in},n}^{(i)} = \begin{cases} Z_{\mathrm{o},n}^{(i)}; & \text{without termination} \\ \mathrm{j}Z_{\mathrm{o},n}^{(i)} \tan \kappa_n^{(i)} h^{(i)}; & \text{terminated by a short-circuit board} \end{cases} \tag{6.32}$$

After collecting the voltage and current modes in Equation (6.31) and putting inside the voltage and current vectors, we obtain the following equation:

$$\underline{V}^{(i)}(0^+) = \mathbf{Z}_{\mathrm{in}}^{(i)} \underline{I}^{(i)}(0^+) \tag{6.33}$$

where $\mathbf{Z}_{\mathrm{in}}^{(i)}$ is a diagonal matrix with the main diagonal entries given in Equation (6.32). We may again collect each of the voltage and current vectors in each ppwg and put them into the super-vectors shown in Equation (6.20). A super-matrix storing the input impedance matrix of each ppwg is then obtained, which is written as

$$\overline{\underline{V}}(0^+) = \overline{\mathbf{Z}}_{\text{in}}\overline{\underline{I}}(0^+) \tag{6.34}$$

$$\overline{\mathbf{Z}}_{\text{in}} = \text{diag}[\mathbf{Z}_{\text{in}}^{(i)}], \quad i = 1, 2, 3, \ldots, P \tag{6.35}$$

Substituting Equation (6.34) into Equation (6.18) and making use of Equation (6.30), we can obtain the input impedance matrix looking into the cavities array in the air region, at $z = 0^-$:

$$\underline{V}(0^-) = \mathbf{Z}_{\text{in}}^{(a)}\underline{I}(0^-) \tag{6.36}$$

$$\mathbf{Z}_{\text{in}}^{(a)} = \mathbf{R}\overline{\mathbf{Z}}_{\text{in}}\mathbf{R}^{\text{H}} \tag{6.37}$$

The voltage and current amplitudes of each space harmonic in the air region satisfy the transmission-line equations, which are written as

$$v_n(z) = \exp(-\text{j}\kappa_n z)a_n + \exp(+\text{j}\kappa_n z)b_n \tag{6.38}$$

$$z_{a,n}i_n(z) = \exp(-\text{j}\kappa_n z)a_n - \exp(+\text{j}\kappa_n z)b_n \tag{6.39}$$

$$\kappa_n = \sqrt{k_0^2\varepsilon_{\text{a}} - \left(k_{xo} + n\frac{2\pi}{d_x}\right)^2} \tag{6.40}$$

The above two equations can be expressed in terms of the matrix equations by putting each of the voltage and current mode amplitudes into vectors, which are given below:

$$\underline{V}(z) = \exp(-\text{j}\mathbf{k}z)\underline{a} + \exp(+\text{j}\mathbf{k}z)\underline{b} \tag{6.41}$$

$$\underline{I}(z) = \mathbf{Y}_a[\exp(-\text{j}\mathbf{k}z)\underline{a} - \exp(+\text{j}\mathbf{k}z)\underline{b}] \tag{6.42}$$

$$\mathbf{k} = \text{diag}(\kappa_n), \quad n = -\infty, \ldots, +\infty \tag{6.43}$$

$$\mathbf{Y}_a = \text{diag}(Y_{a,n}), \quad n = -\infty, \ldots, +\infty \tag{6.44}$$

$$\underline{a} = (a_{-\infty} \quad \cdots \quad a_{+\infty})^{\text{T}} \tag{6.45}$$

$$\underline{b} = \left(b_{-\infty} \quad \cdots \quad b_{+\infty}\right)^{\text{T}} \tag{6.46}$$

Substituting Equations (6.41) and (6.42) into Equation (6.36), the reflection matrix relating the incident and reflected vectors is written as

$$\underline{b} = \Gamma\underline{a} \tag{6.47}$$

$$\Gamma = [\mathbf{Z}_{\text{in}}^{(a)}\mathbf{Y}_a + \mathbf{I}]^{-1}[\mathbf{Z}_{\text{in}}^{(a)}\mathbf{Y}_a - \mathbf{I}] \tag{6.48}$$

Once the reflection matrix is determined, the reflected-wave amplitude of each space harmonic can be readily obtained by Equation (6.47). Furthermore, the electric and magnetic fields in the air region are determined by substituting each voltage and current space-harmonic amplitude into Equations (6.4) and (6.5).

Once the incident and reflected amplitudes of each space harmonic are determined, we can calculate the reflected power in free space. From the definition of Poynting power, we may determine the total reflected power by evaluating the following equation:

$$P_{\mathrm{r}} = \frac{1}{2}\,\mathrm{Real}\int_0^{d_x} E_x^{(\mathrm{r})}(x,0^-)H_y^{(\mathrm{r})}(x,0^-)^{\dagger}\,\mathrm{d}x \tag{6.49}$$

The reflected tangential electric and magnetic fields in free space at the interface $z = 0^-$ can be decomposed into the superpositions of eigenmodes, which are written as

$$E_x^{(\mathrm{r})}(x,0^-) = \sum_{n=-\infty}^{\infty} b_n(0^-)\psi_n(x) \tag{6.50}$$

$$H_y^{(\mathrm{r})}(x,0^-) = \sum_{n=-\infty}^{\infty} Y_{n,a}b_n(0^-)\psi_n(x) \tag{6.51}$$

Substituting Equations (6.50) and (6.51) into Equation (6.49) together with the completeness relation of the eigenfunction

$$\int_0^{d_x} \psi_m(x)\psi_n^{\dagger}(x)\,\mathrm{d}x = \delta_{m,n} \tag{6.52}$$

the total reflected power is written as

$$P_{\mathrm{r}} = \frac{1}{2}\,\mathrm{Real}\left\{\sum_n Y_{n,a}^{\dagger}|b_n(0^-)|^2\right\} \tag{6.53}$$

Problem 6.1

Derive the input–output relations, which are shown in Equations (6.29) and (6.30), of the current amplitudes between the air and waveguide cavities array.

Problem 6.2

Prove Equation (6.53) using the completeness relation in Equation (6.52) together with Equations (6.50) and (6.51).

From the above equation, we know that due to the completeness relation in Equation (6.52) the total reflected power equals to the sum of each space-harmonic power. Additionally, the reflected power of each space-harmonic power can be calculated individually. The reflection angle of the nth space harmonic can be determined by $\theta_n = \sin^{-1}(k_{x,n}/k_{\mathrm{o}})$.

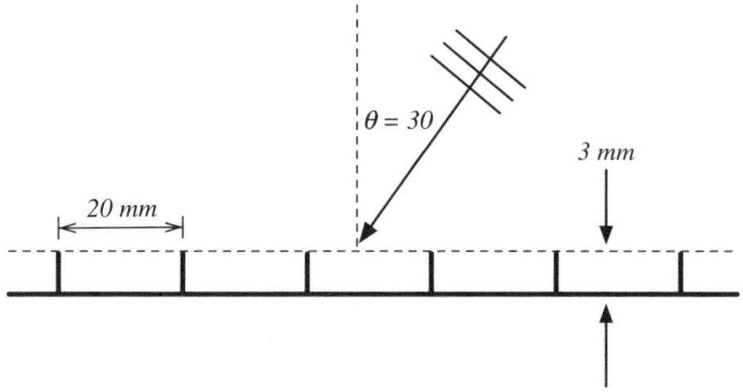

Figure 6.2 Structure configuration of a 1D corrugated metal surface, where the period along the x-axis is 20 mm and the length of corrugation is 3 mm. The thickness of the metallic corrugation is assumed to be zero. The unit cell can be regarded as the ppwg with one end terminated by a metal plate and open at the other end. The periodicity is along the x- axis. The incident angle is $\theta = 30^\circ$, which is counted from the z-axis along the teeth direction

Example 6.1

As shown in Figure 6.2, we first consider a 1D periodic structure made up of a ppwg cavity with width 20 mm and short-circuit at a depth 3 mm. The incident angle is $\theta = 30^\circ$ counted from the z-axis. Figure out the scattering characteristics of the incident plane.

Solution

We follow the mathematical procedures demonstrated previously for calculating the scattering characteristic. Before doing that, we have to check the continuity of tangential electric and magnetic fields at the interface between the 1D corrugated metal surface and the air. The distributions of tangential electric and magnetic fields are shown in Figures 6.3, 6.4, and 6.5, for 3 GHz, 7.49 GHz, and 10.2 GHz, respectively. The fields in the air and cavity sides are individually drawn with dashed and solid lines.

Looking at Figures 6.3–6.5, the continuity of the tangential magnetic fields, in general, is good except for the tangential electric field near the cavity edges. Owing to the edge condition, the tangential electric field is increasing dramatically near the edges, exhibiting a discontinuity in the tangential electric-field distribution, as depicted in the three figures. For such a discontinuity, the partial sum of the Fourier series has a large oscillation near the jump, which is well known as the Gibbs phenomenon.

Additionally, the electric and magnetic fields inside the 1D cavity array in Figures 6.3 and 6.4 are symmetric with respect to the center of the cavity along the x-axis; however, they are totally different in Figure 6.5. Returning to the modal function in the ppwg in Equation (6.3), the fundamental mode TM_0 has no cutoff frequency, while the cutoff wavelength of the nth mode is written as

$$\lambda_c^n = \frac{2w}{n} \tag{6.54}$$

where w is the width of the ppwg.

From the above equation, the cutoff frequency of the first higher-order mode is 7.5 GHz. Thus, only the fundamental mode is excited when the operational frequency is below the cutoff frequency of TM_1. It is the reason why the electric- and magnetic- field distributions are symmetric with respect to the center of the waveguide cavity, $x = 0.5w$. On the other hand, the first higher order mode, which is an anti-symmetry mode, in the waveguide cavity is above cutoff at 10.2 GHz; therefore, the field distributions in Figure 6.5 are not symmetric.

In addition to the field distributions shown previously, we also calculate the surface impedance, defined as the ratio of $E_x(x, z = 0)$ to $H_y(x, z = 0)$ at the interface, in the air and ppwg regions, respectively. The mathematical expressions of the impedances are written as

$$Z_s^{(p)}(x, 0^+) = \frac{\sum_{n=0}^{\infty} V_n^{(p)}(0^+)\phi_n^{(p)}(x)}{\sum_{n=0}^{\infty} I_n^{(p)}(0^+)\phi_n^{(p)}(x)} \tag{6.55}$$

and

$$Z_s^{(a)}(x, 0^+) = \frac{\sum_{n=-\infty}^{\infty} V_n(z)\psi_n(x)}{\sum_{n=-\infty}^{\infty} I_n(z)\psi_n(x)} \tag{6.56}$$

where the superscripts "p" and "a" denote the ppwg and air regions, respectively. The position x belongs the range $[x_i, x_i + w_i]$.

The surface-impedance distributions are shown in Figures 6.6–6.8 corresponding to 3 GHz, 7.49 GHz, and 10.2 GHz, respectively.

As is well known, the input impedance of the fundamental mode TM_0 in a ppwg that is short-circuited at a distance t can be written as

$$Z_{in} = jZ_o \tan k_o t = jZ_o \tan \frac{2\pi t}{\lambda} \tag{6.57}$$

Substituting the structure parameters including the width and depth of the ppwg and the operational frequency into the above equation, the normalized surface impedances (normalized to free-space intrinsic impedance, 120π) are given below.

$$\frac{Z_{in}}{Z_o} = j0.1908, \quad f = 3\,\text{GHz} \tag{6.58}$$

$$\frac{Z_{in}}{Z_o} = j0.5087, \quad f = 7.49\,\text{GHz} \tag{6.59}$$

$$\frac{Z_{\text{in}}}{Z_{\text{o}}} = \text{j}0.7459, \quad f = 10.2\,\text{GHz} \tag{6.60}$$

The normalized surface impedance in Figures 6.6 and 6.7 is individually, on average, close to the result predicted by the above three equations. Regarding the third case, because the first higher order mode is excited, the surface impedance shall include the contribution from the first higher order mode; that is, TM_1. Therefore, the input impedance of the fundamental mode can no longer predict the surface impedance, depicted in Figure 6.8. Moreover, the spatial variation of the surface impedance, in fact, implies the excitation of multi-modes involved in the waveguide.

After verifying the continuity of the tangential electric and magnetic fields over the interface, we are now in a good position to carry out the scattering analysis. As far as the scattering characteristics of a lossless periodic structure are concerned, the power is conserved; that is, the incident power equals the sum of reflected and transmitted power. Since each of the ppwgs is short-circuited at its bottom end and the structure is infinite in extent along the x- and y-axes, the incident wave cannot transmit into the opposite side. Therefore, all the incident power will be returned back to the air. Notably, there are infinite numbers of space harmonics involved in the scattering process; however, most of them are nonpropagating (evanescent) ones and some of them are propagating ones contributing to the real power conservation. Being limited to the finite computational resources, we have to truncate the infinite matrices and vectors into finite ones; however, the chosen space harmonics must cover all the propagating ones and some of the nonpropagating ones to make sure of the continuity of the tangential electric and magnetic fields at the interface. Although not shown here, I have completed the power conservation check and obtained excellent results for ensuring the accuracy of the computer program.

Figure 6.9 demonstrates the scattering characteristics of the 1D corrugated metal surface shown in Figure 6.2. The solid line, dashed line, dashed line with triangle, and solid line with circle represent the reflectances of the -1th-,0th-,$+1$th-, and -2th-order space harmonics, respectively. Apparently, the -2th, -1th, and $+1$th orders have cutoff phenomena and the cutoff frequencies are respectively 10 GHz, 20 GHz, and 30 GHz. Furthermore, the cutoff wavelength of each higher order space harmonic can be figured out by vanishing the z-axis component of the propagation constant in air, which is given below:

$$k_{zn} = \sqrt{k_{\text{o}}^2 - \left(k_{\text{o}} \sin\theta + n\frac{2\pi}{d_x}\right)^2} \tag{6.61}$$

$$\left|k_{\text{o}} \sin\theta + n\frac{2\pi}{d_x}\right| = k_{\text{o}} \tag{6.62}$$

$$\lambda_{\text{c}}^n = d_x \left(\frac{-\sin\theta \pm 1}{n}\right) \tag{6.63}$$

By substituting of period and incident angle into the above equation, we can correctly predict the cutoff frequencies of the three higher order space harmonics given previously.

From the previous analysis, we know that the input impedance of the fundamental mode looking into a ppwg cavity can be regarded as the surface impedance (IBC) of a cavities array with the cavity width much smaller than the operational wavelength. It will become clear later on that such an input impedance can be employed for modeling the IBC for scattering analysis. It is worth noting that such a model is valid before the excitation of the first higher order ppwg mode.

6.1.1 An AMC Surface Made of Corrugated Metal Surface with Quarter-Wavelength Depth

From the previous examples, we know that the IBC over the top of the corrugated metal surface with an electrical small period ($d_x \ll \lambda$) can be approximated by the input impedance of the fundamental mode looking into the ppwg. For the case of TM polarization in a ppwg, the input impedance on the top of the waveguide is

$$Z_{\text{in}} = \text{j}Z_{\text{o}} \tan k_{\text{o}} t \tag{6.64}$$

where t is the depth of the ppwg.

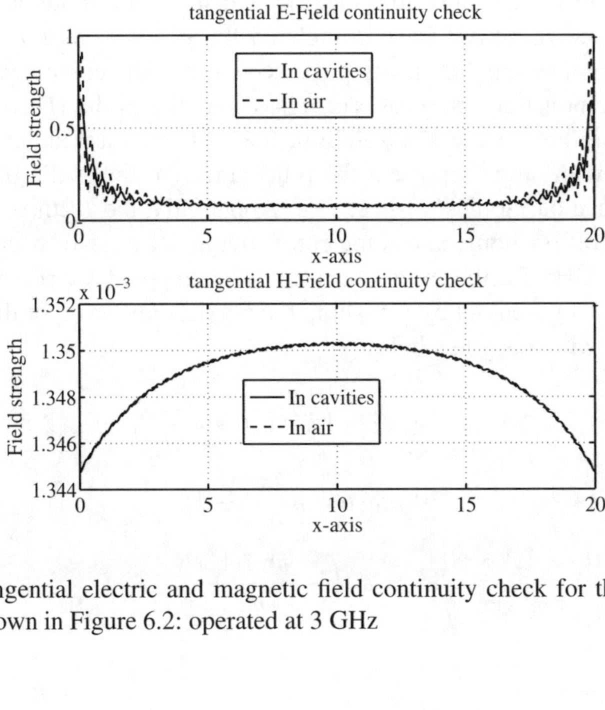

Figure 6.3 Tangential electric and magnetic field continuity check for the 1D corrugated metal surface shown in Figure 6.2: operated at 3 GHz

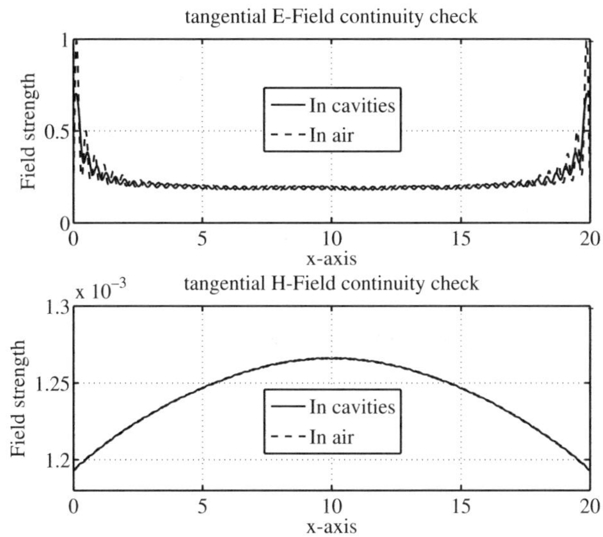

Figure 6.4 Tangential electric and magnetic field continuity check for the 1D corrugated metal surface shown in Figure 6.2: operated at 7.49 GHz

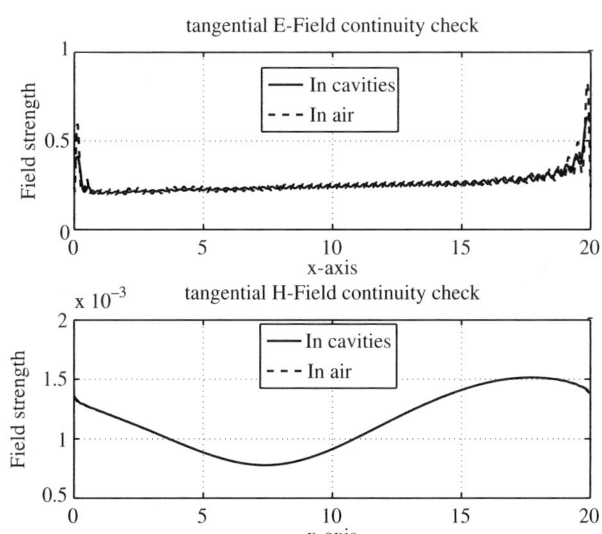

Figure 6.5 Tangential electric and magnetic field continuity check for the 1D corrugated metal surface shown in Figure 6.2: operated at 10.2 GHz

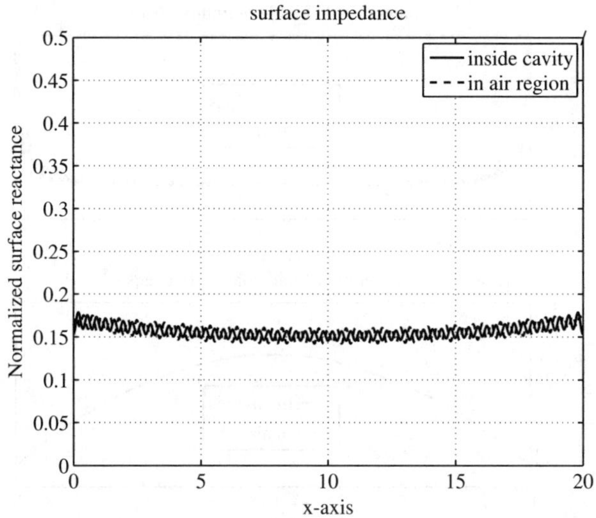

Figure 6.6 Imaginary part of the normalized surface impedance (normalized to 120π) of the 1D corrugated metal surface shown in Figure 6.2: operated at 3 GHz

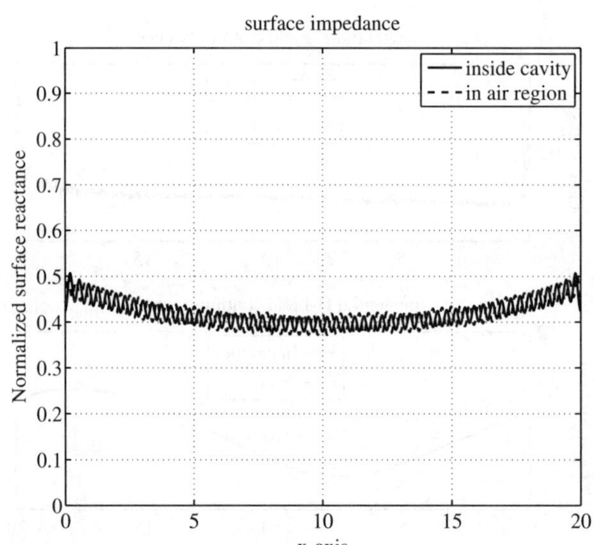

Figure 6.7 Imaginary part of the normalized surface impedance (normalized to 120π) of the 1D corrugated metal surface shown in Figure 6.2: operated at 7.49 GHz

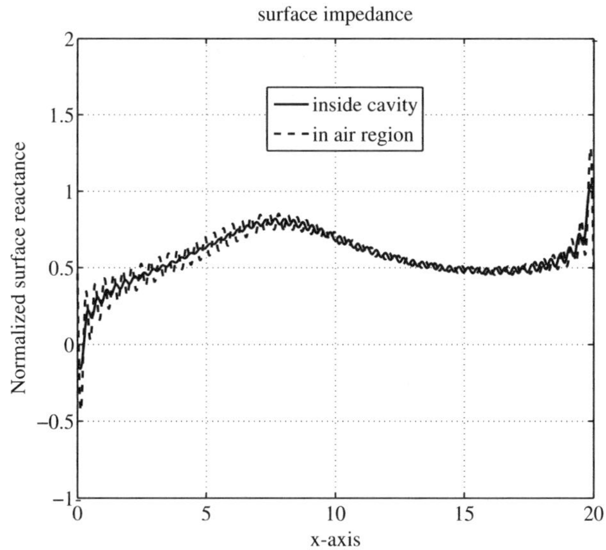

Figure 6.8 Imaginary part of the normalized surface impedance (normalized to 120π) of the 1D corrugated metal surface shown in Figure 6.2: 10.2 GHz

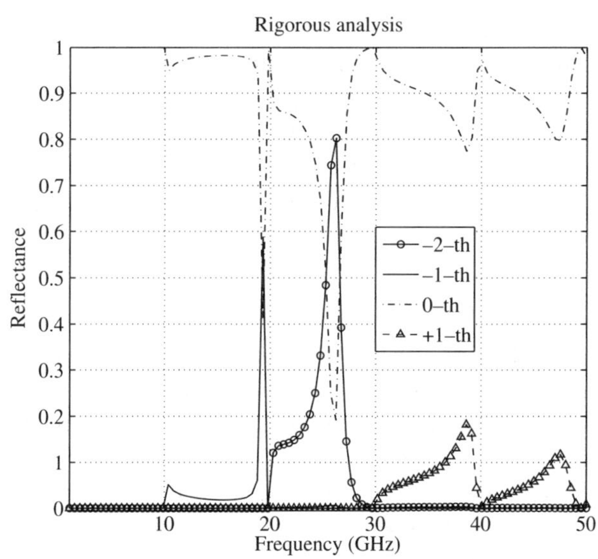

Figure 6.9 Reflectance of the three space harmonics, including the fundamental, -1th, -2th, and $+1$th, of the 1D corrugated metal surface shown in Figure 6.2: incident angle $\theta = 30°$

If a plane wave is normally incident into the corrugated metal surface, the reflection coefficient of the incident plane wave can be written as follows:

$$\Gamma = \frac{Z_{\text{in}} - Z_{\text{o}}}{Z_{\text{in}} + Z_{\text{o}}} = \frac{\text{j} \tan k_{\text{o}}t - 1}{\text{j} \tan k_{\text{o}}t + 1} = \text{e}^{\text{j}\psi} \tag{6.65}$$

The reflection coefficient Γ, in general, is a complex number with its amplitude equal to one. The reflected phase angle can be determined as

$$\psi = \pi - 2k_{\text{o}}t = \pi - 4\pi\frac{t}{\lambda} \tag{6.66}$$

So, if the phase angle equals an odd multiple of π, the reflection coefficient is -1. This implies that the impedance surface is similar to a PEC or SCB condition. On the other hand, if the phase angle equals an even multiple of π, the reflection coefficient is in phase with the incident wave. Such an impedance surface is analogous to a so-called the PMC or OCB condition. Consequently, we may obtain the zero phase angle $\psi = 0$ by setting the corrugation depth equal to a quarter-wavelength, $t = \lambda/4$. Since PMCs do not exist in nature, the above-mentioned method is an option for achieving an AMC. Nevertheless, such an AMC exhibits a high impedance in a narrow band in the vicinity of a quarter-wavelength.

Example 6.2

In this example we consider a 1D periodic structure composed of an alternating PEC and AMC surface; each unit cell contains PEC and AMC surfaces with 50% duty cycle; see Figure 6.10. Specifically, the AMC surface is implemented by a corrugated metal surface with 19 corrugations and a quarter-wavelength in depth. A TM plane wave is normally incident on the surface. The period is 20 mm along the x-axis. The width between two corrugations is 0.5 mm and length of the corrugation is 3 mm. The thickness of the corrugation is negligible in this example.

Solution

Before the elaborate analysis, we have examined the numerical convergence for the reflected power of the fundamental and -1th-order space harmonics against the number of space harmonics that are employed. The variation in the reflected power versus the total number of space harmonics is shown in Figure 6.11. From this figure, we may see that the reflected power converges excellently as the space harmonic number is greater than 40. However, the minimum number of space harmonics are chosen as 200 to obtain an excellent performance in the field continuity.

Figure 6.12 shows the tangential components of the electric (top) and magnetic (bottom) fields at the interface between the air and 1D periodic structure within a unit cell. As was explained in the previous example, the oscillation of the tangential electric field is due to the Gibbs phenomenon occurring at the edge discontinuities. The tangential field components, in general, exhibit excellent agreement. The tangential electric field ranges from $x = 10$ to 20 mm, which locates inside the PEC, and vanishes due to the electromagnetic boundary condition.

Figure 6.13 shows the distribution of the normalized surface impedance (imaginary part) along the x-axis in a period; the impedance surface is a pure imaginary number. From $x = 0$ to 10, this belongs to the region of corrugated metal surface and exhibits a high impedance surface since the length of corrugation (depth of the ppwg with short-circuit termination) is around a quarter-wavelength (operational frequency is 26 GHz). From $x = 10$ to 20, the impedance surface is zero since it is on the metal surface.

Figure 6.14 shows the power spectrum of the reflected space harmonics, including the fundamental, ± 1th, and -2th space harmonics. Because of normal incidence, the ± 1th space harmonics share the same response, as does the ± 2th order. From Equation (6.63), the cutoff frequencies of the $n = \pm 1$ and $n = \pm 2$ are respectively 15 GHz and 30 GHz, which are shown in the figure. Additionally, there are several bands with the near-zero reflected power for the fundamental space harmonic. This indicates that the reflected wave along the incident direction vanishes, while the reflected power is dominated by the higher order space harmonic, such as $n = \pm 1$, which are reflected away from the normal direction ($\theta_{\pm 1} = \pm \sin^{-1} \lambda / d_x \approx \pm 35.23°$). Specifically, around 25 GHz we have a wide bandwidth for minimizing the reflected wave along the normal direction. Such a unique characteristic enables the amazing application of this structure in the radar cross-section (RCS) reduction in microwave and millimeter-wave applications. The same concept, but implementing the AMC surface using the metal patches printed on a grounded dielectric substrate, was published by Hwang and Tsai (2012).

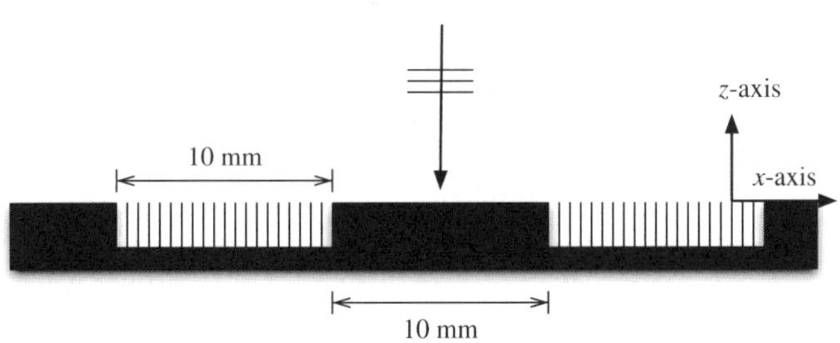

Figure 6.10 A 1D periodic structure consists of alternating PEC and AMC surfaces, where the AMC surface is implemented by a corrugated metal surface. Each unit cell contains a PEC and an AMC surface with 50% duty cycle. The width and depth of the ppwg (between two corrugations) are 0.5 mm and 3 mm, respectively. The period along the x-axis is 20 mm; the width of the AMC surface is 10 mm. The plane wave with TM polarization is normally incident on the structure

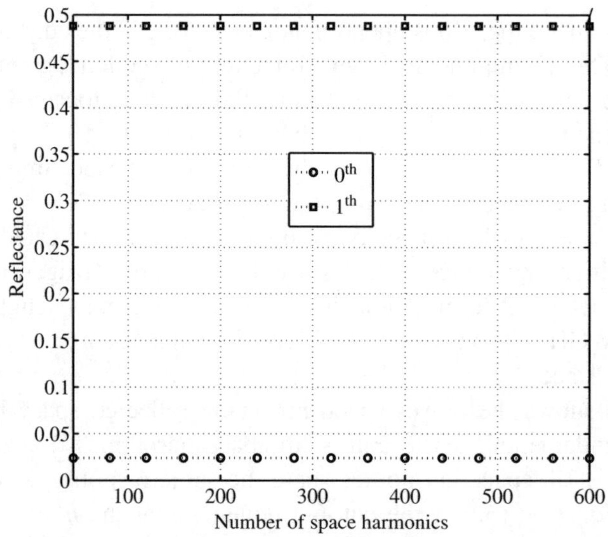

Figure 6.11 Numerical convergence check for the reflected power of the fundamental and −1th-order space harmonics against the number of space harmonics. The structure configuration and parameters are shown in Figure 6.10; the operational frequency is 26 GHz

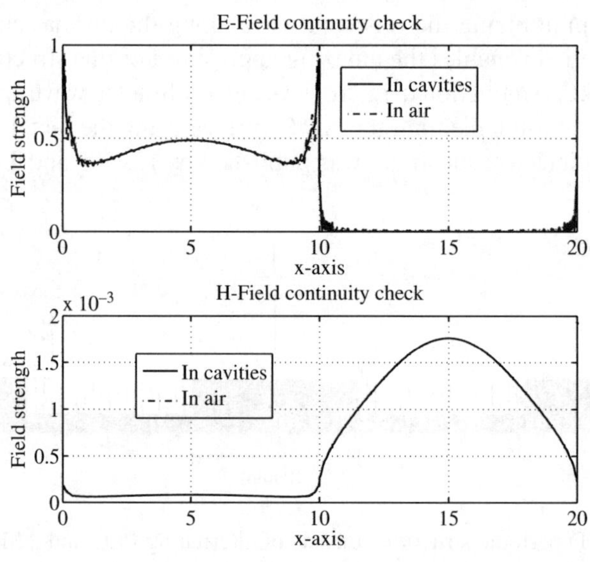

Figure 6.12 Continuity of the tangential components of the electric and magnetic fields at the interface. The structure configuration and parameters are shown in Figure 6.10. The operational frequency is 26 GHz

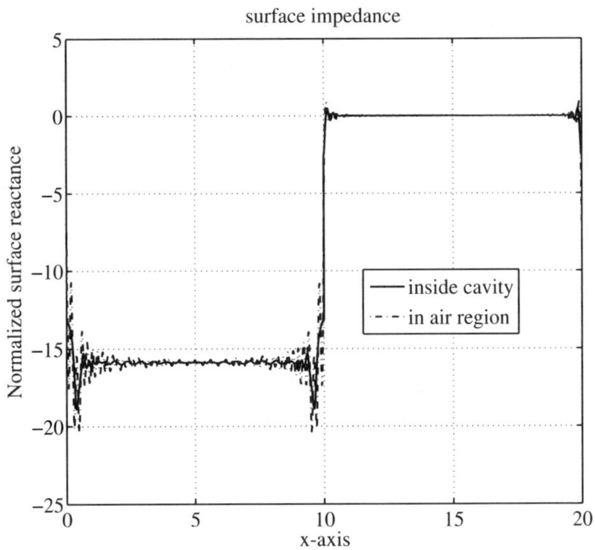

Figure 6.13 Imaginary part of the normalized impedance surface (normalized to 120π at the interface (real part is zero). The structure configuration and parameters are shown in Figure 6.10. The operational frequency is 26 GHz

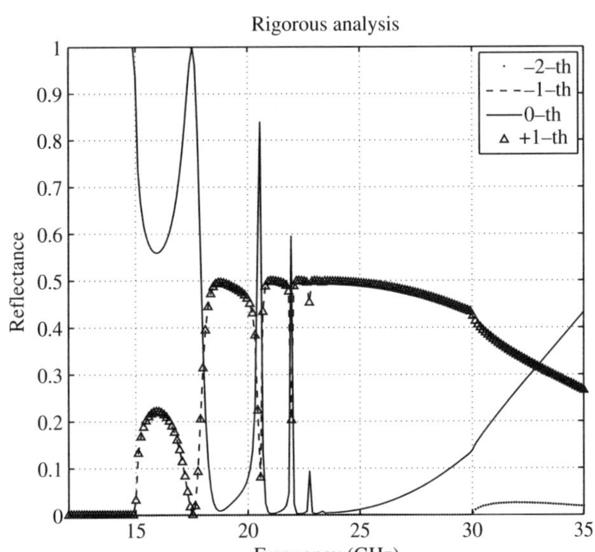

Figure 6.14 Distribution of reflected power for the fundamental, ±1th, and ±2th space harmonics against the operational frequency. The structure configuration and parameters are shown in Figure 6.10. The solid line is for $n = 0$; the dashed line with triangle is for $n = \pm 1$; the dotted line is for $n = \pm 2$

6.2 Periodic Impedance Surface Approach (PISA)

In Section 6.1, the rigorous MMA was developed for carrying out the scattering analysis regarding a 1D periodic structure made of a corrugated metal surface. Significantly, the IBC on the top of the 1D periodic structure composed of a *sub-wavelength* width of ppwg cavities was evaluated and found to be able to be approximated by the input impedance looking into the waveguide cavity. With this approximation, the mathematical analysis can be further simplified by merely solving the electromagnetic fields outside the periodic structure subject to an appropriate IBC; however, it is not necessary to take into account the fields inside the structure.

In this section, I will present the rigorous mathematical procedures for solving the electromagnetic fields in a semi-infinite medium subject to a 1D periodic impedance surface (or boundary condition).

We consider the 1D periodic IBC on the plane at $z = 0$ sketched in Figure 6.15. The medium above the impedance surface is assumed to be the air. A plane wave is impinging on the impedance surface along the principal plane $\phi = 0$ with the elevation angle denoted as θ. If the periodic boundary condition on the plane at $z = 0$ is characterized by $Z_s(x)$, the tangential components of the electric and magnetic field satisfy the following relationship:

$$E_x(x, z = 0) = Z_{\mathrm{s}}^{\mathrm{TM}}(x) H_y(x, z = 0) \tag{6.67}$$

where E_x and H_y are the x-component electric field and y-component magnetic field on the plane at $z = 0$, and $Z_s(x)$ is a periodic function along the x-axis with period d_x. The tangential components of the electric and magnetic fields in the last equation belong to

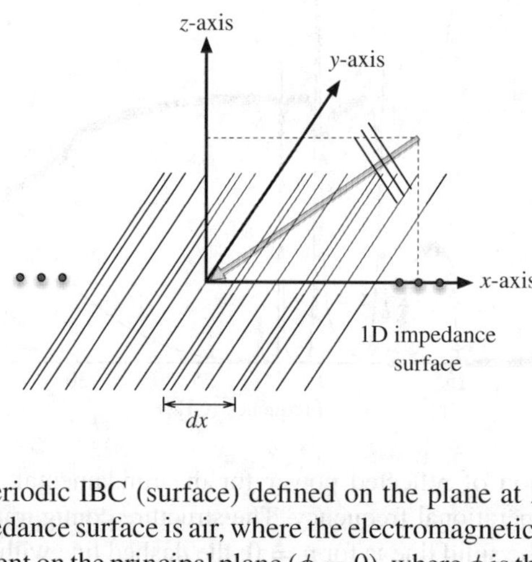

Figure 6.15 A 1D periodic IBC (surface) defined on the plane at $z = 0$. The semi-infinite medium above the impedance surface is air, where the electromagnetic fields are to be resolved. The plane wave is incident on the principal plane ($\phi = 0$), where ϕ is the azimuth angle counted from the x-axis

the TM-polarization wave. On the other hand, the IBC regarding the TE-polarization wave is written as

$$E_y(x, z = 0) = Z_s^{\mathrm{TE}}(x) H_x(x, z = 0) \tag{6.68}$$

Since $Z_s(x)$ is a periodic function, which satisfies $Z_s(x + d_x) = Z_s(x)$, it can be expressed in terms of the Fourier series expansion written as

$$Z_s(x) = \sum_{n=-\infty}^{n=+\infty} z_n \exp\left(\mathrm{j}n\frac{2\pi x}{d_x}\right) \tag{6.69}$$

where z_n is the Fourier coefficient of the nth harmonic.

Let us now consider the TM-polarized waves and the TM periodic impedance surface is considered. Substituting the tangential components of the electric and magnetic fields in Equations (6.4) and (6.5) together with Equation (6.69) into Equation (6.67), we obtain the following equation:

$$\sum_{n=-\infty}^{\infty} V_n(0)\,\mathrm{e}^{-\mathrm{j}k_{x,n}x} = \sum_{p=-\infty}^{p=+\infty} z_p\,\mathrm{e}^{\mathrm{j}p\frac{2\pi x}{d_x}} \cdot \sum_{n=-\infty}^{\infty} I_n(0)\,\mathrm{e}^{-\mathrm{j}k_{x,n}x} \tag{6.70}$$

By combining the two exponential functions on the right-hand side together with the operation $k_{x,n} - 2\pi p/d_x = k_{x,o} + (n - p)2\pi/d_x = k_{x,n-p}$, the above equation becomes

$$\sum_{n=-\infty}^{\infty} V_n(0)\,\mathrm{e}^{-\mathrm{j}k_{x,n}x} = \sum_{p=-\infty}^{p=+\infty}\sum_{n=-\infty}^{\infty} I_n(0) z_p\,\mathrm{e}^{-\mathrm{j}k_{x,n-p}x} \tag{6.71}$$

By replacing the index n by $n + p$, both sides of the above equation have the nth-order basis function in common, which is given below:

$$\sum_{n=-\infty}^{\infty} V_n(0)\,\mathrm{e}^{-\mathrm{j}k_{x,n}x} = \sum_{n=-\infty}^{n=+\infty}\left[\sum_{p=-\infty}^{\infty} z_p I_{n+p}(0)\right]\mathrm{e}^{-\mathrm{j}k_{x,n}x} \tag{6.72}$$

The above equation defines the relationship between the voltage- and current-amplitude of each space harmonic. Moreover, by equating the same Fourier components on both sides, we obtain the following relation:

$$V_n(0) = \sum_{p=-\infty}^{\infty} z_p I_{n+p}(0) \tag{6.73}$$

Notably, the voltage of the nth space harmonic relates to the current of all space harmonics, enabling the occurrence of electromagnetic field coupling among space harmonics.

We first consider a simple case with the periodic impedance surface given below:

$$Z_s(x) = z_o\left(1 + 2\delta\cos\frac{2\pi x}{d_x}\right) \tag{6.74}$$

where δ is the modulation index and z_o is the average surface impedance.

Therefore, Equation (6.73) can be written as

$$V_n(0) = \delta z_o I_{n-1}(0) + z_o I_n(0) + \delta z_o I_{n+1}(0) \tag{6.75}$$

Collecting the index n from positive to negative infinity, we obtain a system of linear equations. After arranging them and putting into a matrix-and-vector form, we get the following matrix equation:

$$\begin{pmatrix} \vdots \\ V_{n=+1}(0) \\ V_{n=0}(0) \\ V_{n=-1}(0) \\ \vdots \end{pmatrix} = \begin{pmatrix} \ddots & \ddots & 0 & 0 & 0 & \ddots \\ \ddots & z_o & \delta z_o & 0 & 0 & 0 \\ 0 & \delta z_o & z_o & \delta z_o & 0 & 0 \\ 0 & 0 & \delta z_o & z_o & \ddots & 0 \\ 0 & 0 & 0 & \ddots & \ddots & \ddots \\ \ddots & 0 & 0 & 0 & \ddots & \ddots \end{pmatrix} \begin{pmatrix} \vdots \\ I_{n=+1}(0) \\ I_{n=0}(0) \\ I_{n=-1}(0) \\ \vdots \end{pmatrix} \tag{6.76}$$

In Equation (6.76), the impedance matrix is a sparse matrix with its main diagonal filling with z_o and the first upper and lower off-diagonal entries filling with δz_0.

With the experience in constructing the impedance matrix, we can extend the sinusoidal variation impedance surface in Equation (6.74) into a more complex case with an arbitrary periodic function, which is able to be expressed in terms of the sum of trigonometric functions shown in Equation (6.69). We know that, for an arbitrary periodic function, the higher order harmonic terms in Equation (6.69), in general, exist. Therefore, the impedance matrix in Equation (6.76) is no longer a sparse matrix; it may be a full matrix. To obtain a succinct presentation of the impedance matrix, a new expression for exhibiting the impedance matrix is given below:

$$\mathbb{Z}_s = \sum_p \text{diag}[z_p\,\text{ones}(1, N-p), -p] \tag{6.77}$$

The expression ones$(1, N - p)$ represents a row vector with its $N - p$ entries filling with unity. Moreover, $\mathbb{Z}_s$ is a square matrix with the elements of $z_p\,$ones$(1, N - p)$ on the $-p$th diagonal. Parameter $p = 0$ is the main diagonal, $p > 0$ means the $|p|$th lower off-diagonal, and $p < 0$ stands for the $|p|$th upper off-diagonal. Once the impedance matrix is constructed, we may further formulate the input–output relation for the voltage and current waves in the semi-infinite uniform medium. The detailed procedures will be described as follows. Once the impedance matrix is constructed and denoted $\mathbb{Z}_s$, Equation (6.76) can be written as

$$\underline{V}(0) = \mathbb{Z}_s \underline{I}(0) \tag{6.78}$$

The voltage and current amplitudes of each space harmonic contain the incident and reflected waves, so that the voltage and current vectors can be decomposed into the sum of the incident and reflected vectors given below:

$$\underline{V}(0) = \underline{a} + \underline{b} \tag{6.79}$$

$$\underline{I}(0) = \mathbf{Y}_a(\underline{a} - \underline{b}) \tag{6.80}$$

where $\mathbf{Y}_a$ is a diagonal matrix with its main entry representing the admittance of each space harmonic, which is $Y_{a,n} = k_{z,n}^{(a)}/\omega\mu_o$. Parameters $\underline{a}$ and $\underline{b}$ are the incident and reflected wave amplitudes in air. Substituting Equations (6.79) and (6.80) into Equation (6.78), the incident and reflected wave vectors can be related by the reflection matrix Γ, which is written as

$$\underline{b} = \Gamma \underline{a} \tag{6.81}$$

$$\Gamma = (\mathbf{I} + \mathbb{Z}_s\mathbf{Y}_a)^{-1}(\mathbb{Z}_s\mathbf{Y}_a - \mathbf{I}) \tag{6.82}$$

where $\mathbf{I}$ is the identity matrix.

Once the incident wave vector is given, the reflected amplitude of each space harmonic can be determined by Equation (6.81). Moreover, the electromagnetic fields in air can also be totally resolved.

Example 6.3

In this example we will revisit a structure similar to the one illustrated in the example in Section 2; however, the PISA will be employed to carry out the scattering analysis. Specifically, the input impedance method mentioned in Section 6.2 will be used for approximating the periodic IBC. Additionally, the *exact* result based on the MMA will also be demonstrated for understanding the feasibility of using such a simple periodic IBC for approximating the complex 1D periodic structure.

Solution

In this example, the widths of the sub-wavelength ppwg are reduced to be 0.25 mm. The duty cycle (defined as the ratio between PEC width and period) and period remain the same, which are 50% and 20 mm. Although not shown here, we have examined the numerical convergence of the reflectance against the number of space harmonics and found that 101 harmonics is enough for obtaining the convergent numerical results (PISA). For obtaining an excellent field continuity over the discontinuity, three evanescent modes in each sub-wavelength waveguide and 240 space harmonics in air are used in the numerical calculation (MMA). From Figures 6.16–6.18, the reflectances of the fundamental and higher order space harmonics against the operational frequency are shown. The solid and dashed lines represent the results obtained using the MMA and PISA, respectively.

In Figure 6.16, the incident angle is near the normal direction with $\theta = 1^\circ$. From Equation (6.63), the cutoff frequency of $(n = -1)$ is 14.7427 GHz. Obviously, the deviation between them is on the relative frequency shift; nevertheless, they almost share the same profile. In the vicinity of the frequencies around 19 GHz, the reflectance of the fundamental space harmonic is decreasing, while the space harmonics $(n = \pm 1)$ reflect most of the incident power.

In addition to the case of near-normal incidence, we also take into account the conditions of oblique incidence, which are shown in Figures 6.17 and 6.18 with incident angles 30° and 60°, respectively. It is apparent that the PISA exhibits excellent fidelity. We may conjecture that the success of the PISA is mainly attributed to the correct estimation of the periodic IBC. For an operation wavelength much larger than that of the waveguide width (sub-wavelength waveguide), only the fundamental waveguide mode is excited and the input impedance looking into the ppwg cavity can indeed correctly approximate the IBC.

Additionally, the reflectances of the fundamental and higher order space harmonics against the incident angle are also calculated using both the MMA and PISA. Figures 6.19, 6.20, and 6.21 show the angular spectrum of reflectance at 15 GHz, 24 GHz and 26 GHz, respectively. They, in general, show a good agreement between the two approaches. Significantly, in Figures 6.20 and 6.21, for an incident angle below around 20° the incident power is reflected into the space harmonics of $(n = -1)$ and $(n = -2)$, while the reflected wave along the incident direction is minimized. For the two frequencies 24 GHz and 26 GHz, the ppwg cavity has a depth close to a quarter-wavelength; therefore, it exhibits a extremely high-impedance surface. From the Huygens principle, for a wave incident near the normal direction, the equivalent radiating sources on PEC and PMC surfaces having the opposite phase angles can cancel out the far field in the vicinity around the normal direction. The results again elucidate that such a class of composite structure consisting of PEC and artificial PMC can reflect the impinging power away from the incident direction, which can serve as an RCS reduction device in microwave and millimeter-wave applications.

6.3 Scattering of Plane Wave by 1D Periodic Impedance Surface: Non-Principal Plane Propagation

In Section 6.2 we compared the scattering properties of a 1D periodic structure composed of an alternating corrugated metal surface and metal plate between two different

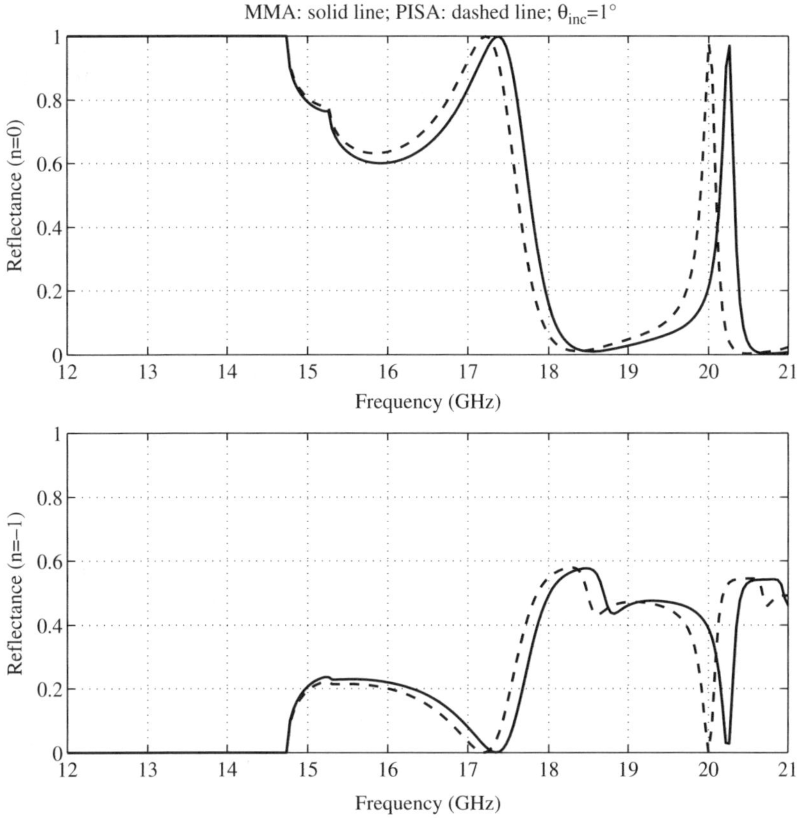

Figure 6.16 Reflectance of the fundamental and −1th-order space harmonics against the frequency (in gigahertz): the 1D periodic structure is shown in Figure 6.10. The width of the sub-wavelength ppwg cavity is 0.25 mm. The duty cycle and period are 50% and 20 mm. The incident angle of the TM-polarized plane wave is 1°. The solid and dashed lines respectively represent the results from the MMA and PISA

approaches: the MMA and PISA. According to the Huygens principle, the field solution in a region V' is completely determined by the tangential fields specified over the surface S' enclosing V'. Therefore, if we are able to determine the exact IBC over an arbitrary 1D periodic structure, the results obtained from the MMA definitely coincide with those obtained from the PISA. Actually, for an arbitrary 1D periodic structure, it is not easy to obtain its exact IBC. Fortunately, for the 1D periodic structure composed of sub-wavelength ppwg cavities, the periodic impedance surface can be appropriately approximated by the input impedance of the fundamental mode in the waveguide. Consequently, the PISA can consistently predict the scattering characteristics within certain error ranges. Significantly, in comparison with the MMA, considerably less computational resources and a relatively short CPU time are needed for the PISA.

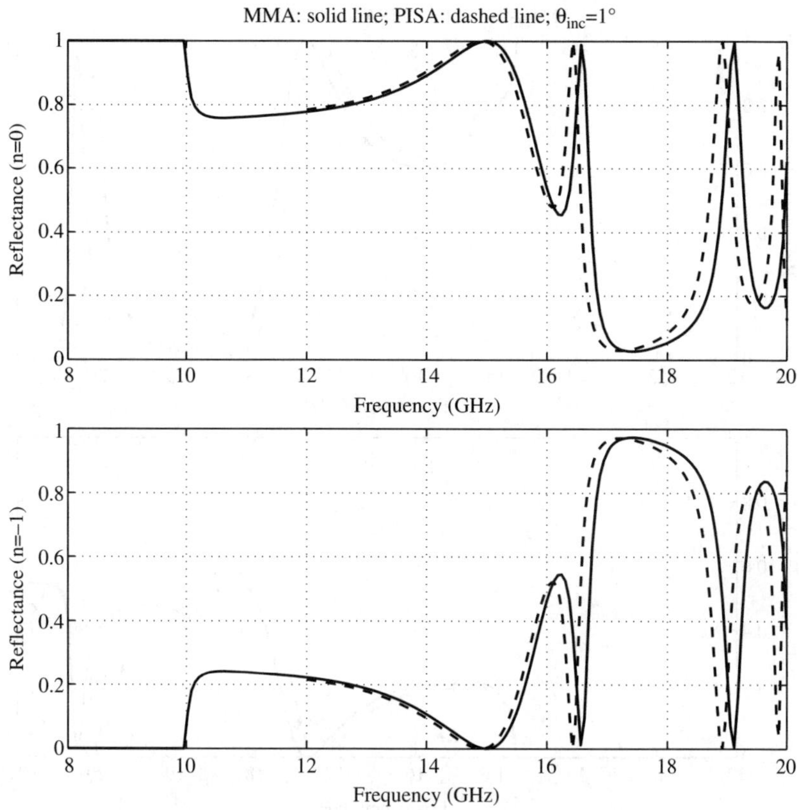

Figure 6.17 Reflectance of the fundamental and −1th-order space harmonics against the frequency (in gigahertz): the 1D periodic structure is shown in Figure 6.10. The width of the sub-wavelength ppwg cavity is 0.25 mm. The duty cycle and period are 50% and 20 mm. The incident angle of the TM-polarized plane wave is 30°. The solid and dashed lines respectively represent the results from the MMA and PISA

In this section we will extend the previous mathematical formulation into one that is able to deal with the condition of non-principal plane incidence ($\phi \neq 0$); see Figure 6.22. Under this situation, the individual polarization can no longer exist ($k_y \neq 0$) and both the TE and TM modes have to be taken into account simultaneously. The IBC relating the tangential electric and magnetic fields over the boundary in Equations (6.67) and (6.68) is modified as

$$\underline{z}_o \times \underline{E}_t(x, y, z = 0) = Z_s(x)\underline{H}_t(x, y, z = 0) \tag{6.83}$$

where $Z_s(x)$ is the periodic impedance surface that is varying periodically along the x-axis and is invariant along the y-axis. The subscript "t" attached to the electric and

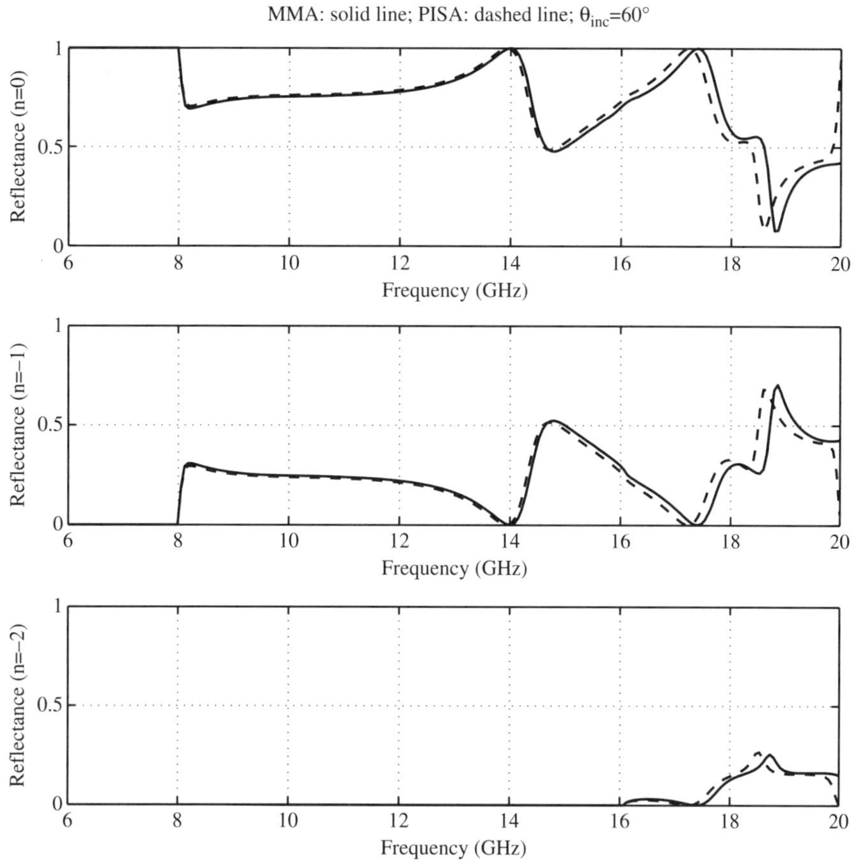

Figure 6.18 Reflectance of the fundamental, −1th- and −2th-order space harmonics against the frequency (in gigahertz): the 1D periodic structure is shown in Figure 6.10. The width of the sub-wavelength ppwg cavity is 0.25 mm. The duty cycle and period are 50% and 20 mm. The incident angle of the TM-polarized plane wave is 60°. The solid and dashed lines respectively represent the results from the MMA and PISA

magnetic field vectors denotes the tangential component. In the presence of periodic nature, the tangential electric and magnetic fields in free space can be written as

$$\underline{z}_o \times \underline{E}_t(x, y, z = 0) = \sum_n [\underline{a}'_n v'_n(0) + \underline{a}''_n v''_n(0)]\, \mathrm{e}^{-\mathrm{j}\underline{k}_{t,n}\underline{\rho}} \tag{6.84}$$

$$\underline{H}_t(x, y, z = 0) = \sum_n [\underline{a}'_n i'_n(0) + \underline{a}''_n i''_n(0)]\, \mathrm{e}^{-\mathrm{j}\underline{k}_{t,n}\underline{\rho}} \tag{6.85}$$

where $\underline{k}_{t,n} = \underline{x}_0(k_{x,0} + n2\pi/d_x) + \underline{y}_0 k_{y,0}$ and $\underline{\rho} = \underline{x}_0 x + \underline{y}_0 y$.

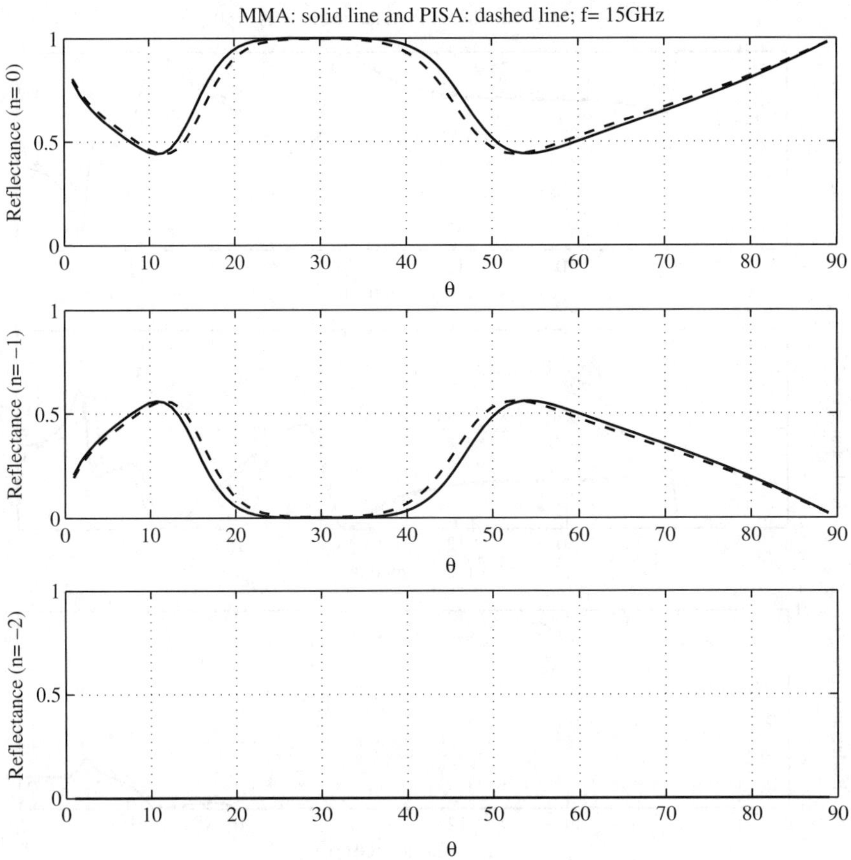

Figure 6.19 Reflectance of the fundamental, −1th-order, and −2th-order space harmonics against the incident angle from the 1D periodic structure shown in Figure 6.10. The solid and dashed lines respectively represent the results from the MMA and PISA. The operation frequency is 15 GHz. The −2th-order space harmonic is below cutoff at this frequency

Substitution of Equations (6.84) and (6.85) into Equation (6.83) by invoking the Fourier series expansion of the periodic function $Z_s(x)$ gives

$$Z_s(x) = \sum_p z_p \exp\left(\mathrm{j}\frac{p2\pi x}{d_x}\right) \tag{6.86}$$

We may obtain the relationship between the voltage and current waves written as

$$\sum_n [\underline{a}'_n v'_n(0) + \underline{a}''_n v''_n(0)]\,\mathrm{e}^{-\mathrm{j}\underline{k}_{t,n}\underline{\rho}} = \sum_p z_p\,\mathrm{e}^{\mathrm{j}\frac{p2\pi x}{d_x}} \sum_n [\underline{a}'_n i'_n(0) + \underline{a}''_n i''_n(0)]\,\mathrm{e}^{-\mathrm{j}\underline{k}_{t,n}\underline{\rho}} \tag{6.87}$$

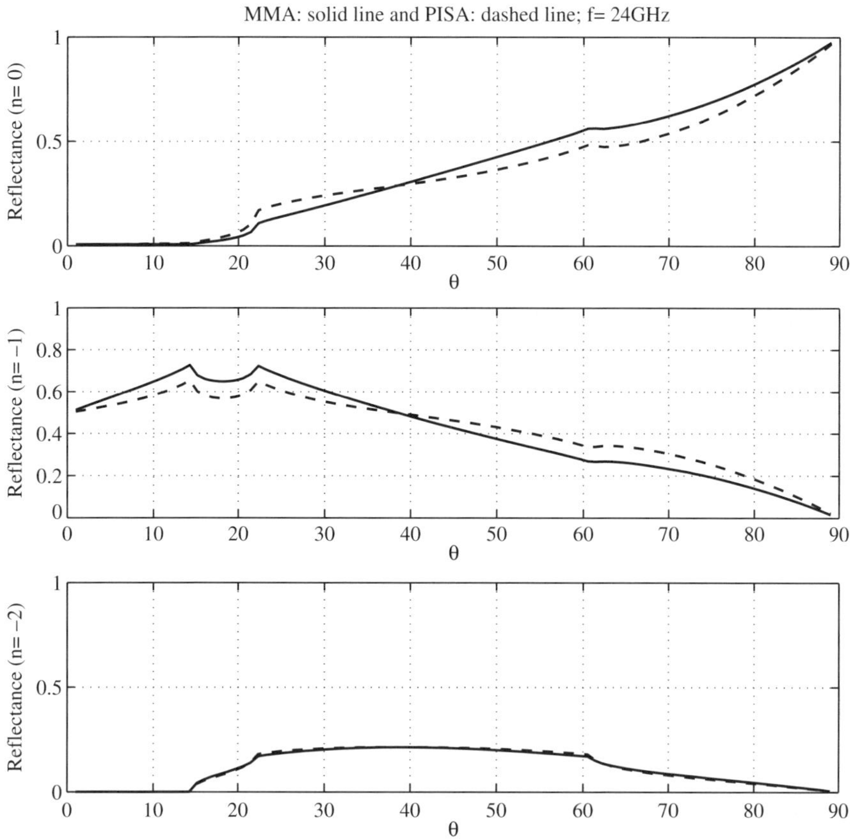

Figure 6.20 Reflectance of the fundamental, −1th-order, and −2th-order space harmonics against the incident angle from the 1D periodic structure shown in Figure 6.10. The solid and dashed lines respectively represent the results from the MMA and PISA. The operation frequency is 24 GHz

Looking at the right-hand side of Equation (6.87), by interchanging the summations and combining the two exponential terms, the above equation can be rewritten as

$$\sum_n [\underline{a}'_n v'_n(0) + \underline{a}''_n v''_n(0)]\,\mathrm{e}^{-\mathrm{j}\underline{k}_{\mathrm{t},n}\underline{\rho}} = \sum_n \sum_p z_p[\underline{a}'_n i'_n(0) + \underline{a}''_n i''_n(0)]\,\mathrm{e}^{-\mathrm{j}\underline{k}_{\mathrm{t},n-p}\underline{\rho}} \tag{6.88}$$

where

$$\underline{k}_{\mathrm{t},n-p} = \underline{x}_{\mathrm{o}} \left[k_{x,0} + (n-p)\frac{2\pi}{d_x}\right] + \underline{y}_0 k_{y,0}$$

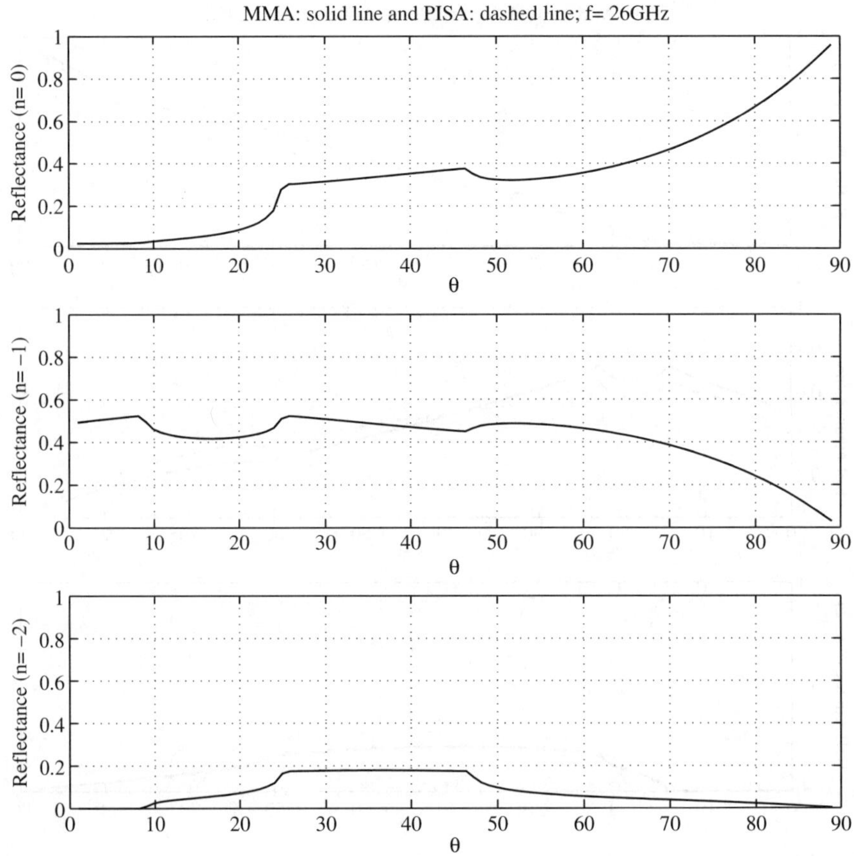

Figure 6.21 Reflectance of the fundamental, −1th-order, and −2th-order space harmonics against the incident angle from the 1D periodic structure shown in Figure 6.10. The solid and dashed lines respectively represent the results from the mode-matching approachMMA and PISA. The operation frequency is 26 GHz

By taking the transformation of variables, Equation (6.88) can be rewritten as

$$\sum_n [\underline{a}'_n v'_n(0) + \underline{a}''_n v''_n(0)] e^{-j\underline{k}_{t,n}\underline{\rho}} = \sum_n \sum_p z_p [\underline{a}'_{n+p} i'_{n+p}(0) + \underline{a}''_{n+p} i''_{n+p}(0)]\, \mathrm{e}^{-j\underline{k}_{t,n}\underline{\rho}} \tag{6.89}$$

Because the Fourier amplitudes of each harmonic must be equal to each other on both sides of Equation (6.89), we obtain the relation

$$\underline{a}'_n v'_n(0) + \underline{a}''_n v''_n(0) = \sum_p z_p [\underline{a}'_{n+p} i'_{n+p}(0) + \underline{a}''_{n+p} i''_{n+p}(0)] \tag{6.90}$$

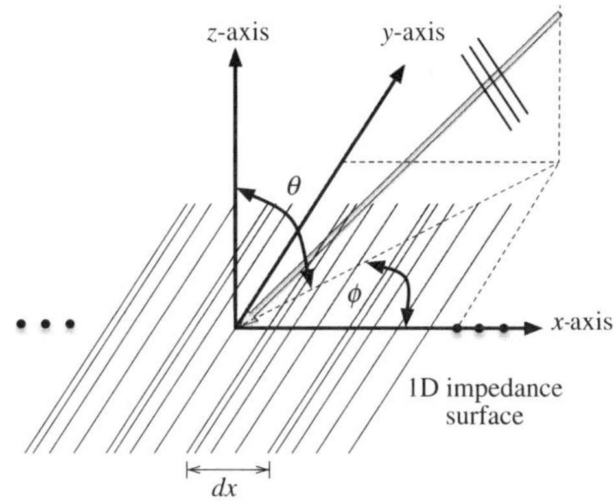

Figure 6.22 A 1D periodic impedance surface on the plane at $z = 0$ incident by a plane wave in the non-principal plane with azimuthal angle ϕ

Incorporating the orthogonality $\underline{a}'_n \cdot \underline{a}''_n = 0$, the relationship between the voltage and current amplitudes can be written as follows:

$$v'_n(0) = \sum_p z_p[\zeta^{(1,1)}_{n,n+p} i'_{n+p}(0) + \zeta^{(1,2)}_{n,n+p} i''_{n+p}(0)] \tag{6.91}$$

$$v''_n(0) = \sum_p z_p[\zeta^{(2,1)}_{n,n+p} i'_{n+p}(0) + \zeta^{(2,2)}_{n,n+p} i''_{n+p}(0)] \tag{6.92}$$

with the definition of four inner products:

$$\zeta^{(1,1)}_{n,n+p} = \underline{a}'_n \cdot \underline{a}'_{n+p} \tag{6.93}$$

$$\zeta^{(1,2)}_{n,n+p} = \underline{a}'_n \cdot \underline{a}''_{n+p} \tag{6.94}$$

$$\zeta^{(2,1)}_{n,n+p} = \underline{a}''_n \cdot \underline{a}'_{n+p} \tag{6.95}$$

$$\zeta^{(2,2)}_{n,n+p} = \underline{a}''_n \cdot \underline{a}''_{n+p} \tag{6.96}$$

Notably, the presence of the cross-coupling terms $\zeta^{(1,2)}_{n,n+p}$ and $\zeta^{(2,1)}_{n,n+p}$ elucidates that the cross-coupling indeed takes place for the non-principal plane incidence.

For each n, we may have a system of linear equation for Equations (6.91) and (6.92). Therefore, by collecting the index n running from negative to positive infinities, we may establish the following two matrix equations:

$$\underline{v}'(0) = \mathbb{Z}^{(1,1)}\underline{i}'(0) + \mathbb{Z}^{(1,2)}\underline{i}''(0) \tag{6.97}$$

$$\underline{v}''(0) = \mathbb{Z}^{(2,1)}\underline{i}'(0) + \mathbb{Z}^{(2,2)}\underline{i}''(0) \tag{6.98}$$

where the impedance matrix $\mathbb{Z}^{(i,j)}$ =

$$\begin{pmatrix} z_0\zeta_{n,n}^{(i,j)} & z_{-1}\zeta_{n,n-1}^{(i,j)} & z_{-2}\zeta_{n,n-2}^{(i,j)} & z_{-3}\zeta_{n,n-3}^{(i,j)} & z_{-4}\zeta_{n,n-4}^{(i,j)} & \ddots & \ddots \\ z_1\zeta_{n-1,n}^{(i,j)} & z_0\zeta_{n-1,n-1}^{(i,j)} & z_{-1}\zeta_{n-1,n-2}^{(i,j)} & z_{-2}\zeta_{n-1,n-3}^{(i,j)} & z_{-3}\zeta_{n-1,n-4}^{(i,j)} & z_{-4}\zeta_{n-1,n-5}^{(i,j)} & \ddots \\ \ddots & \ddots & \ddots & \ddots & \ddots & \ddots & \ddots \\ z_3\zeta_{0,3}^{(i,j)} & z_2\zeta_{0,2}^{(i,j)} & z_1\zeta_{0,1}^{(i,j)} & z_0\zeta_{0,0}^{(i,j)} & z_{-1}\zeta_{0,-1}^{(i,j)} & z_{-2}\zeta_{0,-2}^{(i,j)} & z_{-3}\zeta_{0,-3}^{(i,j)} \\ z_4\zeta_{-1,3}^{(i,j)} & z_3\zeta_{-1,2}^{(i,j)} & z_2\zeta_{-1,1}^{(i,j)} & z_1\zeta_{-1,0}^{(i,j)} & z_0\zeta_{-1,-1}^{(i,j)} & z_{-1}\zeta_{-1,-2}^{(i,j)} & z_{-2}\zeta_{-1,-3}^{(i,j)} \\ \ddots & \ddots & \ddots & \ddots & \ddots & \ddots & \ddots \\ \ddots & \ddots & z_4\zeta_{-n,-n+4}^{(i,j)} & z_3\zeta_{-n,-n+3}^{(i,j)} & z_2\zeta_{-n,-n+2}^{(i,j)} & z_1\zeta_{-n,-n+1}^{(i,j)} & z_0\zeta_{-n,-n}^{(i,j)} \end{pmatrix} \tag{6.99}$$

To obtain a succinct expression, the above equation can be rewritten as

$$\mathbb{Z}^{(i,j)} = \sum_p z_p \operatorname{diag}[\underline{\zeta}_p^{(i,j)}, -p] \tag{6.100}$$

where $\underline{\zeta}_p^{(i,j)}$ is the vector storing the element $\zeta_{m,n}^{(i,j)}$ with the indices (m, n) satisfying $m - n = -p$.

Combining Equations (6.97) and (6.98), we may have a super-matrix equation written as

$$\underline{v}(0) = \mathbb{Z}_s \underline{i}(0) \tag{6.101}$$

where $\underline{v}(0) = [\underline{v}(0)'\underline{v}(0)'']^T$ and $\underline{i}(0) = [\underline{i}(0)'\underline{i}(0)'']^T$. Matrix $\mathbb{Z}_s$ is a partitioned matrix with its (i, j) th entry filled with the sub-matrix $\mathbb{Z}^{(i,j)}$, where $i(j) = 1, 2$.

Substituting the voltage and current wave vectors in free space (given in Chapter 5) into Equation (6.101) yields

$$\mathbb{A}(\underline{a} + \underline{b}) = \mathbb{Z}_s \mathbb{A} \mathbb{Y}_a (\underline{a} - \underline{b}) \tag{6.102}$$

The reflection matrix between the incident ($\underline{a}$) and reflected ($\underline{b}$) vectors can be determined. Once the incident plane wave is given, the reflected amplitude of each space harmonic of TE and TM polarization can be determined.

Problem 6.3

Prove the relationship between the incident and reflected vectors is $\underline{b} = (\mathbb{Z}_s \mathbb{A} \mathbb{Y}_a + \mathbb{A})^{-1} (\mathbb{Z}_s \mathbb{A} \mathbb{Y}_a - \mathbb{A})\, \underline{a}$.

6.3.1 *Guiding Characteristics of Waves Supported by a 1D Periodic Impedance Surface*

In addition to the scattering properties of a 1D periodic impedance surface given in Equation (6.102) and the reflection matrix in Problem 6.3, we may also determine the guiding characteristics of waves supported by such a 1D periodic impedance surface by solving the following equation:

$$\det(\mathbb{Z}_s \mathbb{A} \mathbb{Y}_a + \mathbb{A}) = 0 \tag{6.103}$$

Equation (6.103) is the so-called transverse resonance equation, which determines the relationship among propagation constants along the x- and y-axes, k_x and k_y, and the free-space wavenumber, k_o, for given structure parameters of the 1D periodic impedance surface. For a given phase constant k_y (designated by the incident condition), k_x generally is a complex number, which is characterized as $k_x = \beta_x - j\alpha_x$, with the phase ($\beta$) and attenuation ($\alpha$) constants along the x-axis. In the bound-wave triangle, the attenuation constant is zero except for those in the stop-band region. Outside the bound-wave region, it becomes a *leaky wave* radiating its energy into the surrounding medium. The dispersion relation is usually drawn in the format of a Brillouin diagram similar to those shown in Figures 3.46 and 3.47.

6.4 Scattering of Plane Wave by a Dyadic 2D Periodic Impedance Surface

In this section I will extend the 1D periodic impedance surface into a 2D one; that is, the periodic variations are along both the x- and y-axes. Regarding the scattering and guiding characters of a scalar 2D periodic impedance, the theoretical formulation and numerical calculations have been carried out (Hwang 2000; Hwang and Peng 1999, 2000). Moreover, without loss of generality, the general condition of a 2D dyadic impedance surface will be considered, although such a tensor impedance surface, of course, is hard to obtain for a complex 2D periodic structure.

In this section, the rigorous mathematical formulation for analyzing the scattering characteristics of plane wave by a 2D dyadic impedance surface will be developed. Here, each of the elements in the impedance dyadic is a periodic function along both x and y-axes and can be expressed in terms of the double Fourier series expansion. As to the method for determining the impedance dyadic, this is indeed an essential research topic that is beyond the scope of this chapter.

The dyadic periodic IBC defines the relationship of the tangential electric and magnetic fields at $z = 0$ on the x–y plane, which is given as

$$\underline{z}_0 \times \underline{E}_t(x, y, z = 0) = \overline{\overline{Z}}_s(x, y) \cdot \underline{H}_t(x, y, z = 0) \tag{6.104}$$

where the 2D periodic dyadic impedance surface is defined as

$$\overline{\overline{Z}}_{\rm s} = z_{xx}(x, y)\underline{x}_0\underline{x}_0 + z_{xy}(x, y)\underline{x}_0\underline{y}_0 + z_{yx}(x, y)\underline{y}_0\underline{x}_0 + z_{yy}(x, y)\underline{y}_0\underline{y}_0 \tag{6.105}$$

Substituting the plane-wave solutions incorporating the periodicity along the x- and y-axes into the above equations, we obtain

$$\begin{aligned} &\sum_m \sum_n [\underline{a}'_{mn} v'_{mn}(0) + \underline{a}''_{mn} v''_{mn}(0)]\, \mathrm{e}^{-\mathrm{j}\underline{k}_{tmn}\cdot\underline{\rho}} \\ &\quad = \overline{\overline{Z}}_s(x, y) \cdot \sum_m \sum_n [\underline{a}'_{mn} i'_{mn}(0) + \underline{a}''_{mn} i''_{mn}(0)]\, \mathrm{e}^{-\mathrm{j}\underline{k}_{\rm t,\mathit{mn}}\cdot\underline{\rho}} \end{aligned} \tag{6.106}$$

By taking the dot product of the surface impedance dyadic and $\underline{a}'_{mn}$, we obtain

$$\overline{\overline{Z}}_{\rm s} \cdot \underline{a}'_{mn} = \underline{x}_0(z_{xx}a_{x,mn} + z_{xy}a_{y,mn}) + \underline{y}_0(z_{yx}a_{x,mn} + z_{yy}a_{y,mn}) \tag{6.107}$$

The dot product of the surface impedance dyadic and $\underline{a}''_{mn}$ yields

$$\overline{\overline{Z}}_{\rm s} \cdot \underline{a}''_{mn} = \underline{x}_0(-z_{xx}a_{y,mn} + z_{xy}a_{x,mn}) + \underline{y}_0(-z_{yx}a_{y,mn} + z_{yy}a_{x,mn}) \tag{6.108}$$

Substituting Equations (6.107) and (6.108) into Equation (6.106), the x-component on both sides of Equation (6.106) must be equal, which can be written as

$$\begin{aligned} &\sum_m \sum_n [v'_{mn}(0)a_{x,mn} - v''_{mn}(0)a_{y,mn}]\, \mathrm{e}^{-\mathrm{j}\underline{k}_{\rm t,\mathit{mn}}\cdot\underline{\rho}} \\ &\quad = \sum_m \sum_n [(z_{xx}a_{x,mn} + z_{xy}a_{y,mn})i'_{mn}(0) \\ &\quad + (-z_{xx}a_{y,mn} + z_{xy}a_{x,mn})i''_{mn}(0)]\, \mathrm{e}^{-\mathrm{j}\underline{k}_{\rm t,\mathit{mn}}\cdot\underline{\rho}} \end{aligned} \tag{6.109}$$

By multiplying $\exp(+\mathrm{j}\underline{k}_{t,pq} \cdot \underline{\rho})$ on both sides of Equation (6.109) and taking the integration over a unit cell, we obtain

$$\begin{aligned} &v'_{pq}(0)a_{x,pq} - v''_{pq}(0)a_{y,pq} \\ &\quad = \sum_m \sum_n (z^{xx}_{pq,mn}a_{x,mn} + z^{xy}_{pq,mn}a_{y,mn})i''_{mn}(0) \\ &\quad + (-z^{xx}_{pq,mn}a_{y,mn} + z^{xy}_{pq,mn}a_{x,mn})i''_{mn}(0) \end{aligned} \tag{6.110}$$

with the definition of the inner products given below:

$$z^{xx}_{pq,mn} = \iint \mathrm{e}^{+\mathrm{j}\underline{k}_{\mathrm{t},pq}\cdot\underline{\rho}}\, \mathrm{e}^{-\mathrm{j}\underline{k}_{\mathrm{t},mn}\cdot\underline{\rho}} z_{xx}(x, y)\,\mathrm{d}x\,\mathrm{d}y \tag{6.111}$$

$$z^{xy}_{pq,mn} = \iint \mathrm{e}^{+\mathrm{j}\underline{k}_{\mathrm{t},pq}\cdot\underline{\rho}}\, \mathrm{e}^{-\mathrm{j}\underline{k}_{\mathrm{t},mn}\cdot\underline{\rho}} z_{xy}(x, y)\,\mathrm{d}x\,\mathrm{d}y \tag{6.112}$$

$$z^{yx}_{pq,mn} = \iint \mathrm{e}^{+\mathrm{j}\underline{k}_{\mathrm{t},pq}\cdot\underline{\rho}}\, \mathrm{e}^{-\mathrm{j}\underline{k}_{\mathrm{t},mn}\cdot\underline{\rho}} z_{yx}(x, y)\,\mathrm{d}x\,\mathrm{d}y \tag{6.113}$$

$$z^{yy}_{pq,mn} = \iint \mathrm{e}^{+\mathrm{j}\underline{k}_{\mathrm{t},pq}\cdot\underline{\rho}}\, \mathrm{e}^{-\mathrm{j}\underline{k}_{\mathrm{t},mn}\cdot\underline{\rho}} z_{yy}(x, y)\,\mathrm{d}x\,\mathrm{d}y \tag{6.114}$$

Moreover, by equating the y-component on both sides of Equation (6.106) and taking the inner product with $\exp(+\mathrm{j}\underline{k}_{t,pq}\cdot\underline{\rho})$ on both sides, we obtain the following equation:

$$\begin{aligned} &v'_{pq}(0)a_{y,pq} + v''_{pq}(0)a_{x,pq} \\ &\quad = \sum_m\sum_n (z^{yx}_{pq,mn}a_{x,mn} + z^{yy}_{pq,mn}a_{y,mn})i'_{mn}(0) \\ &\quad +(-z^{yx}_{pq,mn}a_{y,mn} + z^{yy}_{pq,mn}a_{y,mn})i''_{mn}(0) \end{aligned} \tag{6.115}$$

We group the index pair (p, q) and assign them a new index r, and index s is designated as index pair (m, n). Therefore, Equation (6.110) can be rewritten as

$$\begin{aligned} &v'_r(0)a_{x,r} - v''_r(0)a_{y,r} \\ &= \sum_s (z^{xx}_{r,s}a_{x,s} + z^{xy}_{r,s}a_{y,s})i'_s(0) + (-z^{xx}_{r,s}a_{y,s} + z^{xy}_{r,s}a_{x,mn})i''_s(0) \end{aligned} \tag{6.116}$$

Each index r corresponds to a system of linear equations. By collecting these systems of linear equations, we can obtain the following matrix equation:

$$\mathbb{A}_x\underline{v}'(0) - \mathbb{A}_y\underline{v}''(0) = (\mathbb{Z}^{xx}\mathbb{A}_x + \mathbb{Z}^{xy}\mathbb{A}_y)\underline{i}'(0) + (-\mathbb{Z}^{xx}\mathbb{A}_y + \mathbb{Z}^{xy}\mathbb{A}_x)\underline{i}''(0) \tag{6.117}$$

where $\mathbb{A}_x$ and $\mathbb{A}_y$ are diagonal matrices with their entries representing $a_{x,r}$ and $a_{y,r}$, respectively. Variables $\mathbb{Z}^{xx}$, $\mathbb{Z}^{xy}$, $\mathbb{Z}^{yx}$, and $\mathbb{Z}^{yy}$ are full matrices. Vectors $\underline{v}(0)$ and $\underline{i}(0)$ contain the voltage and current amplitudes of each space harmonic.

Similarly, Equation (6.115) can be rewritten as follows:

$$v'_r(0)a_{y,r} + v''_r(0)a_{x,r} = \sum_s (z^{yx}_{r,s}a_{x,s} + z^{yy}_{r,s}a_{y,s})i'_s(0) + (-z^{yx}_{r,s}a_{y,s} + z^{yy}_{r,s}a_{x,s})i''_s(0) \tag{6.118}$$

A similar procedure employed previously is applied here to obtain the matrix equation given below:

$$\mathbb{A}_y\underline{v}'(0) + \mathbb{A}_x\underline{v}''(0) = (\mathbb{Z}^{yx}\mathbb{A}_x + \mathbb{Z}^{yy}\mathbb{A}_y)\underline{i}'(0) + (-\mathbb{Z}^{yx}\mathbb{A}_y + \mathbb{Z}^{yy}\mathbb{A}_x)\underline{i}''(0) \tag{6.119}$$

Combining Equations (6.117) and (6.119), we obtain the following super-matrix equation:

$$\begin{pmatrix} \mathbb{A}_x & -\mathbb{A}_y \\ \mathbb{A}_y & \mathbb{A}_x \end{pmatrix} \begin{pmatrix} \underline{v}'(0) \\ \underline{v}''(0) \end{pmatrix} = \begin{pmatrix} \mathbb{Z}^{xx} & \mathbb{Z}^{xy} \\ \mathbb{Z}^{yx} & \mathbb{Z}^{yy} \end{pmatrix} \begin{pmatrix} \mathbb{A}_x & -\mathbb{A}_y \\ \mathbb{A}_y & \mathbb{A}_x \end{pmatrix} \begin{pmatrix} \underline{i}'(0) \\ \underline{i}''(0) \end{pmatrix} \tag{6.120}$$

$$\underline{v}(0) = \mathbb{A}^{-1}\overline{\overline{\mathbb{Z}}}\mathbb{A}\underline{i}(0) \tag{6.121}$$

where the super-matrix $\mathbb{A}$ and $\overline{\overline{\mathbb{Z}}}$ are defined as

$$\begin{aligned} \overline{\overline{\mathbb{Z}}} &= \begin{pmatrix} \mathbb{Z}^{xx} & \mathbb{Z}^{xy} \\ \mathbb{Z}^{yx} & \mathbb{Z}^{yy} \end{pmatrix} \\ \mathbb{A} &= \begin{pmatrix} \mathbb{A}_x & -\mathbb{A}_y \\ \mathbb{A}_y & \mathbb{A}_x \end{pmatrix} \\ \underline{v}(0) &= \begin{pmatrix} \underline{v}'(0) \\ \underline{v}''(0) \end{pmatrix} \\ \underline{i}(0) &= \begin{pmatrix} \underline{i}'(0) \\ \underline{i}''(0) \end{pmatrix} \end{aligned} \tag{6.122}$$

Substituting of the incident-wave vector ($\underline{a}$) and reflected-wave vector ($\underline{b}$) into (6.121), we obtain:

$$\underline{a} + \underline{b} = \mathbb{A}^{-1}\overline{\overline{\mathbb{Z}}}\mathbb{A}\mathbb{Y}_a(\underline{a} - \underline{b}) \tag{6.123}$$

By solving the relationship between the incident and reflected wave vectors, we obtain the equation

$$\underline{b} = \Gamma\underline{a} \tag{6.124}$$

where the reflection matrix Γ is

$$\Gamma = (\mathbb{A}^{-1}\overline{\overline{\mathbb{Z}}}\mathbb{A}\mathbb{Y}_a + \mathbf{I})^{-1}(\mathbb{A}^{-1}\overline{\overline{\mathbb{Z}}}\mathbb{A}\mathbb{Y}_a - \mathbf{I}) \tag{6.125}$$

where $\mathbf{I}$ is the identity matrix.

Finally, once the incident wave vector is given, the reflectance of each space harmonic (m, n) can be determined from Equation (6.125).

References

Hessel, A. and Oliner, A. A. (1965) A new theory of Wood's anomalies on optical gratings. *Applied Optics* **4**, 1275–1298.

Hwang, R.-B. (2000) Scattering characteristics of two-dimensionally periodic impedance surface. *IEEE Transactions on Antennas and Propagation* **48**, 1521–1527.

Hwang, R.-B. and Peng, S.-T. (1999) Guidance characteristics of two-dimensionally periodic impedance surface. *IEEE Transactions on Microwave Theory and Techniques* **47**, 2503–2511.

Hwang, R.-B. and Peng, S.-T. (2000) Guided waves on 2D periodic structures and their relation to planar photonic band gap structures. *IEICE Transactions on Electronics* **E83-C**, 705–712.

Hwang, R.-B. and Tsai, Y.-L. (2012) Reflection characteristics of a composite planar AMC surface. *AIP Advances* **2**, 012128. DOI: 10.1063/1.3682352.

Leontovich, M. A. (1948) *Investigation of Propagation of Radiowaves, Part II*, Printing House of the Academy of Sciences, Moscow.

7

Exotic Dielectrics Made of Periodic Structures

Investigation on the electromagnetic properties for artificial dielectrics is absolutely not a new topic. To mention a few, metallic rods or strips immersed in a host dielectric medium to mimic a natural dielectric material were implemented by Kock (1948), Cohn (1949), and Morita and Cohn (1949); the static approach, such as conformal mapping, had been developed to evaluate the effective dielectric constant of 2D periodic structures consisting of conducting-strip with negligible thickness (Collin 1991). In this chapter we will focus our attention on an artificial dielectric exhibiting a periodic nature; the theoretical framework developed in the previous chapters will be employed. Additionally, the macroscopic viewpoint of an *effective medium* in connection with the eigenwave properties in a periodic medium will be elucidated. Finally, the *effective refraction index* of the eigenwave in an exotic[1] dielectric made of a periodic structure will be demonstrated.

7.1 Synthetic Dielectrics Using a 2D Dielectric Columns Array

Figure 7.1 presents a synthetic dielectric composed of a 2D dielectric columns array immersed in a uniform dielectric medium. Here, we assume that the length of each dielectric rod is infinite in extent; the propagation constant along that direction, say the z-axis, is not zero, however. From the theoretical background illustrated in the previous chapters, we know that for $k_z \neq 0$ the 3D boundary-value problem for the vectorial electromagnetic fields has to be resolved rather than the scalar one.

[1] An "exotic" dielectric, in fact, is different from the well-known artificial dielectric because of its complex phase-relation diagram.

Periodic Structures: Mode-Matching Approach and Applications in Electromagnetic Engineering, First Edition.
Ruey-Bing (Raybeam) Hwang.

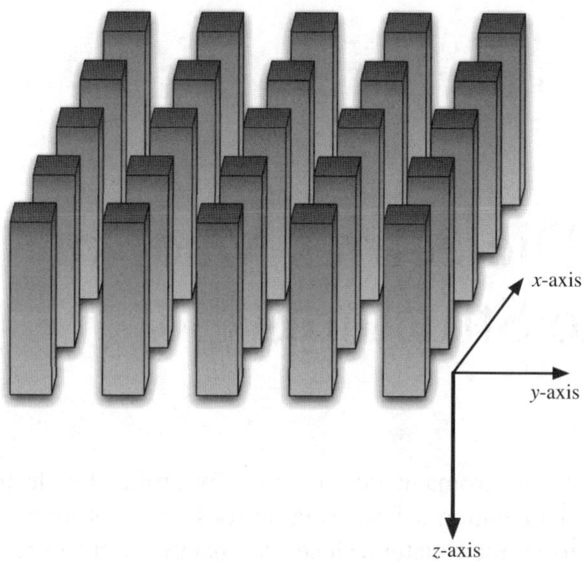

Figure 7.1 A synthetic dielectric material composed of a 2D dielectric columns array immersed in a uniform dielectric medium

The single polarization mode (E_z or H_z) can no longer exist independently. The resultant dispersion relation cannot be simply recognized as the E_z or H_z mode, so they are termed E_z-like or H_z-like modes hereafter.

7.1.1 Description of the Example

The example to be demonstrated is a synthetic dielectric made of dielectric square rods arranged in a 2D square lattice and immersed in a uniform dielectric medium (Figure 7.2). The periods along both the x- and y-axes are denoted as d; the dimension of the square column is a. The propagation constant along the z-axis is k_z, which can be non-vanished. Here, we assume that the wave propagates along the x–y plane, which is called *in-plane propagation*. The non-vanished k_z is given by the excitation condition of the incident wave.

As was described in Section 4.2, the generalized eigenvalue equation shown there can be applied for solving the phase relation among the k_x, k_y, and k_o for a given k_z.[2] Additionally, the propagation characteristic of the eigenwave along the y-axis can also be determined by solving the same generalized eigenvalue equation subject to known

[2] For correctly applying the equations given in Section 4.2.1, you have to interchange the variables y and z there, because the rod length is along the y-axis, as shown in Figure 4.29.

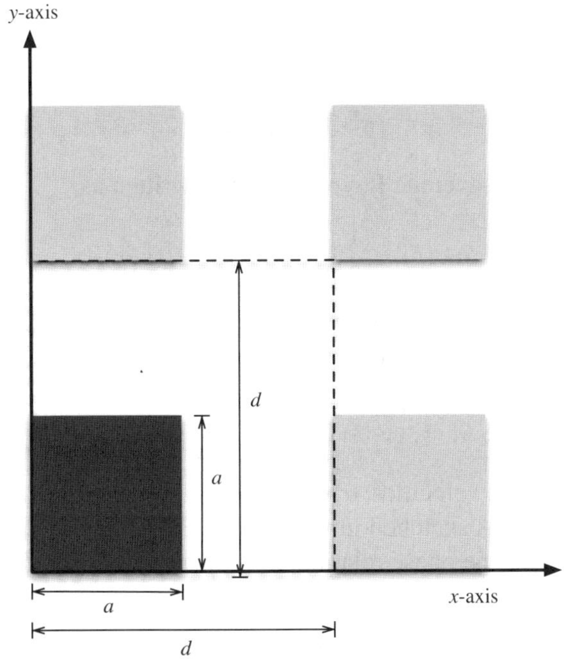

Figure 7.2 A synthetic dielectric material composed of 2D dielectric columns array immersed in a uniform dielectric medium: top view. The structure is arranged in a 2D square lattice with period d along the x- and y-axes; the dimension of the square dielectric column is denoted as a. The relative dielectric constant of the dielectric rod and the surrounding medium are ε_{s} and ε_{u}, respectively

k_x and k_z; the propagation constant $k_y = \beta_y - \mathrm{j}\alpha_y$ is generally a complex number with phase and attenuation constants β_y and α_y, respectively.

7.1.2 Phase-Relation Diagram of a Uniform Dielectric Medium

Before discussing the refractive index of synthetic dielectrics, we will first review that of a commonly used dielectric medium. In a uniform medium, if the plane wave has $k_z = 0$, its phase (or k) relation satisfies the equation: $k_x^2 + k_y^2 = k_{\mathrm{o}}^2\varepsilon_{\mathrm{u}}$, where k_x and k_y are the propagation constants along the x- and y-axes, and k_{o} is the free-space wavenumber; parameter ε_{u} is the relative dielectric constant of the uniform medium.

The phase relation forms a circle with respect to the two axes k_x and k_y with a radius $k_{\mathrm{o}}\sqrt{\varepsilon_{\mathrm{u}}}$. Incidentally, the wave, distinct from a particle, has three types of velocity: *phase* velocity, *group* velocity, and *energy* velocity. Furthermore, the phase velocity is the propagation velocity of an equal-phase surface, defined as $v_{\mathrm{p}} = \frac{\omega}{k}$; the group velocity is that of a wave packet, defined as $\underline{v}_{\mathrm{g}} = \frac{\partial\omega}{\partial\underline{k}}$, wherein $\underline{k}$ is the propagation vector. The energy velocity is defined as the propagation velocity of electromagnetic field energy (Sakoda 2004), written as

$$v_e = \frac{S(\underline{k}, \underline{r})}{U(\underline{k}, \underline{r})} \tag{7.1}$$

where $S(\underline{k}, \underline{r})$ is the time-average Poynting vector, written as

$$S(\underline{k}, \underline{r}) = \frac{1}{2}\Re[\underline{E}(\underline{k}, \underline{r}) \times \underline{H}^{\dagger}(\underline{k}, \underline{r})] \tag{7.2}$$

where the symbol "†" means the complex conjugate. Moreover, parameter $U(\underline{k}, \underline{r})$ is the electromagnetic-field energy density, defined as

$$U(\underline{k}, \underline{r}) = \frac{1}{4}[\varepsilon_o\varepsilon_u|\underline{E}(\underline{k}, \underline{r})|^2 + \mu_o|\underline{H}(\underline{k}, \underline{r})|^2] \tag{7.3}$$

Alternatively, the three velocities, particularly on their directions, can be understood by a graphical method, as sketched in Figure 7.3. Here, they share the same velocity because of a circular shape phase relation, as indicated in this figure.

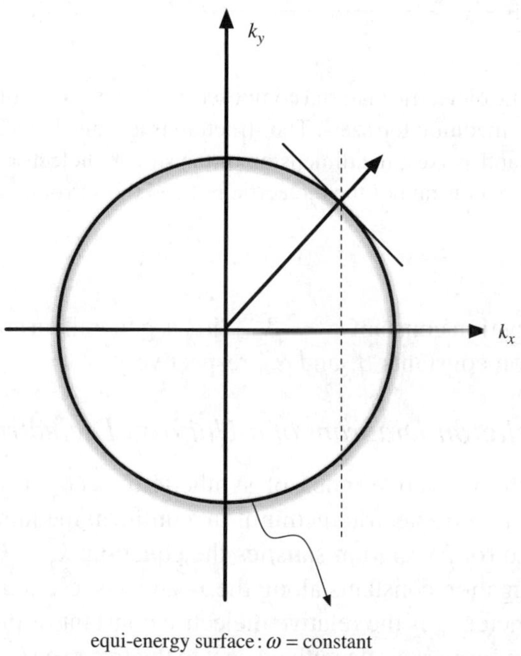

Figure 7.3 Phase-relation diagram of a plane wave in a uniform dielectric medium (the circle with the equal angular frequency is denoted an equi-energy surface); for a prescribed direction of propagation, the group velocity direction is along that perpendicular to the equi-energy surface

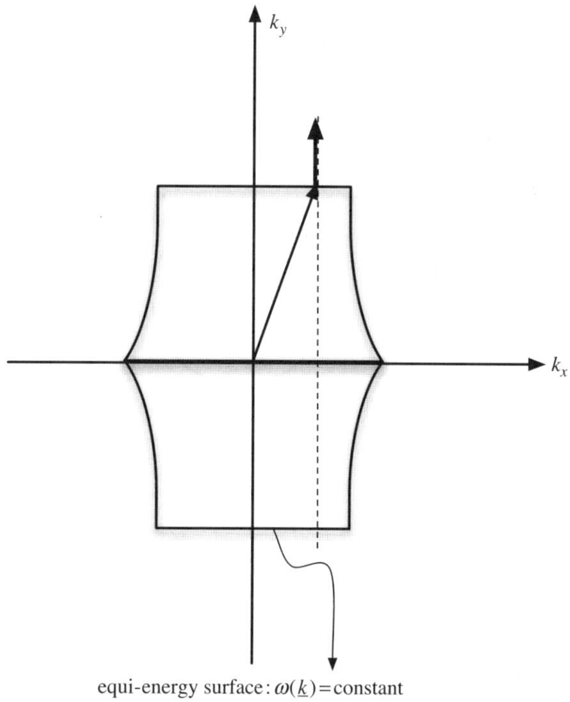

Figure 7.4 A cartoon diagram for the phase-relation of an eigenwave in a synthetic medium; for a prescribed direction of propagation, the group velocity direction is along that perpendicular to the equi-energy surface

Different from a uniform medium, a synthetic dielectric may have a complex equi-energy diagram similar to that shown in Figure 7.4. Apparently, for a given k_x, the phase velocity is the one pointing from the origin to the point on the surface with the prescribed k_x; however, the group velocity has a direction normal to the surface at that point, and the energy velocity is parallel to the group velocity. Generally speaking, in a medium having periodic variation, the group (or energy) velocity does *not* coincide with the phase velocity, while the energy velocity is parallel to the group velocity (Sakoda 2004). In the following numerical example, we will demonstrate some phase diagrams, which are obtained from rigorous MMA, with complex equi-energy surfacea to clarify the point that we address here.

7.2 Refractive Index of a 2D Periodic Medium

Returning to Figure 7.2, we synthesize a composite dielectric using a 2D square lattice. The detailed structure parameters are given below. The relative dielectric constant of the dielectric column and uniform medium are $\varepsilon_s = 10.2$ and $\varepsilon_u = 1.0$. The dimensions of square rod and period are $a = 0.5$ and $d = 1.0$. All the dimensions are in

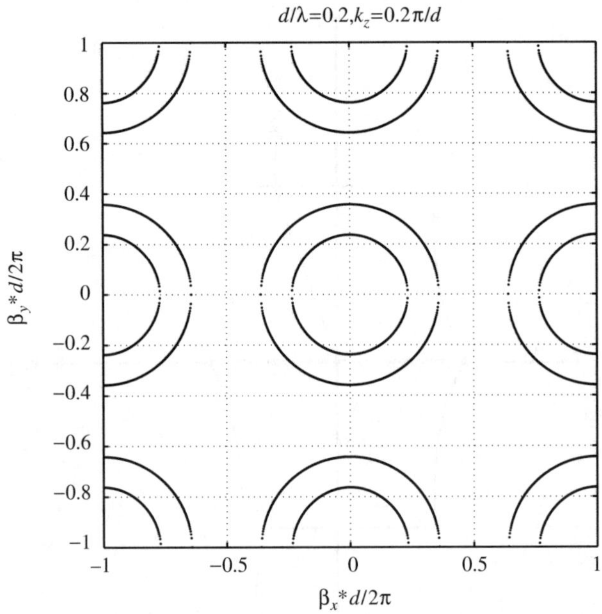

Figure 7.5 Phase-relation diagram (equi-energy surface) of an eigenwave in 2D periodic dielectric medium shown in Figure 7.2: $d/\lambda = 0.2$, $k_z d/2\pi = 0.1$; the relative dielectric constants of the dielectric rod and uniform medium are $\varepsilon_s = 10.2$ and $\varepsilon_u = 1.0$. The dimensions of the square rod and period are $a = 0.5$ and $d = 1.0$. All the dimensions are in arbitrary units

arbitrary units. We follow the theoretical analysis given in Chapter 4 for building up the generalized eigenvalue equation and solve the eigenvalue and associated eigenvector numerically. Here, we consider the condition of *in-plane* propagation; however, the propagation constant along the z-axis is not zero. It is designated by the excitation condition; the following numerical calculations are carried out under the condition $k_z d/2\pi = 0.1$.

Figure 7.5 demonstrates the phase-relation diagram at a fixed normalized frequency $d/\lambda = 0.2$ (so it is called the equi-energy surface). In view of the periodicity along the x- and y-axes, the curves for the space harmonics are all displaced by $2\pi/d$ from each other. Since the horizontal and vertical axes are normalized to $2\pi/d$, the space harmonics are displaced by an integer number. Looking at the first Brillouin zone ($|\beta_x d/2\pi| \leq 0.5$ and $|\beta_y d/2\pi| \leq 0.5$), the two circles are the phase diagram due to the E_z*-like* and H_z*-like* modes. In the presence of k_z, both types of polarization modes certainly exist. However, they seem not to couple to each other under this situation. Since they exhibit circular patterns, their radii allow us to know the *effective* refractive index n_{eff}, which is obtained by dividing the radius by d/λ. From this figure, the effective refractive indices of the two polarization eigenwaves are close to

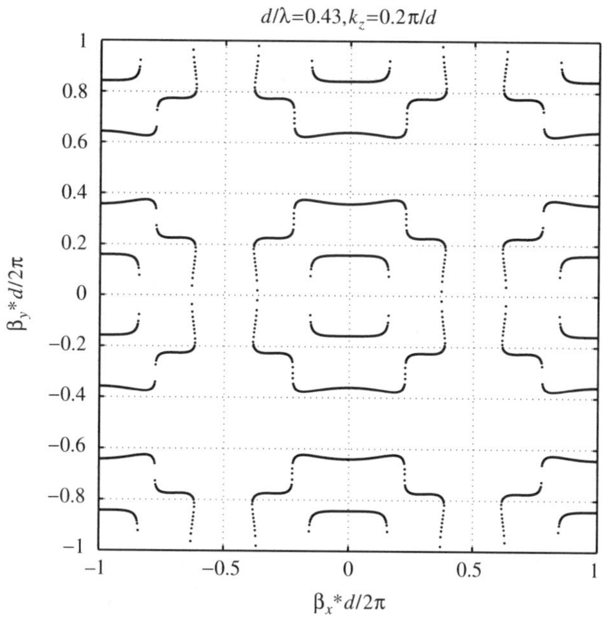

Figure 7.6 Phase-relation diagram (equi-energy surface) of an eigenwave in a 2D periodic dielectric medium shown in Figure 7.2: $d/\lambda = 0.43$, $k_z d/2\pi = 0.1$; the relative dielectric constants of the dielectric rod and uniform medium are $\varepsilon_s = 10.2$ and $\varepsilon_u = 1.0$. The dimensions of the square rod and period are $a = 0.5$ and $d = 1.0$. All the dimensions are in arbitrary units

1.0 and 2.0. Intuitively, if the composite dielectric considered here is regarded as a bulk medium (electrical large approximation), the effective refractive index may be lowered by averaging it with a low refractive medium, such as air in this example. In comparison with the arithmetic mean, which is $n_{avg} = (12\frac{1}{4} + 1\frac{3}{4})^{1/2} \approx 1.94$ in this case, we know that, in fact, n_{eff} can be lower than the *visualized* refractive index. The above results can provide us an excellent option for synthesizing a *foam* medium with a refractive index near that of free space, especially in microwave and millimeter-wave applications.

As the normalized frequency is increased to $d/\lambda = 0.43$, which is beyond the first stop-band of the two polarized modes, the equi-energy surface is no longer a circle for both polarized modes. They become complex shapes[3] depicted in Figure 7.6. Referring to this figure, the equi-energy surface is almost enclosed by straight lines, which means that the direction of the group velocity is around the normal direction as the ending point of the vector $\underline{k}_t$ locates within each straight-line segment.

[3] Notably, the eigenwave in a periodic medium is contributed by Floquet's solutions, which can be regarded as the superposition of plane waves.

Interestingly, for the smaller one which is like a square, the dimension of the square is smaller than the operation normalized frequency $d/\lambda = 0.43$; therefore, we have $n_{\text{eff}} \approx 0.365$, which is smaller than that of free space. It may be conjectured that if a line source is placed inside such a medium with finite dimensions along the x- and y-axes, the outgoing wave transmitted (or refracted) into the air outside the 2D periodic structure will tend to be perpendicular to the interface. Two types of artificial dielectrics with effective refractive index smaller than unity have been designed (Hwang *et al.* 2009, 2012), fabricated, and measured. A line source implemented at the center of the meta-material-based dielectric block excites the beam radiating into free space with a quasi-plane wavefront along the $\pm x$- and $\pm y$-axes, respectively. As a consequence, four directional beams were obtained and can be employed as a spatial beam splitter.

Another two figures plotted at normalized frequencies $d/\lambda = 0.44$ and $d/\lambda = 0.45$ are shown in Figures 7.7 and 7.8. They almost retain a similar shape; however, a much more complex pattern appears in Figure 7.8. This may be conjectured to be in connection with the complex electromagnetic coupling between higher order space harmonics. Although not shown here, we have continuously increase the normalized frequency and found that the much more complex phase-relation diagrams appear to be contributed by the complex coupling caused by the higher order space harmonics.

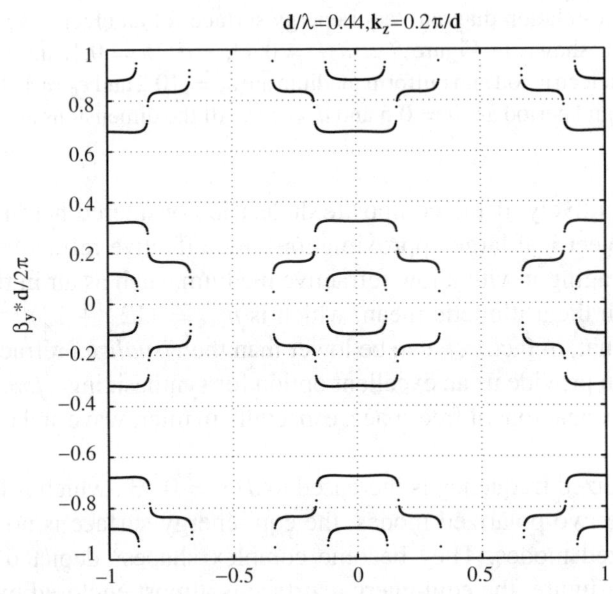

Figure 7.7 Phase-relation diagram (equi-energy surface) of an eigenwave in a 2D periodic dielectric medium shown in Figure 7.2: $d/\lambda = 0.44$, $k_z d/2\pi = 0.1$; the relative dielectric constants of the dielectric rod and uniform medium are $\varepsilon_s = 10.2$ and $\varepsilon_u = 1.0$. The dimensions of square rod and period are $a = 0.5$ and $d = 1.0$. All the dimensions are in arbitrary units

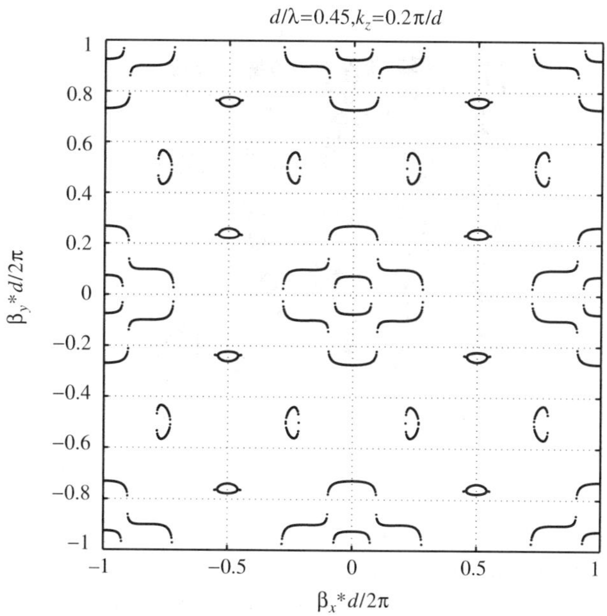

Figure 7.8 Phase-relation diagram (equi-energy surface) of an eigenwave in a 2D periodic dielectric medium shown in Figure 7.2: $d/\lambda = 0.45$, $k_z d/2\pi = 0.1$; the relative dielectric constants of the dielectric rod and uniform medium are $\varepsilon_s = 10.2$ and $\varepsilon_u = 1.0$. The dimensions of square rod and period are $a = 0.5$ and $d = 1.0$. All the dimensions are in arbitrary units

7.2.1 Conclusion

From the numerical results shown previously, the synthetic dielectric made of the 2D periodic structure exhibits an angular- and frequency-dependent effective refractive index, which can be written as $n_{\text{eff}}(k_o, k_x, k_y)$ for a given k_z. The effective refractive index with respect to arbitrary phase constants k_x and k_y can be read from the above figures subject to a certain normalized frequency. However, if we want to know the frequency-dependence of the refractive index, we have to stack up those equi-energy plots for a certain range of frequencies with a small frequency step. Therefore, a 3D plot and a sophisticated computer graphics technique are needed to generate and view the graph. Unfortunately, it is hard to present them in this book.

Although the effective medium approach – which extracts the effective permittivity and permeability from the scattering parameters, including the reflectance and transmittance of an artificial dielectric (or meta-material) slab, by carrying out the scattering analysis with respect to a specific incident angle – was extensively used to determine the electrical properties of an artificial dielectric, their uncertainty, which is caused by the excitation of higher order space harmonics, cannot be resolved.

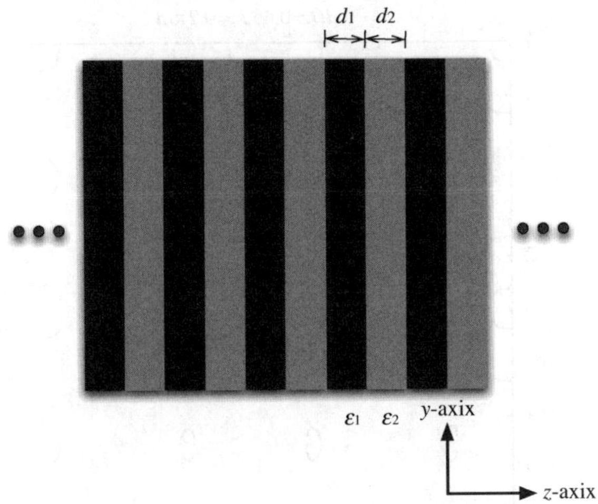

Figure 7.9 An artificial dielectric made of 1D periodic dielectric layers; in this figure, each of the unit cells contains two parallel-plane dielectric mediums characterized by ε_1 and ε_2 with widths d_1 and d_2, respectively. The structure can be viewed as an infinite stack (along the z-axis) of alternating dielectric slab 1 and slab 2. Each of the slabs is infinite in extent along the transverse (x–y) plane

Conversely, the modal-solution approach developed in this book can rigorously determine the frequency- and angular-dependent refractive index without any approximation once the finite truncation of the matrices is sufficient for achieving the numerical convergence.

Last but not the least, the 2D periodic structure not only allows us to synthesize a specific refractive index, but also provides us with a feasible means for manipulating the propagation of electromagnetic waves (energy flow).

7.3 An Artificial Dielectric Made of 1D Periodic Dielectric Layers

Let us now return to a relatively simple scheme of an artificial dielectric made of 1D periodic dielectric layer. Figure 7.9 depicts an artificial dielectric made of 1D periodic dielectric layers; in this figure, each of the unit cells contains two[4] parallel-plane dielectric media denoted as "medium 1" and "medium 2," respectively. The relative dielectric constants of the two media are designated as ε_1 and ε_2 with widths d_1 and d_2, respectively. The structure can be viewed as an infinite stack (along the z-axis) of alternating slab 1 and slab 2. Each of the slabs is infinite in extent along the transverse (x–y) plane.

[4] The unit cell, of course, can be designed to have much more dielectric slabs if needed.

In view of the infinite periodic medium with period $d = d_1 + d_2$, along the z-axis, the eigenvalue equation of the Bloch wave in the periodic medium is written as (see Section 5.1 for the detailed mathematical procedure):

$$\cos \kappa d = \cos k_{z,1}d_1 \cos k_{z,2}d_2 - Q \sin k_{z,1}d_1 \sin k_{z,2}d_2 \tag{7.4}$$

$$Q = \frac{1}{2}\left(\frac{Z_1}{Z_2} + \frac{Z_2}{Z_1}\right) \tag{7.5}$$

$$k_{z,i} = \sqrt{k_o^2 \varepsilon_i - k_t^2} \tag{7.6}$$

where κ is the propagation constant of the Bloch wave along the z-axis. Parameter k_t is the propagation constant along the transverse plane, which is prescribed by the excitation condition, and is written as

$$k_t^2 = k_x^2 + k_y^2 \tag{7.7}$$

where Z_i is the characteristic impedance in the ith medium, which is $\omega\mu_o/k_z^{(i)}$ for TE polarization and $k_z^{(i)}/\omega\varepsilon_o\varepsilon_i$ for TM polarization.

7.3.1 Effective Refractive Index of the 1D Dielectric Periodic Medium

Through Equation (7.4), we can obtain the propagation constant of the Bloch wave along the z-axis, which is given as

$$\kappa = \frac{1}{d}\cos^{-1}(\cos k_{z,1}d_1 \cos k_{z,2}d_2 - Q \sin k_{z,1}d_1 \sin k_{z,2}d_2) \tag{7.8}$$

Notice that if the absolute value of the right-hand side in Equation (7.4) is greater than unity, κ is a complex number, which means that the eigenwave is in the stop-band; otherwise, κ is a real number and the eigenwave is in the pass-band. The *effective* refractive index n_{eff} of the composite dielectric can be written as

$$n_{\text{eff}}(k_o, k_x, k_y) = \frac{1}{dk_o}\cos^{-1}(\cos k_{z,1}d_1 \cos k_{z,2}d_2 - Q \sin k_{z,1}d_1 \sin k_{z,2}d_2) \tag{7.9}$$

The above equation indicates that the *effective* refractive index is seen along the z-axis for the eigenwave preserving its transverse component of phase constants, k_x and k_y. The refractive index n_{eff} will change in accordance with the change in k_x and k_y.

7.3.2 Effective Wave-Impedance of the 1D Dielectric Periodic Medium

In addition to the eigenvalue being determined by Equation (7.4), its eigenvector with its components representing the voltage and current amplitudes at the input terminal of the unit cell can also be obtained. The ratio of voltage- to current-amplitude is called the *Bloch impedance*, mentioned in Chapter 5:

$$Z_B = \frac{t_{1,2}}{\lambda - t_{1,1}} \tag{7.10}$$

$$t_{1,1} = \cos k_{z,1}d_1 \cos k_{z,2}d_2 - \frac{Z_1}{Z_2} \sin k_{z,1}d_1 \sin k_{z,2}d_2 \tag{7.11}$$

$$t_{1,2} = -jZ_2 \sin k_{z,2}d_2 \cos k_{z,1}d_1 - jZ_1 \sin k_{z,1}d_1 \cos k_{z,2}d_2 \tag{7.12}$$

Therefore, Z_B can be regarded as the *effective* wave-impedance Z_{eff}.

With the effective refractive index as well as the wave impedance, the medium can be represented by a new transmission-line specified by the two parameter: $\kappa = k_o n_{eff}$ and Z_{eff}. Consequently, the plane wave scattering by such a composite dielectric with a prescribed thickness can be determined by considering the network shown in Figure 7.10.

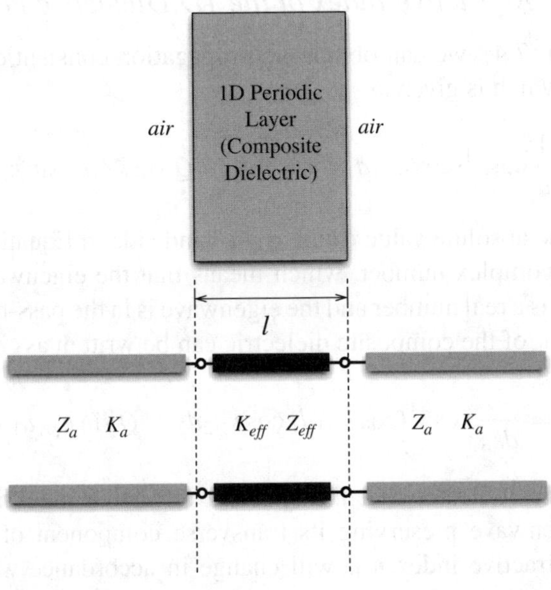

Figure 7.10 Transmission-line network representation of the scattering of a plane wave by an artificial dielectric made of 1D periodic dielectric layers; the two semi-infinite transmission-lines with propagation constant and impedance κ_a and Z_a represent the fields in air, while the finite length characterized by κ_{eff} and Z_{eff} is that in the 1D periodic medium. The length of the effective transmission-line is denoted l

Problem 7.1

With the effective refractive index and the wave impedance, determine the effective permittivity and permeability of the composite dielectrics for TE- and TM-polarization waves, respectively.

7.4 Conclusion

In fact, the above-mentioned theory can also be applied in a much more complex structure; for example, an artificial dielectric consisting of metallic circular disks or metallic strips or columns. Once the equivalent lumped-circuit model of each dielectric layer having metallic inclusions is determined, the eigenvalue equation can be developed and the effective impedance and refractive index can be obtained for establishing an effective transmission-line. It is noted, however, that such a class of structures are 2D or 3D; however, we treated their behavior in only a 1D fashion. To obtain the 2D or 3D behavior of 2D or 3D periodic structures, it is necessary to deal with the vector boundary-value problem and matrix generalized eigenvalue problem, as elucidated in Section 7.2.

References

Cohn, S. B. (1949) Analysis of the metal strip delay structures for microwave lenses. *Journal of Applied Physics* **20**, 257–262.

Collin, R. E. (1991) *Field Theory of Guided Waves*, 2nd edn, IEEE Press, New York, NY.

Hwang, R.-B., Liu, H.-W., and Chin, C.-Y. (2009) A metamaterial-based E-plane horn antenna. *Progress in Electromagnetics Research* **93**, 275–289.

Hwang, R.-B., Hsu, N.-C., and Chin, C.-Y. (2012) A spatial beam splitter consisting of a near-zero refractive index medium. *IEEE Transactions on Antennas and Propagation* **60**, 417–420.

Kock, W. E. (1948) Metallic delay lenses. *Bell Systems Technical Journal* **27**, 58–82.

Morita, T. and Cohn, S. B. (1956) Microwave lens matching by simulated quarter-wave transformers. *IEEE Transactions on Antennas and Propagation* **4**, 33–39.

Sakoda, K. (2004) *Optical Properties of Photonic Crystals*, Springer.

Index

Periodic Structures: Mode-Matching Approach and Applications in Electromagnetic Engineering, First Edition.
Ruey-Bing (Raybeam) Hwang.